Current Research in Biopsychology

John P. J. Pinel

THE UNIVERSITY OF BRITISH COLUMBIA

ALLYN AND BACON
Boston London Toronto Sydney Tokyo Singapore

Copyright © 1991 by Allyn and Bacon
A Division of Simon & Schuster
160 Gould Street
Needham Heights, MA 02194

All rights reserved. No part of the material protected by this copyright notice may be reproduced or utilized in any form or by any means, electronic or mechanical, including photocopying, recording, or by any information storage and retrieval system, without written permission from the copyright owner.

ISBN 0-205-13000-3

Printed in the United States of America

10 9 8 7 6 5 4 3 2 1   95 94 93 92 91

ACKNOWLEDGEMENTS

We are grateful to the publishers, editors, and authors who graciously permitted this facsimile printing of their articles.

To my teachers and students —
for what they have taught
me.

About The Author

John P.J. Pinel is a professor in the Department of Psychology at the University of British Columbia in Vancouver, B.C., Canada. He is the author of well over 100 research reports, review articles, and book chapters, and is well-known for his innovative research on a variety of biopsychological topics: including learning, memory, epilepsy, defensive behavior, and drug tolerance. In recognition of his research accomplishments, he has been made a fellow of the American Psychological Association, the American Psychological Society, and the International Society for Research on Aggression; and he is a recent winner of the prestigious Killam Memorial Senior Fellowship.

Professor Pinel's contributions to biopsychology have not been restricted to the laboratory; he is an award-winning teacher, and his recent textbook Biopsychology (Allyn & Bacon, 1990) has won him wide recognition—Biopsychology has been acclaimed for its clear, insightful, forward-looking coverage of the field of biopsychology. He has the rare ability to speak directly and personally to students through his writing, with just the right blend of humor and enthusiasm on one hand and scholarship on the other.

In this book, Professor Pinel introduces 24 recently published journal articles that are at the very cutting edge of biopsychological research. These exciting articles and Pinel's colorful commentary are an educational treat.

TABLE OF CONTENTS

Preface -- vii

To the Student --- x

1. BRAIN: DEVELOPMENT, DAMAGE, AND RECOVERY

Article 1   Brain Development, Plasticity, and Behavior
Bryan Kolb (1989) ------------------------- 1

Article 2   Reorganization of Retinotopic Cortical Maps in Adult Mammals After Lesions of the Retina
Jon H. Kaas, Leah A. Krubitzer, Yuzo M. Chino, Andy L. Langston, Edward H. Polley, and Norman Blair (1990) ------ 15

Article 3   Open Microsurgical Autograft of Adrenal Medulla to the Right Caudate Nucleus in Two Patients with Intractable Parkinson's Disease
Ignacio Madrazo, René Drucker-Colín, Victor Díaz, Juan Martínez-Mata, César Torres, and Juan José Becerril (1987) -- 23

2. THE NEURAL BASES OF PERCEPTION

Article 4   Knowledge Without Awareness: An Autonomic Index of Facial Recognition by Prosopagnosics
Daniel Tranel and Antonio R. Damasio (1985) ------------------------------------- 31

Article 5   A Face-Responsive Potential Recorded from the Human Scalp
D. A. Jeffreys (1989) --------------------- 37

Article 6   Vision Guides the Adjustment of Auditory Localization in Young Barn Owls
Eric I. Knudsen and Phyllis F. Knudsen (1985) ---------------------------------- 51

3. THE BIOPSYCHOLOGY OF MOTIVATED BEHAVIOR

Article 7   Studies of Instrumental Behavior With Sexual Reinforcement in Male Rats (*Rattus norvegicus*): II. Effects of Preoptic Area Lesions, Castration, and Testosterone
Barry J. Everitt and Pamela Stacey (1987) ----------------------------- 59

Article 8   Escalation of Feline Predation Along a Gradient From Avoidance Through "Play" to Killing
Sergio M. Pellis, Dennis P. O'Brien, Vivien C. Pellis, Philip Teitelbaum, David L. Wolgin, and Susan Kennedy (1988) 77

Article 9   Gastric Emptying Changes Are Neither Necessary Nor Sufficient for CCK-Induced Satiety
Kent L. Conover, Stephen M. Collins, and Harvey P. Weingarten (1989) ---------- 99

4. SLEEP AND CIRCADIAN RHYTHMS

Article 10   Sleep on the Night Shift: 24-Hour EEG Monitoring of Spontaneous Sleep/Wake Behavior
Lars Torsvall, Torbjörn Åkerstedt, Katja Gillander, and Anders Knutsson (1989) ---------------------------------- 111

Article 11 Transplanted Suprachiasmatic Nucleus Determines Ciradian Period
Martin R. Ralph, Russel G. Foster, Fred C. Davis, and Michael Menaker (1990) ----------------------------------123

5. THE BIOPSYCHOLOGY OF LEARNING

Article 12 Female Visual Displays Affect the Development of Male Song in the Cowbird
Meredith J. West and Andrew P. King (1988) ----------------------------------131

Article 13 Caenorhabditis elegans: A New Model System for the Study of Learning and Memory
Catherine H. Rankin, Christine D.O. Beck, and Catherine M. Chiba (1990) ----------------------------------139

Article 14 Social Influences on the Selection of a Protein-Sufficient Diet by Norway Rats (*Rattus norvegicus*)
Matthew Beck and Bennett G. Galef, Jr. (1989) ----------------------------------147

6. THE NEURAL BASES OF MEMORY

Article 15 Human Amnesia and Animal Models of Amnesia: Performance of Amnesic Patients on Tests Designed for the Monkey
Larry R. Squire, Stuart Zola-Morgan, and Karen S. Chen (1988) -------------159

Article 16 Hippocampus and Memory for Food Caches in Black-Capped Chickadees
David F. Sherry and Anthony L. Vaccarino (1989) ------------------------175

Article 17 Spatial Selectivity of Rat Hippocampal Neurons: Dependence on Preparedness for Movement
Tom C. Foster, Carl A. Castro, and Bruce L. McNaughton (1989) -----------------191

Article 18 Organizational Changes in Cholinergic Activity and Enhanced Visuospatial Memory as a Function of Choline Administered Prenatally or Postnatally or Both
Warren H. Meck, Rebecca A. Smith, and Christina L. Williams (1989) ----------199

7. DRUGS, BRAIN, AND BEHAVIOR

Article 19 Alcohol Inhibits and Disinhibits Sexual Behavior in the Male Rat
James G. Pfaus and John P.J. Pinel (1989) ----------------------------211

Article 20 Buprenorphine Suppresses Cocaine Self-Administration by Rhesus Monkeys
Nancy K. Mello, Jack H. Mendelson, Mark P. Bree, and Scott E. Lukas (1989) ----------------------------------223

Article 21 Lesions of the Nucleus Accumbens in Rats Reduce Opiate Reward but Do Not Alter Context-Specific Opiate Tolerance
John E. Kelsey, William A. Carlezon, Jr., and William A. Falls (1989) -----------231

8. CEREBRAL LATERALITY AND COGNITIVE PROCESSES

Article 22 The Corpus Callosum Is Larger with Right-Hemisphere Cerebral Speech Dominance
John O'Kusky, Esther Strauss, Brenda Kosaka, Juhn Wada, David Li, Margaret Druhan, and Julie Petrie (1988) ------243

Article 23 Reading With One Hemisphere
Karalyn Patterson, Faraneh Vargha-Khadem, and Charles E. Polkey (1989) ----------------------------------253

Article 24 Mental Rotation of the Neuronal Population Vector
Apostolos P. Georgopoulos, and Joseph T. Lurito, Michael Petrides, Andrew B. Schwartz, and Joe T. Massey (1989) -283

Preface

These are exciting times for students and teachers of biopsychology; significant biopsychological discoveries are being reported almost daily. This book, Current Research in Biopsychology, is designed to introduce students of biopsychology to some of the most interesting and important of these current findings; it contains 24 outstanding biopsychology journal articles, all published since 1985. Current Research in Biopsychology is designed for use as a supplementary text in undergraduate biopsychology courses (variously titled Biopsychology, Physiological Psychology, Brain and Behavior, Psychobiology, Behavioral Neuroscience, Behavioral Neurobiology).

Contrary to first impressions, Current Research in Biopsychology is not just another book of neuroscience readings. It is a book of "biopsychology" readings that has been carefully tailored to the specific needs and interests of biopsychology students. The following are some of the general approaches and good intentions that guided its preparation.

Emphasizing Behavioral Research

In many biopsychology textbooks, the coverage of neurophysiology, neurochemistry, and neuroanatomy subverts the coverage of behavioral research. This prejudice is particularly obvious in the books of readings commonly used as supplements in biopsychology courses: Many of the readings in these books have precious little direct connection to biopsychological research and issues. In contrast, this book gives biopsychology top billing. It recognizes that neuroscience is a team effort and that the unique contribution made by biopsychologists to this team effort is their behavioral expertise and their interest in the neural bases of behavior—it contains only those neuroscience articles whose behavioral orientation clearly qualifies them as biopsychological.

Focusing on Primary Research Reports

Students of biopsychology receive general overviews of biopsychological research from both their textbooks and their instructors. Current Research in Biopsychology is intended to complement these two indirect sources of research coverage by introducing students to some of the best in recent biopsychology research reports. It is based on the premise that there is no satisfactory substitute for the primary biopsychological research article in the education of biopsychology students.

Increasing the Coverage of Human Research

In recent years, the field of biopsychology has become more human oriented. Current Research in Biopsychology reflects this trend; it includes several studies of human patients and of animal models of human neuropsychological disorders. One of its major themes is that major advances in our understanding of brain-behavior relations are often the result of the convergence of research involving human and nonhuman subjects.

Focusing on the Cutting Edge

Current Research in Biopsychology focuses on the best in recent biopsychological research—all the articles in it have been published since 1985, most since 1989—and all describe major new breakthroughs, ideas, or trends. The focus on current research is designed to complement the more historical treatment provided by most general biopsychology textbooks.

Selecting Articles of Particular Interest to Students

I believe that the principles of good science are best taught in contexts that are of particular interest to the students—too often they are not. All of the articles in this book meet the highest scientific standards, but, in addition, they are all clear concise descriptions of research on topics that, in my experience, are proven student favorites.

Helping Students Learn

The articles in this book were all written by biopsychologists for biopsychologists and other neuroscientists. Accordingly, to fully appreciate these research articles, the beginning biopsychology student may occasionally need some help. I supply this help in five different forms:

Student-Oriented Introductions to Each Article
The introductions that I have written to each article are designed to generate interest, to provide any background material that students will need to understand the article and to tell students in clear nontechnical language what was done and why. The introductions pave the way.

Glossaries
A glossary is included with each article so that students have ready access to the meanings of any new terms that they might encounter. The glossaries are also useful for study and review.

Essay Study Questions
The essay study questions that follow each article are designed to encourage students to think about principles and conceptual issues.

Multiple-Choice Study Questions
Multiple-choice study questions are included to help students review the main points of each article and to prepare themselves for examinations. Answers are included at the end of the text.

Food-for-Thought Questions
The food-for-thought questions that are included with each article are designed to provoke original thought. These questions are excellent topics for classroom discussion.

Topic Areas

In selecting the 24 articles for this book, I tried to choose articles that covered as many different biopsychology topic areas as possible. As a result, I tried to select articles that are integrative: that combine approaches, techniques, and theories from more than one topic area. The following table provides an overview of the breadth and emphases of the chosen 24.

Acknowledgements

First and foremost, I thank the authors of the articles in this book and the publishers of these articles for giving me permission to reprint them here. Many of my colleagues also deserve thanks for recommending articles to me: Michael Baum, Kent Berridge, Robert Blanchard, Brian Bland, David Booth, Rod Cooper, Michael Corcoran, Verne Cox, B.G. Galef, Bill Greenough, Charles Malsbury, Morris Moscovitch, Antonio Nunez, David Olton, George Paxinos, Sergio Pellis, Ron Racine, Neil Roland, Tim Schallert, Doug Wahlsten, Neil Watson, Harvey Weingarten, Ian Whishaw, Don Wilkie, Roy Wise, Steve Woods, Eran Zaidel, and Stuart Zola-Morgan. I am also grateful to my publisher, Allyn and Bacon, and in particular to Diane McOscar, Diana Murphy, Elaine Ober, and Sandi Kirshner, who were largely responsible for designing, producing, and marketing this book. Finally, I would like to thank Liz McCririck, Rose Tamdoo, Bev Charlish, and my son Greg for their valuable clerical support.

Table 1. Topics Covered by the 24 Articles in This Collection

ABBREVIATED TITLES OF ARTICLES	Development	Brain Damage and Recovery	Perception	Motivation	Learning	Memory and Amnesia	Sleep and Circadian Rhythms	Psychopharmacology	Cognition and Language	Brain Laterality	Ethoexperimental Approach	Neuropsychology	Psychophysiology	Human Subjects	Nonhuman Primate Subjects	Avian Subjects	Case Studies	Clinical Implications	Single-Cell Recording
1. Brain Development, Plasticity, and Behavior	x	x			x							x	x					x	x
2. Reorganization of Retinotopic Cortical Maps in Adult Mammals		x	x															x	x
3. Autograft of Adrenal Medulla to the Caudate Nucleus of Patients With Parkinson's Disease		x										x	x	x				x	
4. Knowledge Without Awareness		x	x			x						x	x	x				x	
5. Face-Responsive Potential Recorded from the Human Scalp		x											x	x				x	
6. Vision Guides the Adjustment of Auditory Localization in Young Barn Owls	x		x								x					x			
7. Studies of Sexual Reinforcement in Male Rats		x		x															
8. Feline Predation Along a Gradient				x							x								
9. Gastric Emptying Changes and CCK-Induced Satiety				x							x								
10. Sleep on the Night Shift: 24-Hour EEG Monitoring							x						x	x					
11. Transplanted Suprachiasmatic Nucleus Determines Circadian Period		x					x												
12. Female Visual Displays Affect the Development of Male Song in the Cowbird	x			x	x						x					x			
13. Caenorhabditis elegans: A New Model System for the Study of Learning and Memory					x														
14. Social Influences on the Selection of a Protein-Sufficient Diet by Norway Rats	x			x	x						x								
15. Performance of Amnesic Patients on Tests Designed for the Monkey		x				x						x		x	x			x	
16. Hippocampus and Memory for Food Caches in Black-Capped Chickadees		x				x					x					x			
17. Spatial Selectivity of Rat Hippocampal Neorons						x													x
18. Cholinergic Activity and Enhanced Memory After Early Choline Administration	x					x		x										x	
19. Alcohol Inhibits and Disinhibits Sexual Behavior in the Male Rat				x				x										x	
20. Buprenorphine Suppresses Cocaine Self-Administration by Rhesus Monkeys				x				x							x			x	
21. Lesions of the Nucleus Accumbens Reduce Opiate Reward but Not Context-Specific Opiate Tolerance		x		x		x												x	
22. Corpus Callosum is Larger with Right-Hemisphere Cerebral Speech Dominance									x	x		x		x				x	
23. Reading With One Hemisphere									x	x		x		x			x	x	
24. Mental Rotation of the Neuronal Population Vector			x						x						x		x		x

The Next Edition

As I write these words, Current Research in Biopsychology is not yet in press, but I am already searching for new articles for its next edition—I am planning to update Current Research in Biopsychology frequently to maintain its focus on the cutting edge of biopsychological research. I would greatly appreciate receiving your recommendations, and I would be pleased to acknowledge them in the preface of the next edition. Please send your recommendations to me at the Department of Psychology, University of British Columbia, Vancouver, B.C., Canada V6T 1Y7.

To the Student

If you have already skimmed the preface and the table of contents of this book, you will already know that it contains 24 research articles recently published in scientific journals by some of the world's best biopsychologists. For many of you, this book may constitute your first direct exposure to the primary products of biopsychology research—most of your previous exposure will have been second or third hand through your teachers and general textbooks. I can understand why you now might be regarding this book with a feeling of apprehension—it is hard to imagine how reading 24 articles that were written by scientists for other scientists could be anything other than a frustrating experience for a student. But relax, this book comes with a guarantee: I guarantee that you will find Current Research in Biopsychology to be clear, informative, and interesting—a truly worthwhile and enjoyable educational experience.

You may wonder how I would dare issue such an unqualified guarantee—and in writing yet. I feel confident in doing so because the 24 articles in this book all passed three separate tests. In addition to meeting the highest standard of scientific merit, each article is clearly and concisely written with a minimum of scientific jargon, and each article focuses on a topic, that, in my experience as a teacher, is likely to be of great interest to you. Today's biopsychologists are doing amazing things, and the articles in this book describe some of the most amazing—you are about to enter a world of brain transplants, microscopic worms that are capable of learning, birds with amnesia, drugs that improve memory, and people with only half a brain. Furthermore, if you do happen to encounter some minor difficulties along the way, I will be there to help you. To guide you through each article, I have written a stage-setting introduction, a glossary of technical terms, essay study questions, multiple-choice study questions, and food-for-thought questions.

Because my job as a producer and teacher of biopsychological research has meant so much to me, I have done everything that I can to make this introduction to the world of biopsychological research a positive one for you. I do hope that you enjoy it and that it whets your appetite for more biopsychology. If you are so inclined, please write to me at the Department of Psychology, University of British Columbia, Vancouver, B.C., Canada, V6T 1Y7. I welcome your comments, suggestions, and questions.

John P.J. Pinel

John P.J. Pinel is a graduate of McGill University and a Professor of Psychology at the University of British Columbia in Vancouver, B.C., Canada. He is the author of well over 100 research reports, review articles, and book chapters on a variety of biopsychological topics: including learning, memory, epilepsy, defensive behavior, and drug tolerance. In recognition of his research accomplishments, Professor Pinel has been made a fellow of both the American Psychological Association and the American Psychological Society, and he is a recent winner of the prestigious Killam Memorial Senior Fellowship, but his contributions to psychology have not been restricted to the laboratory. He is an award-winning teacher, and his recent textbook Biopsychology (Allyn and Bacon, 1990) has been acclaimed both for its clear, insightful, forward-looking coverage of the field of biopsychology and for its effective blend of humor, enthusiasm, and scholarship. Professor Pinel claims that he pauses from his rigorous writing schedule only long enough for the occasional afflatus.

ARTICLE 1

Brain Development, Plasticity, and Behavior

B. Kolb

Reprinted from American Psychologist, 1989, September, 1203-1212

Kolb and his collaborators—most notably Whishaw and Sutherland—have recently published a series of experiments that have delved into the adverse effects of early brain damage on behavioral and neural development. In this article, Kolb summarizes the results and implications of these experiments.

By blending the results of human neuropsychological case studies with the results of laboratory experiments, Kolb makes a strong case for the need to qualify the Kennard doctrine: the widely held view that recovery from brain damage is greatest when the damage occurs early in life. Kolb begins by pointing out that the Kennard doctrine has little to say about why the deficits produced by brain damage lesions occurring at one age are sometimes qualitatively different from the deficits produced by similar brain damage occurring at another age. The human nervous system continues to develop for several months after birth, and Kolb argues that the particular effects of early brain damage depend to a substantial degree on the stage of neural development that is in progress at the time that the damage occurs. The Kennard doctrine also fails to explain how the development of the brain and its recovery from brain damage are influenced by experience. For example, it cannot explain Kolb and Elliot's (1981) observation that rats with frontal cortex lesions experienced fewer deficits as adults if they were raised in an enriched homecage environment.

The most serious challenges to the Kennard doctrine come from the numerous demonstrations of cases in which greater recovery is not associated with earlier brain lesions. For example, Kolb and Whishaw (1981) found that frontal cortex lesions permanently impaired the performance of a variety of species-typical behaviors (e.g., the defensive burying of dangerous inanimate objects) in adult rats whether the lesions were created when the rats were 7 days old or when they were young adults. In contrast, the recovery of the ability of the same two groups of rats to perform various tests of learning (e.g., Morris water maze) was consistent with the Kennard doctrine; the recovery was considerably greater in the rats that had received their lesions when they were 7 days old. Clearly, the Kennard doctrine is too simple; it can account for some instances of recovery from brain damage but not others.

Recently, Kolb and Whishaw found evidence that recovery from early cortical damage might be mediated by dendritic arborization (i.e., by the growth of dendritic branches) in the remaining healthy cortex. In general, behavioral recovery from brain damage was found to be greatest in those experimental conditions that favored dendritic arborization.

Science Watch

Brain Development, Plasticity, and Behavior

Bryan Kolb   *University of Lethbridge*

ABSTRACT: Damage to the infant brain is associated with a complex array of behavioral and anatomical effects. Recent research is leading to a new understanding of the nature of, and mechanisms underlying, recovery from brain damage.

Nearly one half million people will suffer traumatic brain injury in the United States alone this year. When one adds the people who will suffer a stroke, develop dementing disorders, or suffer from other types of brain dysfunctions such as mental retardation, cerebral palsy, or epilepsy, it becomes clear that there are a large number of people who have permanent behavioral abnormalities that may include disorders of movement, perception, or memory; loss of language; and the alteration of social behavior and personality. Thus, whereas the study of brain–behavior relationships was once restricted largely to physiological psychologists, the development of neuropsychology has moved the study of brain and behavior into the mainstream of psychology to involve significant numbers of human experimental and clinical psychologists. One problem, however, is that most of the basic work in neuroscience is largely divorced from psychology and is inaccessible to the bulk of psychologists who quite rightly have difficulty in seeing the direct relevance of this work to psychological issues. My goal in this article is to review recent work on the nature of brain development and plasticity and its relation to the understanding of behavior.

There are numerous approaches to the study of brain–behavior relationships. The first is to study how normal mature brains work. This can be done either by examining the morphological and physiological correlates of behavior or by studying the structure of cognitive processes and making predictions about how the brain must be processing information. Studies of morphological changes during learning provide an example of the former type, and psychophysical experiments provide examples of the latter. A second approach is to study the behavioral correlates of brain dysfunction, with the goal of making predictions about normal function. This has been the principle method of neuropsychology for over 100 years. A third approach is to study the manner in which the brain and behavior normally develop, with the hope of gaining insight into both how the brain comes to produce behavior and how one might gain control of the processes of development. In the latter case it is proposed that it might be possible to re-initiate developmental processes to repair injury. Indeed, because both fish and amphibian brains can do this after injury and some birds annually regrow structures necessary for song each spring, it is possible that under certain conditions, hormonal events may re-initiate neural growth in mammals. A final approach is to alter the brain during development in order to see how the anatomical and behavioral organization changes. This allows an opportunity not only to look at the processes involved in brain development but also to try to determine what rules can predict when restitution of function is likely to occur and what anatomical changes might correlate with behavioral recovery. Furthermore, this approach allows one to look at the nature of localization of functions in the brain, which is an issue that has fascinated psychologists since the time of Gall. It is this final approach that I wish to examine in detail, and I will begin with an illustrative example.

Consider the following case histories of two young women. The first, P.B., is a 22-year-old business school graduate who was struck by a car and suffered a serious head injury, requiring emergency surgery to repair her skull and to relieve the pressure from subdural bleeding. It was necessary to remove a large portion of her right posterior temporo-parietal cortex. After the accident she had a left visual field defect but was able to return to her job as a typist/clerk. Upon neuropsychological examination six years after the injury, she obtained an average IQ score, although she was relatively better at verbal tests than those requiring manipulation of pictorial information. She had particular difficulty drawing and remembering pictorial information, including faces. Her motor skills were good, and although she initially had difficulty reading because of the visual loss, she overcame this handicap and could read as well as IQ-matched controls. The second case, S.S., is an 18-year-old woman who had a difficult birth and forceps delivery and began having epileptic seizures at 14 years of age. Neurological examination revealed a right parietal cyst; this was removed surgically, and the seizures were arrested. She was an average student in school but had difficulty in 12th grade, especially with English and mathematics. Her neuropsychological assessment at age 18 revealed an average IQ, but she was relatively better at pictorial tests than verbal ones, which is in direct contrast to P.B. Furthermore, she had a poor vocabulary score considering her education, IQ, and socioeconomic group, and she had a difficult time with arithmetic. She also had difficulty in repeating sequences of movements shown to her by the examiner, especially those of the face, and had difficulty on tests that are typically sensitive to frontal-lobe injury. In contrast, she had no difficulty on tests of drawing or visual memory. In short, P.B. and S.S. had similar brain damage, but at different ages, and the consequences could not have been much more different. I note, parenthetically, that because P.B. had a closed head injury, one might expect some nonspecific damage, such as tearing of connections or bruising elsewhere in the brain, in addition to her focal lesions. Her symptoms were typical of patients with vascular lesions in adulthood, however, and there was some evidence of nonspecific damage on tests of interhemispheric transfer (see Kolb & Whishaw, 1985, Chapter 16, for more examples).

Several questions arise from these two cases. If functions are localized in the cortex, why are the symptoms different when the damage included the same tissue? Why did S.S. have symptoms typical of frontal-lobe injury when there is no evidence of any damage to her frontal lobe? Was the age at which brain damage was sustained responsible for her behavioral differences? Why was there a permanent loss of functions in both cases, even with years of recovery? How did the function of the remaining tissue in the two brains change after the injuries? Were the changes the same? I return to these questions later.

Brain Development

One of the wonders of human development is the manner in which the human brain, which consists of over 100 billion neurons, can develop so quickly from just a few initial neural cells. According to Cowan (1979), during the time the brain is growing in utero it must be generating neurons at a rate of more than 250,000 neurons per minute. Furthermore, once the neurons are "born," they must move to their correct locations and form connections, which have been estimated at up to 15,000 per neuron.

The gross development of the human brain is summarized in Figure 1, but these general morphological changes provide little insight into the details, most of which have been discovered in the last two decades by studies on laboratory animals, especially rats and monkeys. It is now known that the development of the cortex in any species occurs in several stages (e.g., Cowan, 1979; Rakic, 1988). These include cell proliferation, cell migration, cell differentiation, dendritic and axonal growth, cell and axonal death, and gliogenesis. I consider these stages briefly.

Like all mammalian brains, the human brain begins as a hollow tube and gradually develops the features of

---

I wish to thank J. Vokey and L. DeLude for comments on an earlier version of this article.

Correspondence concerning this article should be addressed to Bryan Kolb, Department of Psychology, University of Lethbridge, Lethbridge, Alberta, Canada T1K 3M4.

the adult brain. The hollow area in the tube forms the ventricular system, and the cells of the brain are generated along the ventricular wall and then migrate out to their proper locations. As the brain develops, the newly formed cells must travel farther and farther to reach their final locations. As might be predicted, the precise timing of the development and migration of cells to different cortical regions varies with the particular area in question. Once cells find their correct location in the cortex, they develop the characteristics of the cell type that they are to be (e.g., stellate or pyramidal cell) and begin to grow their dendrites and axons and to form synapses. One particularly interesting aspect of neural development is that the brain overproduces neurons, possibly by a factor of two, and the extra cells are lost by a process of cell death. Similarly, a large proportion of the cortical synapses are lost during development, perhaps as many as 50%. This cell and synaptic loss is probably not random, although the controlling factors are still unknown. Curiously, it has been suggested that a failure of cell death or synaptic loss may lead to retardation or contribute to the emergence of developmental disorders, possibly even schizophrenia (e.g., Feinberg, 1982).

It has been possible to determine the timetable for many of these stages by labeling cells with various tracers. For example, thymidine is a compound that is incorporated into cells only during cell division. If a radioactive isotope is attached to the thymidine, the radioactivity will be detectable later only in those cells that were exposed to the thymidine during their mitosis. Cells born before or after this time will not be labeled. By labeling cells at different points in development, it is possible not only to chart the time of birth of cells but also to track their route during migration (see Figure 2). Thus, the thymidine technique has shown that cells that form the innermost layers of the cortex are born first, followed by those in the external layers. One consequence of this arrangement is that newly produced cells must migrate through the existing layers to reach their correct locations. It is also known that all cells forming a particular layer in a particular region of the cortex proliferate and migrate at the same time. Thus, brief prenatal events (e.g., drugs, toxins, or stress) that jeopardize developing cells could lead to the development of a brain without a particular cell group, the anomaly depending upon the precise timing of the prenatal event.

Detailed studies of brain development have shown that in most altricial mammals such as rats, cats, monkeys, or humans the stages of cell proliferation and migration are largely prenatal and much of the development of neuropil (axons and dendrites) and cell death are postnatal. Thus, most mammals are born with practically a full complement of neurons, and few, if any, neurons are born postnatally. One exception is the hippocampus, which continues to develop neurons throughout the life of some species. In general, however, if the brain is damaged after cell proliferation has ceased, it is obvious that any compensation will have to be accomplished by changes in the remaining cells. The fact that the growth

Figure 1
The Development of the Human Brain

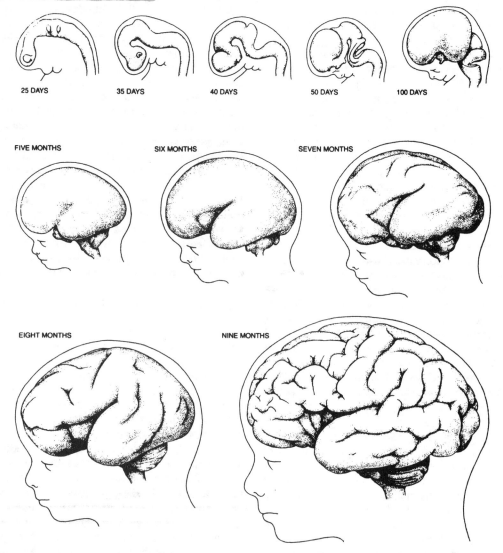

Note. Adapted from "The Development of the Brain" by W. M. Cowan. In *The Brain,* (p. 59), 1979, San Francisco: Freeman. Copyright 1979 by W. H. Freeman. Adapted by permission.

of neuropil and the loss of cells is postnatal is important for it is obvious that the extrauterine environment could have a direct influence on these processes.

Consider the following analogy. If one were to make a statue, it would be possible to do it either by starting with grains of sand and glueing them together to form the desired shape or by starting with a block of stone and chiseling the unwanted pieces away. The brain uses the latter procedure. The "chisel" in the brain could be of several forms including genetic signal, environmental stimulation, gonadal hormones, stress, and so on. Similarly, the same processes are likely to affect the development of dendrites, axons, and synapses. Cell death and neuropil development do not end in infancy but rather may continue well into adolescence. For example, it appears that cell death continues in the human frontal lobe until about 16 years of age (Huttenlocher, 1979). I should note here that cell death continues at a greatly slowed pace throughout adult life, but it is unclear what role environmental events may play in this process.

One example of the effect of environmental stimulation on brain development comes from the work of Janet Werker and Richard Tees (1984). They studied the ability of infants to discriminate phonemes taken from widely disparate languages such as English, Hindi, and Salish. Their results showed that infants can discriminate speech sounds of different languages without previous experience, but there is a decline in this ability, over the first year of life, as a function of specific language experience. One might speculate that neurons in the auditory system that

Figure 2
Migration of Cells

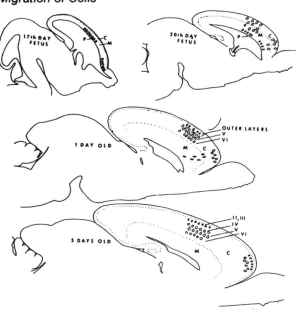

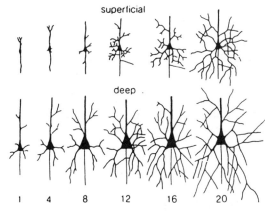

Note. The top panel is adapted from "Cell Migrations to the Isocortex of the Rat" by S. P. Hicks and C. J. D'Amato, 1968, *Anatomical Record, 160,* p. 621. Copyright 1968 by Allen R. Liss, Inc. Adapted by permission.

The top panel is a diagram of labeled cells that originated in the 17th (ovals) or 20th (black) intrauterine day in the rat brain. Notice that migration continues after birth.

The bottom panel is adapted from "Cellular Differentiation: Development of Dendritic Arborizations Under Normal and Experimentally Altered Conditions" by M. Berry, 1980, *Neurosciences Research Program Bulletin, 20,* p. 456. Copyright 1980 by MIT Press. Adapted by permission.

The bottom panel is a schematic illustration of the rate of acquisition of basal and apical dendrites of the pyramidal cells of the superficial and deep layers of the cerebral cortex of the rat.

are not stimulated early in life may be selected against and die, although there are other explanations.

The postnatal stages of brain development provide not only an explanation for how early experience could affect later behavior but also why brain damage at different ages might have very different effects and lead to different cerebral organizations.

Behavioral Effects of Early Brain Damage

In the 1930s Margaret Kennard was studying the effects of cortical lesions on motor performance in monkeys and reported the provocative finding that lesions in infant monkeys had less severe effects on behavior than similar lesions occurring in adulthood. This claim led to the development of the Kennard doctrine (1936), which stated that the earlier one suffers brain damage, the less severe the behavioral loss. This view was reinforced by Lennenberg (1967), who reviewed the effects of cortical lesions on language in children and concluded that left hemisphere damage in the first few years of life allowed substantial recovery of language processes, presumably because of some sort of cortical reorganization. By the mid 1970s, the Kennard doctrine was being challenged on the basis of studies using a variety of laboratory animals including monkeys, cats, and rats. For example, Patricia Goldman and her colleagues began to find that early cortical injury in monkeys was not always advantageous; the outcome depended on when behavior was assessed, the type of behavioral test employed, and the sex of the animal (e.g., Goldman, 1974). Indeed, even Kennard's experiments recently failed replication when redone using more modern behavioral methods (Passingham, Perry, & Wilkinson, 1983). Similarly, although the initial studies done by me and my colleagues on the effects of cortical lesions on infant rats supported the Kennard doctrine, we fell victim to the general principle of science that the more one studies something, the smaller the effects become (e.g., Kolb, 1987)!

Studies using children have also failed to support either Kennard's or Lennenberg's findings. Although there clearly is recovery of language after early left hemisphere lesions to the language areas and there is evidence that the language zones can shift either to the other hemisphere, if the damage is in the first two years of life (or within the left hemisphere, if the damage occurs between two and five years), children with early brain injuries suffer significant cognitive loss. For example, Woods (1980) found that the IQs of children with brain damage in the first year of life were well below average (WISC-R = ~85) as well as below those of children who suffered brain damage later in life. Similar results have been found by others, and it now appears that the period of severe IQ loss may extend as late as four to five years of age. In sum, there is little support for Kennard's original conclusions. Nonetheless, we have now identified a variety of factors that influence the outcome of early brain injury, and there is some evidence that a limited version of Kennard's principle may have some validity.

Behavioral Recovery in Rats

Over the past 15 years my colleagues and I have removed virtually every region of the rat's neocortex and have devised a neuropsychological test battery for the rat that is conceptually similar to that used for people. In contrast to the "rat studies" of the 1930s–1960s, which assumed that a rat should be used for just a single experiment, our test battery assumes that the best way in which to study behavior in any species is to administer multiple measures of many aspects of behavior including both learned and species-typical behaviors (e.g., Kolb, 1984; Whishaw, Kolb, & Sutherland, 1983), as summarized in Table 1.

Contrasting models. To simplify the following discussion, I will focus on two of the preparations that we have used: (a) the hemidecorticate preparation and (b) the bilateral frontal preparation. In the hemidecorticate preparation, the entire neocortical mantle of one hemisphere is removed at different ages. This procedure is especially interesting in that it parallels the surgical procedure used in the treatment of children with major injuries restricted to one hemisphere, for which there is virtually no empirical basis in the primate literature (but see Villablanca, Burgess, & Sonnier, 1984, for parallel studies in kittens). In the bilateral frontal preparation, the frontal cortex that is analogous to the prefrontal and anterior cingulate region of primates is removed at different ages. We chose this preparation because the bulk of the work on infant lesions in primates has been done in animals with frontal lesions and because more is known about the frontal cortex of the rat than any other region of the rodent cortex.

We have varied the age at lesion and found the behavioral and anatomical effects to vary with age at insult (Kolb, Holmes, & Whishaw, 1987; Kolb, 1987; Kolb & Tomie, 1988). Such a finding would be expected given the different stages of development that would be interrupted (e.g., see Figure 2). Thus, lesions on the first day of life, which we call postnatal day 1 (P1), perturb a brain in which cell migration is not yet complete and in which neuropil development has barely begun. In contrast, lesions on postnatal day 5 or 10 (P5, P10) affect a brain in which there is active development of neuropil. In human terms, the P10 lesions would be well into the first year, or even later.

Following adult rat hemidecortication there is a loss, or impairment, in a wide variety of behaviors. For example, there is a significant impairment in control of the limbs contralateral to the removal, a reduced ability to respond to stimuli contralateral to the lesion, and a deficit in most learning tasks requiring visuospatial guidance. Neonatal hemidecortication produces parallel deficits although the earliest operates (P1) were less impaired than adult operates. This advantage was not present in rats operated at P10. An example will illustrate.

Richard Morris (1981) devised an ingeneous test of spatial navigation in the rat in which an animal is placed in a large tank of water. The water is made opaque by a small amount of skimmed milk powder, and there is a hidden platform that the rat must locate to escape from the water (see Figure 3). Because rats are excellent swimmers, they need little training to learn the location of the platform on the basis of visual cues in the environment. Rats hemidecorticated as adults are impaired at this task, however, even if preoperatively trained. Although P1 hemidecorticates are also impaired when tested as adults, they are significantly superior to rats with P10 or adult lesions (see Figure 4). Parallel results are obtained from other measures, such as of motor abilities. Thus, for the hemidecorticate there is evidence that something like the Kennard principle is operating.

The effect of frontal lesions is different, however. Rats with bilateral frontal lesions at P1 are not only worse at most behavioral tasks than animals with lesions at P5, P10, P25, or adulthood, but they are also impaired on tasks at which the P10 and older animals are not. In short, they are much like those children with early lesions who have low IQs and poor recovery. The Morris water task again provides a good example. In contrast to rats with restricted lesions in other neocortical regions, rats with frontal lesions are slow to master this task. Rats with lesions at P1 are truly incapable of completing this task and never learn where the platform is. Surprisingly, however, rats with lesions at P10 are virtually indistinguishable

Table 1
Behavioral Assessment of the Rat: A Partial Summary of Features of Behavior for Examination

Measure	Specific feature
1. Appearance	Body weight, core temperature, eyes, feces, fur, genitals, muscle tone, pupils, responsiveness, saliva, teeth, toenails, vocalizations
2. Sensory and sensorimotor behavior	Response to stimuli of each sensory modality presented both in home cage and in novel place such as open field
3. Posture and immobility	Behavior when spontaneously immobile, immobile without posture or tone; tonic immobility or animal hypnosis; environmental influences on immobility
4. Movement	General activity, movement initiation, turning, climbing, walking, swimming, righting responses, limb movements in different activities such as reaching or bar-pressing, oral movements such as in licking or chewing, environmental influences on movement
5. Species-typical behaviors	All species-typical behaviors such as grooming, food hoarding, foraging, sleep, maternal or sexual behavior, play, and burying
6. Learning	Operant and respondent conditioning, and learning sets, especially including measures of spatial learning, avoidance learning, and memory (for details, see Whishaw et al., 1983)

Figure 3
Illustration of the Morris Water Task

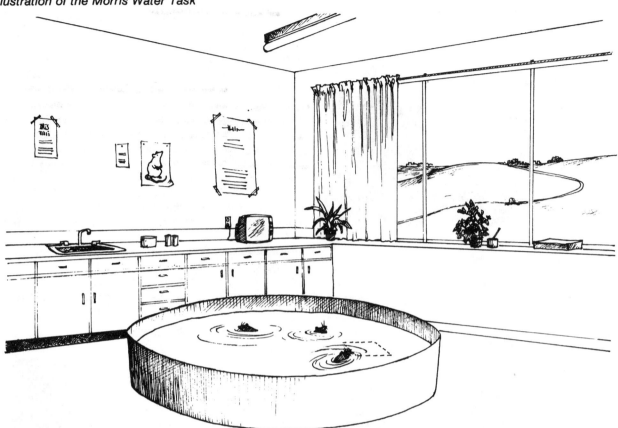

Note. Adapted from "Plasticity in the Neocortex: Mechanisms Underlying Recovery From Early Brain Damage by B. Kolb and I. Q. Whishaw, 1989, *Progress in Neurobiology, 32,* p. 242. Copyright 1989 by Pergamon Press. Adapted by permission.
The rat's task is to locate a submerged, hidden platform by using visuospation cues available in the room.

from control animals. If the lesions are made later in life, the deficit appears, although we do not know the exact age at which it occurs. Nonetheless, in contrast to rats with hemidecortications, in which the earliest lesions allowed the best recovery, in the frontal preparation there is a window of time around 10 days of age in which the Kennard principle holds. Similar results can be shown for other behavioral tests (e.g., Kolb, 1987).

The contrasting effects of hemidecortication and frontal lesions are important because they have provided a behavioral marker that we can use to look for an anatomical correlate. Thus, the task is to find an anatomical change that occurs in P1 hemidecorticates and P10 frontals and that correlates with the Kennard effect, and an absence of this change (or possibly the onset of other changes) that correlates with the poor behavioral performance of the P1 frontal animals. Before describing such a correlate, I will consider some factors that influence recovery.

Factors Influencing Recovery

It has long been known that in different people the severity of symptoms resulting from the same brain damage varies considerably. This is presumably because of differences in a number of variables such as handedness, IQ, and personality. We have searched for modulating factors other than age that might influence the effect of neonatal (or adult) lesions in rats and have identified a large number, including sex, environmental experiences, size of brain lesion, nature of the behavioral test used, age at behavioral testing, and the level of endogenous cortical norepinephrine (e.g., Kolb & Elliott, 1987; Kolb & Sutherland, 1986; Kolb & Whishaw, 1981; Sutherland, Kolb, Whishaw, & Becker, 1982). I will briefly describe two examples.

Behavioral test. It is typical in neuropsychology to use tests of learned habits to assess behavior. When we began our studies, we also emphasized these measures, but as we expanded our tests, we were surprised to find little correspondence between the performance on tests of learned behaviors and measures of species-typical behavior such as hoarding or nest building. An example will serve to illustrate. John Pinel and his colleagues (e.g., Pinel & Treit, 1983) devised a test of natural avoidance behavior in rats in which a noxious stimulus, such as an electrified prod, was introduced into a rat's living quarters.

Figure 4
Comparison of the Effects of Hemidecortication or Frontal Lesions on Performance on the Morris Water Task

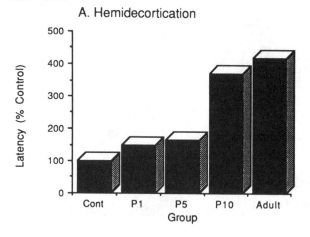

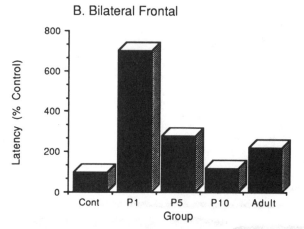

Note. The data are graphed as a percentage of control performance. All animals were tested as adults. Con = control. Ad = adult. P1 = surgery on postnatal day 1. P5 = surgery on postnatal day 5. P10 = surgery on postnatal day 10.

Pinel found that rats, being curious, approached this stimulus but once they were "stung" by it, and recovered from the immediate startle of the experience, they began to bury the stimulus. This highly reliable behavior proved simple to quantify with measures such as the depth of sawdust piled upon the electric prod. When we used Pinel's test with adult rats with frontal lesions, we found that they failed to bury the prod at all, although they clearly learned that the prod was noxious because no animal was shocked a second time; they simply avoided it. When rats with P7 lesions were tested on this task, they too failed to bury the prod. In contrast to the adult operates, however, the P7 animals showed nearly normal performance on several tests of learned behaviors, including delayed response, active avoidance, and spatial reversal learning. Further study showed that the animals that were nearly normal on a variety of tests of learned habits were as impaired at a variety of species-typical tests as were animals with adult lesions (e.g., Figure 5). Thus, it is evident that there is limited generality to the Kennard effect, even when it is present on some tests.

Environmental effects. Because it is now well accepted that environmental stimulation has major effects upon the neocortex (e.g., Greenough, 1986), it is reasonable to predict that some environmental conditions may influence the extent to which lesions influence behavior. In particular, we predicted that if animals with early lesions were housed in complex environments in which they were given considerable opportunity to move about and to explore a frequently changing world, they would show better recovery than those who were raised in standard laboratory cages. This proved to be the case: In contrast to the large impairments in isolated rats, enriched animals with P1 or P5 frontal lesions were significantly better at nearly all behaviors that we assessed, and in some cases the P5 animals were nearly as proficient as

Figure 5
Comparison of the Effects of Frontal Lesions at Postoperative Day 7 on the Performance of the Morris Water Task and the Pinel Burying Task

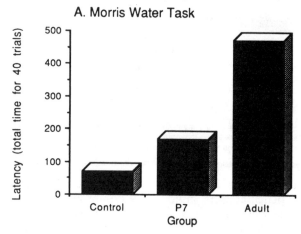

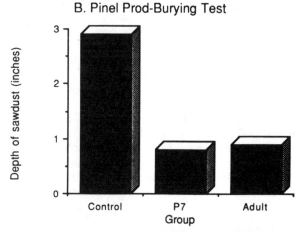

Note. The same animals who showed sparing of function on the learned task showed no sparing on the species typical test.

control animals (Kolb & Elliott, 1987). For example, when tested in the Morris water task the animals with P5 lesions performed as well as our P10 animals. Furthermore, those with P1 lesions, although still seriously impaired, performed as well as adult operates. In sum, we again showed that the Kennard effect is flexible and is not simply a function of the time of surgery, but may also interact with other factors.

Anatomical Effects of Early Brain Injury

As we began to search for anatomical correlates of our behavioral findings, we were immediately struck by one observation: Animals with early lesions had visibly smaller brains, even when lesion size was held constant (e.g., Kolb, Sutherland, & Whishaw, 1983). This led us to measure the brain and its parts in order to find the cause of this small brain syndrome. Although we found many changes in the brain, we were most impressed with the changes in the thickness of the remaining cortex. This was intriguing because these changes correlated with behavior. For example, hemidecortication at P1 led to a thicker cortex than similar lesions at P10. The problem was to find the cause of the variability in cortical thickness. There seemed to be several obvious places to look.

First, others have shown that unilateral cortical lesions, including hemidecortication, lead to the development of abnormal connections. Perhaps lesions at different ages produce major changes in cortico-cortical or cortico-subcortical connections. An increase or decrease in the number of connections would not only affect the number of fibers in the cortex but also the number of synapses. This could be reflected in changes in cortical thickness. Second, perhaps the early lesions differentially affect cell death such that the cortex is thicker or thinner depending on the number of neurons present. Third, there could be a difference in the development of myelin in the cortex. Because myelin takes considerable space, this too could affect cortical thickness. Fourth, it is reasonable to suppose that the lesions affect neuropil development, independent of gross changes in connections. Thus, there could be significant changes in the extent of dendritic or axonal arborization, and these changes could contribute to cortical thickness. Finally, because it is known that strokes in humans lead to changes in vascular flow and in metabolic activity, it is reasonable to suppose that these too are changed by lesions in rats and may be changed differently at different ages.

We began by looking at the connections of the cortex. There has been a revolution in techniques of neural tracing in the past decade. When injected in very small quantities (e.g., .05 µl) into the brain, various compounds (amino acids, proteins, or dyes) are transported either to the cell body from the terminal field of an axon (retrograde transport) or to the terminal field from the cell body (anterograde transport). Using such techniques, it has proven possible to demonstrate numerous, previously unknown, connections in the brain. Thus, there is a new understanding of the principles of cortical connectivity, and function (e.g., Pandya & Yeterian, 1985). In addition, this new technology has provided an opportunity to look for changes in normal connectivity, changes that might underly behavioral recovery. Although we were able to replicate many of the peculiarities in rewiring that others already had observed (e.g., Kolb & Whishaw, 1989), as well as finding many others, these changes could not account for the large changes in cortical thickness, especially in the frontal operates. The biggest changes in cortical connectivity were seen in new connections in the animals with the earliest lesions, but these animals also had the thinnest cortices. Furthermore, it seemed unlikely that factors that affected behavior, such as environment or sex, had any effect on the development of anomolous connections.

Our examination of cell numbers was equally unrevealing for there were simply no differences in the number of cells in a given column of cortical tissue. In retrospect, this finding might have been anticipated because there are no differences across mammalian species either, even though there are up to twofold differences in cortical thickness in different species (Rockel, Hiorns, & Powell, 1980). Thus, if mice and humans have the same number of cells in cortices that are wildly different in thickness, we ought not be surprised when treatments fail to affect the numbers within a species. Furthermore, within a species the number of cells in different cortical areas is constant even though cortical thickness varies considerably. It appears as though there is a genetic signal for the number of cells in a column of cortical tissue. It is the neuropil that varies from area to area and species to species.

When we examined myelin, we did find abnormalities, but they were idiosyncratic from animal to animal and seemed unlikely to allow us to make reliable correlations with behavior. Finally, we began to study the neuropil, and it was here that we began to find correlations with our behavioral data.

Dendritic arborization. The neuropil of cortical neurons is composed of the dendritic and axonal branches of neurons. The dendritic arbor can be visualized by using a Golgi stain in which a small percentage of cortical neurons and their dendrites and unmyelinated axons are stained. Although the reason for the incomplete staining of cortical neurons is still unknown, this property makes it possible to draw neurons and all their processes, and then to quantify them in a number of ways. This procedure is not adequate for axons, however, in part because only the unmyelinated portions are well stained and in part because axons are fine and difficult to identify. Axonal staining must be done using a different procedure. To date, we have studied only dendritic arbor, and so my discussion will be restricted to this.

Our major finding was that changes in the dendritic arbor correlate with changes in behavior. Thus, hemidecortication at P1 and frontal lesions at P10 both increase dendritic arbor in the intact cortex relative to littermate control animals (Kolb & Whishaw, 1989). In contrast, frontal lesions at P1 decrease dendritic arbor in the intact cortex (Figure 6). In other words, in those in-

Figure 6
Effects of Frontal Lesions at P1 and P10

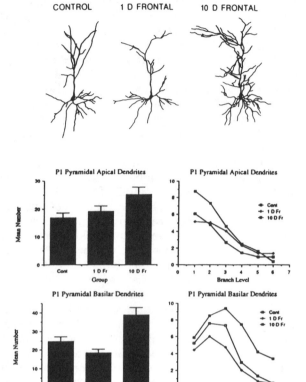

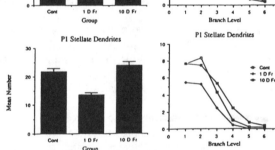

Note. The top panel shows examples of layer 11/111 pyramidal cells from the parietal cortex of a normal rat and a rat with a P1 frontal lesion in one hemisphere and a P10 frontal lesion in the other hemisphere.

The bottom panel shows the quantification of the differences in dendritic arbor of layer 11/111 pyramidal cells and stellate cells in parietal cortex. The apical dendrites are those that extend directly upward on the main dendritic shaft, whereas the basilar dendrites are those that originate from the cell body. The total number of branches is summarized on the left, and the distribution of the differences in different branches is summarized on the right.

stances in which we have evidence of at least a limited Kennard effect we have evidence of increased dendritic arbor. Similarly, in the case of worsened behavior, we have decreased dendritic arbor. Furthermore, we know from the work of others, as well as from our own preliminary observations, that dendritic arbor is affected by environmental conditions and by norepinephrine levels, and both factors influence the Kennard effect. As might be expected, however, there are some difficulties that must be considered.

First, the changes in dendritic arborization are not found everywhere in the cortex. Why not? Second, although the increase in dendritic arbor in the P10 frontals is equivalent to that in the P1 hemidecorticates, the cortical thickness is by no means comparable. The P10 frontals still have a significant thinning of the cortex relative to control animals. Why? Third, are the same factors influencing the changes in dendritic arbor in the hemidecorticate and frontal animals? Fourth, how do changes in dendritic arbor translate into changes in behavior? It is reasonable to argue that if there is an increase in dendritic space, there will be a corresponding increase in the number of synapses. Indeed, this appears to be true in other experimental conditions that affect dendritic arbor, such as environmental enrichment. If there are increased synapses, then we must wonder *what* is synapsing? In sum, we believe that we have made meaningful anatomical correlations of behavioral recovery, but we do not yet understand the mechanism. Future work will have to determine both the mechanism and the variables that affect it.

Generalizations and Conclusions

It is now possible to reach several conclusions regarding the nature of brain plasticity and recovery from early brain injury. First, it is clear that early brain injury has different anatomical and behavioral effects depending on the nature of the injury, the age of the brain at injury, and the presence of various modulating factors such as the environment or the levels of endogenous transmitters or hormones. Second, it is difficult to make simple generalizations about whether functions are spared after early lesions. Different measures of behavior yield different results. Third, although the relationship between the anatomical and behavioral changes is complex, there appears to be a strong correlation between dendritic arborization and behavioral recovery.

The rat has proved to be a convenient and useful model for use in studying the effects of early cortical lesions. It is reasonable to wonder, however, if this work has any generalizability to humans. I believe it does. First, we have already seen in our comparison of two cases with right parietal injury that early lesions in humans can produce paradoxical behavioral effects. Of particular interest in this regard is the brain of Case S.S., the woman with the birth injury. Figure 7 illustrates her computerized tomography (CT) scan, and it is immediately evident that the entire right hemisphere is smaller than the left, even though the lesion is clearly restricted to the parietal cortex. This result is strikingly similar to what we have seen in our rats. Furthermore, because the right frontal lobe is normally somewhat larger than the left (Le May, 1982), the magnitude of the effect on the right hemisphere may even be larger than it appears. Second, we have had an opportunity to study a small sample of patients who had perinatal injuries to their frontal lobes and to compare these patients, as adults, to other adults who had frontal lobe injuries acquired in adulthood. The results are unequivocal: frontal-lobe injury in the first year leads to a

Figure 7
Drawing From a CT-Scan of the Brain of Case S.S. Who Had a Birth-Related Injury to the Right Posterior Cortex

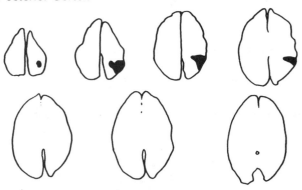

lowered IQ and poor performance on all tests sensitive to frontal lobe injury in adulthood. Frontal-lobe injury in later childhood has a somewhat better outcome, but these people all have significant social difficulties. These results are reminiscent of two findings in our rats: (a) The earliest lesions produce the worst behavioral outcome; and (b) recovery is much better on tests of learned behaviors than on tests of species-typical behavior (Kolb & Whishaw, 1981).

To conclude, it is evident that cortical lesions produce different behavioral and anatomical effects depending on the age at which the brain is damaged. These observations have now provided us with a model to make predictions concerning the processes underlying restitution of function and to search for ways to control the effects of brain damage not only in our laboratory animals but also, someday, in humans. One of the major challenges in the future will be to learn how to control the anatomical changes so that we can gain control over the behavioral outcomes of early brain injury.

REFERENCES

Berry, M. J. (1980). Cellular differentiation: Development of dendritic arborizations under normal and experimentally altered conditions. *Neurosciences Research Program Bulletin, 20,* 451–461.
Cowan, W. M. (1979). The development of the brain. In *The brain* (pp. 56–69). San Francisco: Freeman.
Feinberg, I. (1982). Schizophrenia: Caused by a fault in programmed synaptic elimination during adolescence. *Journal of Psychiatric Research, 17,* 319–334.
Goldman, P. S. (1974). An alternative to developmental plasticity: Heterology of CNS structures in infants and adults. In D. G. Stein, J. J. Rosen, & N. Butters (Eds.), *Plasticity and recovery from brain damage* (pp. 149–174). New York: Academic Press.
Greenough, W. T. (1986). What's special about development? Thoughts on the bases of experience-sensitive synaptic plasticity. In W. T. Greenough & J. M. Juraska (Eds.), *Developmental neuropsychology* (pp. 387–408). New York: Academic Press.
Hicks, S. P., & D'Amato, C. J. (1968). Cell migrations to the isocortex of the rat. *Anatomical Record, 160,* 619–634.
Huttenlocher, P. R. (1979). Synaptic density in human frontal cortex—Developmental changes and effects of ageing. *Brain Research, 163,* 195–205.
Kennard, M. A. (1936). Age and other factors in motor recovery from precentral lesions in monkeys. *Journal of Neurophysiology, 1,* 477–496.
Kolb, B. (1984). Functions of the frontal cortex of the rat: a comparative review. *Brain Research Reviews, 8,* 65–98.
Kolb, B. (1987). Recovery from early cortical damage in rats: I. Differential behavioral and anatomical effects of frontal lesions at different ages of neural maturation. *Behavioural Brain Research, 25,* 205–220.
Kolb, B., & Elliott, W. (1987). Recovery from early cortical lesions in rats: II. Effects of experience on anatomy and behavior following frontal lesions at 1 or 5 days of age. *Behavioural Brain Research, 26,* 47–56.
Kolb, B., Holmes, C., & Whishaw, I. Q. (1987). Recovery from early cortical lesions in rats: III. Neonatal removal of posterior parietal cortex has greater behavioral and anatomical effects than similar removals in adulthood. *Behavioural Brain Research, 26,* 119–137.
Kolb, B., & Sutherland, R. J. (1986). A critical period for noradrenergic modulation of sparing from neocortical parietal cortex damage in the rat. *Neuroscience Abstracts, 12,* 322.
Kolb, B., Sutherland, R. J., & Whishaw, I. Q. (1983). Abnormalities in cortical and subcortical morphology after neonatal neocortical lesions in rats. *Experimental Neurology, 79,* 223–244.
Kolb, B., & Tomie, J. (1988). Recovery from early cortical damage in rats: IV. Effects of hemidecortication at 1, 5, or 10 days of age on cerebral anatomy and behavior. *Behavioural Brain Research, 28,* 259–274.
Kolb, B., & Whishaw, I. Q. (1981). Neonatal frontal lesions in the rat: Sparing of learned but not species typical behavior in the presence of reduced brain weight and cortical thickness. *Journal of Comparative and Physiological Psychology, 95,* 863–879.
Kolb, B., & Whishaw, I. Q. (1985). *Fundamentals of human neuropsychology* (2nd ed.). New York: Freeman.
Kolb, B., & Whishaw, I. Q. (1989). Plasticity in the neocortex: Mechanisms underlying recovery from early brain damage. *Progress in Neurobiology, 32,* 235–276.
Le May, M. (1982). Morphological aspects of human brain asymmetry. *Trends in Neurosciences, 5,* 273–275.
Lennenberg, E. (1967). *Biological foundations of language.* New York: John Wiley.
Morris, R. G. M. (1981). Spatial localization does not require the presence of local cues. *Learning and Motivation, 12,* 239–260.
Pandya, D. N., & Yeterian, E. H. (1985). Architecture and connections of cortical association areas. In A. Peters & E. G. Jones (Eds.), *Cerebral cortex: Vol. 4. Association and auditory cortices* (pp. 3–61). New York: Plenum.
Passingham, R. E., Perry, V. H., & Wilkinson, F. (1983). The long-term effects of removal of sensorimotor cortex in infant and adult rhesus monkeys. *Brain, 106,* 675–705.
Pinel, J. P. J., & Treit, D. (1983). The conditioned burying paradigm and behavioral neuroscience. In T. E. Robinson (Ed.), *Behavioral approaches to brain research* (pp. 212–234). New York: Oxford University Press.
Rakic, P. (1988). Specification of cerebral cortical areas. *Science, 241,* 170–176.
Rockel, A. J., Hiorns, R. W., & Powell, T. D. S. (1980). The basic uniformity in structure of the neocortex. *Brain, 103,* 221–244.
Sutherland, R. J., Kolb, B., Whishaw, I. Q., & Becker, J. (1982). Cortical noradrenaline depletion eliminates sparing of spatial learning after neonatal frontal cortex damage in the rat. *Neuroscience Letters, 32,* 125–130.
Villablanca, J. R., Burgess, J. W., & Sonnier, B. J. (1984). Neonatal cerebral hemispherectomy: A model for postlesion reorganization of the brain. In S. Finger & C. R. Almli (Eds.), *Recovery from brain damage* (Vol. 2, pp. 179–210). New York: Academic Press.
Werker, J. F., & Tees, R. C. (1984). Cross language speech perception: Evidence for perceptual reorganization during the first year of life. *Infant Behavior and Development, 7,* 49–63.
Whishaw, I. Q., Kolb, B., & Sutherland, R. J. (1983). Analysis of the behavior of the laboratory rat. In T. E. Robinson (Ed.), *Behavioral approaches to brain research* (pp. 141–211). New York: Oxford University Press.
Woods, B. T. (1980). The restricted effects of right-hemisphere lesions after age one: Wechsler test data. *Neuropsychologia, 18,* 65–70.

Glossary

altricial - species in which the young are born before they are well developed; humans are altricial

anterograde transport - transport of materials in a neuron forward from the cell body to the axon terminals

bilateral - both sides

cingulate - a region of human cortex that is situated near the top of the brain inside the deep fissure that separates the two hemispheres

cyst - a fluid-filled growth

frontal lobe - one of the four lobes of the human cerebral hemisphere

Gall - the man who incorrectly proposed that people's special skills and personality traits are related to bumps on the surface of their skulls

gliogenesis - the growth of glial cells, the cells that provide nutritive and structural support for neurons in the central nervous system

hemidecorticate - an animal that has had the cortex removed from one of its cerebral hemispheres

hippocampus - a subcortical neural structure of the temporal lobes; the hippocampus plays an important role in memory, and it is one of the few structures in which new neurons continue to be formed in adulthood

in utero - in the uterus

mitosis - the process by which one cell divides to form two complete daughter cells

morphological - structural

Morris water task - a test of rat spatial learning ability; after only a few trials, rats learn to swim directly to an invisible platform beneath the surface of a large, murky milk bath

neuropil - a general term for the neural fibers (i.e., axons and dendrites) in neural tissue

neuropsychology - the study of the behavioral deficits experienced by patients with brain damage

parietal lobe - one of the four lobes of each human cerebral hemisphere; it is situated at the top of the brain, near the back of the head

phonemes - individual speech sounds

postnatal - after birth

prod-burying test - a test of defensive burying in the rat; a rat shocked by a stationary shock prod attached to the wall of its chamber subsequently buries the prod by spraying bedding material at it with its forepaws

psychophysical - pertaining to the study of the relation between the physical properties of a stimulus and the perceptions that the stimulus produces

pyramidal cell - a type of neuron that is found in the human cortex; it is a relatively large neuron that is named after the characteristic triangular shape of its cell body

retrograde transport - transport of materials in a neuron back from the axon terminals to the cell body

species-typical - referring to behaviors that are performed in virtually the same way by all those members of a given species that are of the same age and sex (e.g., nest building, courting, defensive burying)

stellate cell - a type of neuron that is found in the human cortex; it is named after its star-shaped cell body (stellate means star-shaped)

temporo-parietal - involving both the parietal and temporal lobes

thymidine - a compound that is incorporated into the genetic material of cells as they undergo mitotic cell division; thus, if an injection of radioactive thymidine is administered to an infant animal, the location of radioactivity in its brain once it reaches adulthood reveals the location of those neurons that were undergoing division at the time of the injection

tracer - any dye or radioactive substance that binds to a particular molecule or class of molecules and thus reveals their location; tracers are used to study the location of molecules in the brain

Essay Study Questions

1. Kolb describes four approaches to the study of brain-behavior relations? What are they? Which approach does Kolb use?

2. Compare the cases of P.B. and S.S. What points does Kolb illustrate with these cases?

3. Briefly describe the events of neurodevelopment, from cell proliferation to cell death.

4. Describe how radioactive thymidine is used to study neurodevelopment.

5. Describe the Werker and Tees (1984) experiment on the development of an infant's ability to discriminate among the speech sounds of different languages. (It may interest you to know that I served as one of the adult control subjects in this series of experiments; I was amazed to learn that my ability to discriminate foreign speech sounds was less than that of a baby.)

6. Evaluate the Kennard doctrine on the basis of recent research findings.

7. What evidence suggests that it is important to distinguish between the development of learned behaviors and the development of species-typical behaviors?

8. What evidence is there that enriched environments can facilitate recovery from brain damage?

9. Kolb believes that dendritic arborization is an important correlate of recovery from brain damage. What evidence does he offer to support his view?

Multiple-Choice Study Questions

1. Both P.B. and S.S. had damage to their right
    a. frontal cortex.
    b. hippocampus.
    c. eye.
    d. parietal cortex.
    e. both a and d

2. A paradoxical aspect of neurodevelopment is that about half the developing neurons
    a. initially grow two axons.
    b. have no axons.
    c. have bushy dendrites.
    d. die.
    e. divide.

3. The term <u>neuropil</u> refers to
    a. dendrites.
    b. axons.
    c. cell bodies.
    d. all of the above
    e. both a and b

4. A developmental psychobiologist wanted to find out when during development the neurons of the rat visual cortex are created; she injected rats at different stages of neural development with
    a. radioactive thymidine and inspected slices of their brains when they reached maturity.
    b. radioactive thymidine and then immediately inspected slices of their brains.
    c. thymidine and exposed them to radioactivity when they reached maturity.
    d. radioactive thymidine and examined the resulting cell death when they reached maturity.
    e. thymidine and immediately evaluated the effects of radioactivity on cell death in the visual cortex.

5. The experiment of Werker and Tees on the development of phoneme discrimination indicated that

    a. the infants of North American natives have the innate ability to discriminate the phonemes of only their own native dialects.

    b. young infants have the capacity to discriminate between the phonemes of foreign languages, and this ability improves with experience.

    c. young infants have the capacity to discriminate between the phonemes of all languages, but they lose the ability to discriminate phonemes that are not spoken in their presence.

    d. children acquire the ability to recognize the phonemes of their language by repeatedly hearing them.

    e. children acquire the ability to recognize the phonemes of their language through positive reinforcement.

6. Kolb focuses his discussion on his experiments in which the subjects were

    a. rats with transected spinal cords.
    b. hemidecorticate rats.
    c. rats with unilateral frontal lesions.
    d. rats with bilateral frontal lesions.
    e. both b and d

7. Which of the following does Kolb consider to be a good test of rat species-typical behavior?

    a. the Morris water test
    b. the prod-burying test
    c. the Kennard test
    d. the radioactive thymidine test
    e. the arborization test

The answers to the preceding questions are on page 290.

Food-For-Thought Questions

1. According to Kolb, the brain is formed in the same way that a statue is chiseled from granite. Explain and discuss.

2. Many people seem to approach questions regarding the development of a behavioral ability by asking, "How much of it is innate, and how much of it is the result of experience?" However, most developmental psychobiologists have long appreciated that all behaviors are a product of the interaction of genetic programs and experience, and that it is counterproductive to try to apportion development between the two. Asking, "How much of a behavior is innate and how much is the result of experience?" is like asking, "How much of a melody comes from the pianist and how much comes from the piano?" Discuss.

ARTICLE 2

Reorganization of Retinotopic Maps in Adult Mammals After Lesions of the Retina

J.H. Kaas, L.A. Krubitzer, Y.M. Chino, A.L. Langston, E.H. Polley, and N. Blair
Reprinted from Science, 1990, Volume 248, 229-231.

Although the plasticity of the developing visual system is widely recognized, evidence seemed to suggest that the distribution of retinal projections into the visual cortex of adult mammals is immutable (i.e., unchangeable). However, the recent experiment by Kaas and his collaborators demonstrated that the adult visual cortex retains some of its plasticity.

As others had discovered before them, Kaas and his collaborators found that small monocular lesions (i.e., lesions to one eye) to the retina of an adult cat produced no reorganization of neural input into the visual cortex. Most neurons of the visual cortex receive input from corresponding areas of both retinas, and thus destroying a small part of one retina created an area of neurons in the visual cortex that received input from only the intact retina, but there was no reorganization of input. However, the story was entirely different when all input to an area of adult visual cortex was eliminated by destroying a small area of one retina and then removing the other eye. Amazingly, 2 to 6 months after this binocular lesion, the neurons in the area of the cat visual cortex whose input had been totally eliminated responded to visual stimuli presented to the intact portion of the remaining retina. Following the adult binocular lesions, the visual projections from the retina to the cortex had somehow reorganized themselves so that cortical neurons whose pathways of neural input had been totally eliminated were now receiving input from other areas of the retina. Each of the neurons whose input had been eliminated was found to have a receptive field in the intact retina near the boundary of the lesion. Complete reorganization did not occur when the retinal lesions were greater than 10° in diameter.

These results suggest that the basic retinotopic organization of the visual system remains plastic throughout the lifespan of mammals. This adult plasticity may permit adult organisms to partially recover from certain kinds of visual system damage, and it may be the basis for our ability to learn and retain certain visual skills.

Reorganization of Retinotopic Cortical Maps in Adult Mammals After Lesions of the Retina

JON H. KAAS,* LEAH A. KRUBITZER, YUZO M. CHINO, ANDY L. LANGSTON, EDWARD H. POLLEY, NORMAN BLAIR

The organization of the visual cortex has been considered to be highly stable in adult mammals. However, 5° to 10° lesions of the retina in the contralateral eye markedly altered the systematic representations of the retina in primary and secondary visual cortex when matched inputs from the ipsilateral eye were also removed. Cortical neurons that normally have receptive fields in the lesioned region of the retina acquired new receptive fields in portions of the retina surrounding the lesions. The capacity for such changes may be important for normal adjustments of sensory systems to environmental contingencies and for recoveries from brain damage.

ARE THE MAPS OF VISUAL SPACE IN visual cortex capable of reorganization in adult mammals? As in other mammals, the visual cortex of cats contains several retinotopic representations of the visual field, including those in areas 17 and 18 (1). Such systematic representations of peripheral receptor arrays also characterize somatosensory and auditory cortex (2). Under normal circumstances, these sensory maps develop in a highly consistent manner in individuals of the same species. However, development of these topological maps can be altered by abnormal sensory inputs, including those produced by sensory deprivation and damage to the peripheral sensory sheet (3, 4). Thus, the nature of the input from the receptor sheet partly determines the ultimate organization of developing sensory maps. In the visual system, sensory manipulations such as monocular deprivation, induced strabismus, and unilateral defocusing of the image can alter cortical organization (3). However, these manipulations affect cortical organization mainly or only within a critical developmental period extending a few months postnatally in cats or several years in humans (3). Thus, evidence supports the view that the organization of visual cortex remains highly stable after initial development, and there has been little reason to suppose that basic features of retinotopic maps can change in adults.

In contrast to the visual system, recent experiments on somatosensory cortex indicate that the organization of sensory maps can be modified even in adults (4, 5). For example, if part of the normal representation of the hand in primary somatosensory cortex is deprived of its normal source of activation by cutting a peripheral nerve, the cortical representation reorganizes over a period of hours to weeks so that neurons in the deprived zone of cortex acquire new receptive fields on other parts of the hand. Such adult plasticity implies that previously existing connections in the brain are capable of changing in synaptic effectiveness so that new receptive fields and new representational organizations can emerge in cortex. Such changes could be important in normal adjustments of the brain to alterations in the sensory environment, as well as in compensations for peripheral and central damage to the nervous system. Because the potential for such reorganization would seem to exist in other sensory fields, we investigated the possibility of adult plasticity in visual cortex with an experimental approach that has been used successfully for the somatosensory system.

Parts of areas 17 and 18 of the visual cortex were deprived of a normal source of activation by placing lesions 5° to 10° in diameter just above the area centralis in the retina of one eye of adult cats (6). By itself this procedure produced no notable change in retinotopic organization when tested in one cat. Most cortical neurons are binocularly activated and thus have two retinotopi-

---
J. H. Kaas and L. A. Krubitzer, Department of Psychology, Vanderbilt University, Nashville, TN 37240.
Y. M. Chino and A. L. Langston, College of Optometry, University of Houston, Houston, TX 77204.
E. H. Polley and N. Blair, Departments of Anatomy and Ophthalmology, University of Illinois Medical Center, Chicago, IL 60680.

*To whom correspondence should be addressed.

Copyright © 1990 by the American Association for the Advancement of Science

cally matched receptive fields. Hence, recordings made after a retinal lesion simply demonstrated restricted regions of cortex where neurons had receptive fields only in the intact eye. The monocular lesion, therefore, merely revealed an effect of removing one of two sources of activation, rather than any basic reorganization (7). However, when restricted zones of cortex were totally deprived of normal sources of visual activation by placing a lesion in one eye and removing the other eye, dramatic changes in the retinotopic organization of areas 17 and 18 were produced. Neurons in the deprived zone of cortex acquired new receptive fields representing inputs from retinal locations around the margins of the lesion.

To allow time for cortical reorganization to occur, most of our recordings were made 2 to 6 months after the retinal lesion and the enucleation of the other eye. In each experiment, microelectrode recordings were made from neurons in an array of closely spaced electrode penetrations within and around the deprived cortex (8). Outside the zone of altered cortex in areas 17 and 18, neurons had receptive fields of normal locations and sizes. Thus, in the explored region of cortex, rows of recording sites extending mediolaterally from area 17 to area 18 produced rows of receptive fields systematically displaced from the last, forming a progression within the contralateral lower visual quadrant toward the zero vertical meridian as the border of areas 17 and 18 was reached, and back again for sites in area 18. Within the zone of altered cortex, neurons were activated by visual stimuli and had receptive fields of normal sizes. However, the receptive fields of these neurons were displaced from the region of the retinal lesion to adjacent parts of the retina (Fig. 1). Thus, for mediolateral rows of recording sites into the region of deprived cortex, receptive fields progressed from locations just temporal to the scotoma or "blind spot" produced by the lesion to the margin of the scotoma. Then, the progression of receptive fields ceased as the deprived cortex was reached. Receptive fields remained on the temporal side of the scotoma for several successive recording sites over 2 mm of cortex. Next, receptive fields jumped to the opposite side of the scotoma and remained stationary for several recording sites; they then resumed their normal progression for recording sites outside the deprived zone. In addition, some recording sites (Fig. 1, sites d/e in row 4) had two receptive fields, one on each side of the scotoma. The responsiveness of neurons with new receptive fields was not notably abnormal (9). Both area 17 and area 18 were altered in this way, and comparable results were obtained in four cats with retinal lesions of 5° to 10°.

An example of how progressions of receptive fields for rows of recording sites differed in normal and reorganized cortex is shown in Fig. 2. In normal cortex, receptive field centers shift systematically as recording sites progress across the retinotopic representation in area 17. In contrast, receptive fields for recording sites over a considerable tangential distance in cortex can have nearly the same receptive field center in reorganized cortex.

In two other cats, larger retinal lesions of 10° to 15° in diameter produced a larger zone of deprived cortex. In these cases, neurons near the margin of the deprived zone of cortex had displaced receptive fields, but neurons in a 2- to 3-mm-wide center of the deprived visual zone of cortex were unresponsive to visual stimuli. Thus, large zones of deprived cortex may not completely reorganize.

The present results (Fig. 1) indicate that portions of retinotopic cortical maps as large as 4 to 8 mm and encompassing 5° or more of the visual field can reorganize such that neurons within this cortex acquire receptive fields in new locations. Reorganization over such distances could result from changes in

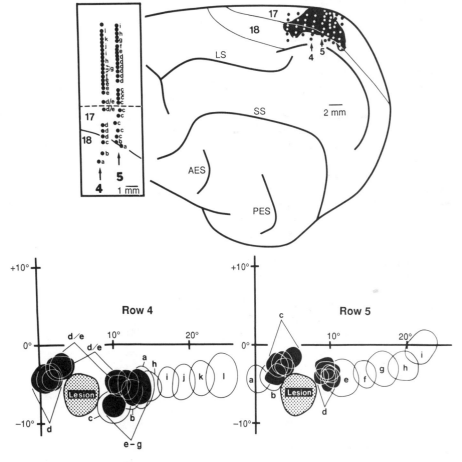

Fig. 1. Evidence for cortical reorganization from one case. Partial results are shown from a total of 121 recording sites, of which 55 produced abnormally located receptive fields and were judged to be within altered cortex. Shown are normal (outlined) and displaced (black) receptive fields for neurons at recording sites in areas 17 and 18 in a cat with a lesion of the retina of the right eye and enucleation of the left eye. Receptive fields were hand-plotted with projected, moving bars of light on a tangent screen or hemisphere (7). Some of the electrode penetrations producing mediolateral rows of recording sites are indicated by dots on a dorsolateral view of the brain on the upper right. The black region approximates the extent of the cortex where neurons had abnormally located receptive fields. Recording sites obtained from electrode penetrations of rows 4 and 5, including successive recording sites in penetrations extending down the medial wall of the cerebral hemisphere in area 17, are lettered in the box, a to l and a to i, respectively. The solid line in the box marks the estimated border of areas 17 and 18, and the dashed line marks the dorsal edge of the medial wall. The receptive fields for the lettered recording sites for each row are indicated in the contralateral hemifield depicted below. The zero vertical and zero horizontal meridians are marked in degrees of visual angle, and the projection of the retinal lesion (scotoma) is shown. Abbreviations: AES, anterior ectosylvian sulcus; LS, lateral sulcus; PES, posterior ectosylvian sulcus; and SS, suprasylvian sulcus. The normal organizations of areas 17 and 18 in cats are given in (1).

17

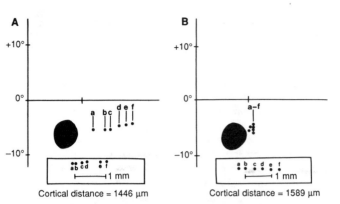

Fig. 2. Receptive field centers for recording sites in normal (**A**) and reorganized (**B**) parts of area 17. (A) Receptive field centers (a to f) systematically progress toward peripheral vision for a row of recording sites (a to f, box) moving away from the border of areas 17 and 18 in a part of area 17 that had normal retinotopic organization. (B) A similar row of recording sites in the part of area 17 that was deprived of normal activation by the retinal lesion produced an accumulation of receptive fields with nearly identical centers next to the blind area produced by the retinal lesion. In (A) and (B) horizontal and vertical meridians through the center of gaze are marked in degrees of visual angle in the visual hemifield contralateral to recording sites. Recording sites progress away from the border of areas 17 and 18 in area 17 (they are reversed from actual order in the left cerebral hemisphere for ease of matching the progressions of receptive fields and recording sites). The projection of the retinal lesion into the visual hemifield is in black.

the effectiveness of synapses within the arbors of thalamocortical axons of previously existing inputs (10). Comparable results have been obtained by Heinen and Skavenski (11) from part of area 17 of one monkey. Cortex with neurons initially unresponsive to visual stimuli after bilateral lesions of the fovea later contained neurons responsive to visual stimuli. Results from visual cortex are similar to those obtained from somatosensory cortex of monkeys; removing the inputs from part of the hand produces a zone of altered cortex where neurons achieve new receptive fields of normal sizes in other parts of the hand (4, 5). Furthermore, removing inputs from more than half of the hand produces a larger zone of deprived cortex where complete reactivation does not occur (12).

These results are important for at least two reasons. First, in certain ocular diseases in humans, lesions are commonly found in the retinas of both eyes, and retinotopic reorganization of visual cortex could result when lesions in the two eyes correspond to the same locations in visual space (13). Second, the present results, together with those from the somatosensory system, imply that basic neuronal properties such as receptive field location are maintained in a dynamic state in sensory-perceptual systems of adult mammals. Such adult plasticity may be important, not only in recoveries from brain damage and adjustments to other impairments, but also in our abilities to maintain, alter, and improve sensorimotor and perceptual skills.

REFERENCES AND NOTES

1. R. J. Tusa, L. A. Palmer, A. C. Rosenquist, *J. Comp. Neurol.* **177**, 213 (1978); *ibid.* **185**, 657 (1979); A. C. Rosenquist, in *Cerebral Cortex*, A. Peters and E. G. Jones, Eds. (Plenum, New York, 1985), vol. 3, pp. 81–117.
2. M. M. Merzenich and J. H. Kaas, *Prog. Psychobiol. Physiol. Psychol.* **9**, 1 (1980).
3. J. A. Movshon and R. C. Van Sluyters, *Annu. Rev. Psychol.* **32**, 477 (1981); S. M. Sherman and P. D. Spear, *Physiol. Rev.* **62**, 738 (1982); R. G. Boothe, V. Dobson, D. Y. Teller, *Annu. Rev. Neurosci.* **8**, 495 (1985).
4. J. H. Kaas, M. M. Merzenich, H. P. Killackey, *Annu. Rev. Neurosci.* **6**, 325 (1983); J. T. Wall, *Trends Neurosci.* **11**, 549 (1988).
5. M. M. Merzenich *et al.*, *Neuroscience* **10**, 639 (1983); D. D. Rasmusson, *J. Comp. Neurol.* **205**, 313 (1982); J. T. Wall and C. G. Cusick, *J. Neurosci.* **4**, 1499 (1984); M. B. Calford and R. Tweedale, *Nature* **332**, 446 (1988); S. A. Clark, T. Allard, W. M. Jenkins, M. M. Merzenich, *ibid.*, p. 444.
6. A single photocoagulation lesion was made with an Argon blue-green laser (Argon Medical, Athens, TX) (spot size, 500 μm; intensity, 2.5 W; duration, 5 s or longer) in the superior and nasal retina of one eye in cats anesthetized intramuscularly with ketamine hydrochloride (20 mg/kg) and xylazine (4 mg/kg). Within a few days, the contralateral eye was enucleated under Nembutal anesthesia (35 mg/kg, intraperitoneally) by standard procedures and under aseptic conditions. The animals were treated with antibiotics and maintained for 2 to 6 months before recordings were made.
7. Although cats have ocular dominance columns and some neurons in layer IV respond exclusively to one or the other eye, most neurons can be activated by either eye [for example, D. H. Hubel and T. N. Wiesel, *J. Physiol.* (London) **160**, 106 (1962); *J. Neurophysiol.* **28**, 229 (1965)].
8. Recordings were largely from neuron clusters 800 to 1200 μm from the surface in the hemisphere contralateral to the lesioned eye. Penetrations were typically placed 200 to 300 μm apart in mediolateral rows about 1 mm apart. Recording sites in penetrations along layers in cortex of the medial wall were typically 100 μm apart. Recording methods were standard and have been described in detail elsewhere [J. H. Kaas and R. W. Guillery, *Brain Res.* **59**, 61 (1973); Y. M. Chino, M. S. Shansky, W. L. Jankowski, F. A. Banser, *J. Neurophysiol.* **50**, 265 (1983); Y. M. Chino, W. H. Ridder, E. P. Czora, *Exp. Brain Res.* **72**, 264 (1988)]. Briefly, in one series of experiments, four cats were anesthetized with an initial dose of urethane (100 mg per 100 g of body weight, intraperitoneally), and the eye was mechanically stabilized. In a second series, two cats were initially immobilized with ketamine hydrochloride (20 mg/kg) and anesthetized with Fluothane and a mixture of 70% nitrous oxide and 30% oxygen. During the experiment, Fluothane was replaced by the intravenous infusion of Surital. These cats were paralyzed with an intravenous mixture of gallamine triethiodide (5 mg/kg per hour), d-tubocurarine (0.5 mg/kg per hour), atropine, and saline and were artificially respired. Both procedures produced similar results. The expired CO_2, blood pressure, electrocardiogram, and electroencephalogram were continuously monitored. Cycloplegia was maintained with 10% Neo-Synephrine and atropine sulfate, and the corneas were protected with gas-permeable contact lenses of appropriate curvature. The optic disc and the area centralis were projected onto a translucent plastic hemisphere or tangent screen. Much of the visual cortex of the dorsal surface of the brain was exposed, and the brain was protected in a chamber filled with silicone fluid. The exposed area was photographed with a scale to later record mapping sites and the sequence of mapping during the experiment. The surface pattern of blood vessels was used as a reference for establishing the position of each penetration. Recordings were made with low-impedance tungsten microelectrodes from clusters of neurons, and projected bars of light served as stimuli. Successive recording depths along penetrations down the medial wall were measured from the surface and from small electrolytic marker lesions placed along penetrations.
9. The response characteristics of recorded neurons were not quantitatively determined, but responses to moving bars of light appeared to be similar for neurons in and outside of the deprived zone of cortex.
10. Terminal arbors of thalamocortical axons in areas 17 and 18 of cats range from 1 to 3.5 mm in tangential width of distributions [A. L. Humphrey, M. Sur, D. J. Uhlrich, S. M. Sherman, *J. Comp. Neurol.* **233**, 159 (1985); *ibid.*, p. 190. Tangential axon collaterals of cortical neurons [C. D. Gilbert and T. N. Wiesel, *Nature* **280**, 120 (1979)] could also play a role.
11. S. J. Heinen and A. A. Skavenski, *Invest. Ophthalmol. Visual Sci.* **29** (suppl.) 23 (1988).
12. J. T. Wall and J. H. Kaas, *Brain Res.* **372**, 400 (1986).
13. For example, patients who develop bilateral macular degeneration typically experience a permanent central scotoma of varied sizes. These patients use an eccentric retinal locus outside the scotoma for fixation, suggesting that such retinal areas act like a newly developed fovea for resolving spatial details [G. K. von Noorden and G. Mackensen, *Am. J. Ophthalmol.* **53**, 642 (1962); G. T. Timberlake *et al.*, *Invest. Ophthalmol. Visual Sci.* **27**, 1137 (1986); G. T. Timberlake, E. Peli, E. A. Essock, R. A. Augline, *ibid.* **28**, 1268 (1987)].

3 October 1989; accepted 31 January 1990

Glossary

arbors - branches of neuronal fibers

area 17 - primary visual cortex; signals from the retina via the thalamus enter the cortex in area 17; area 17 is retinotopically organized; area 17 is at the very back of the cerebral hemispheres

area 18 - an area of visual cortex that is just lateral to area 17 and receives input from it; like area 17, area 18 is retinotopically organized

area centralis - the central area of the retina

binocular - involving two eyes

contralateral - on the other side of the body

critical developmental period - the period, usually early in life, when a particular treatment must be administered if it is to influence the development of the organism

enucleation - removal of an eye

mediolateral - from the midline of the brain (or other structure) out to one or both of its sides; from a medial area to a lateral area

microelectrodes - electrodes whose points are so fine that signals can be recorded from individual neurons or from small groups of neurons

monocular - referring to one eye

ocular - referring to the eyes

postnatally - after birth

receptive field - each neuron in the visual system has a receptive field, which is the area of the visual field within which an appropriate stimulus can influence the firing of that neuron; from a neuroanatomical perspective, a visual neuron's receptive field indicates that area of the retina from which it receives its input

retinotopic - refers to the layout of some of the areas of visual cortex (e.g., areas 17 and 18); a retinotopic cortical area receives neural input in a pattern that reflects the layout of the retina; in effect, the surface of a retinotopic cortical area is a map of the retina

scotoma - an area of blindness in the visual field that is created by damage to a part of the visual system

somatosensory - refers to the sensory system that mediates the perception of sensory signals from the skin and internal organs

strabismus - refers to any case in which there is a deviation in the direction of focus of the two eyes; because the two eyes do not focus on the same spot in the visual field, the images from the two eyes of a person with strabismus cannot be merged by the visual system into a unitary view of the world; strabismus is commonly referred to as crossed eyes

temporal - toward the side of the brain; away from the midline

thalamocortical - referring to neural projections from the thalamus to the cortex

topological - refers to any cortical area whose input is organized so that it maps the surface of the receptor; for example, some areas of visual cortex are retinotopic

unilateral - on one side of the body

visual quadrant - the four areas of the visual field that are created when horizontal and vertical lines are drawn through the focus point; these horizontal and vertical lines are referred to as the horizontal and vertical zero meridians

Essay Study Questions

1. What treatments can influence the retinotopic organization of developing mammals?

2. What evidence suggested to Kaas and his collaborators that the visual cortex of mammals might maintain its plasticity into adulthood?

3. Destroying a portion of one retina does not completely eliminate neural input to the corresponding area of mammalian visual cortex. Why? Does it induce retinotopic reorganization?

4. Describe the method used by Kaas and his collaborators to induce the reorganization of the adult cat retinal projections.

5. Describe the linear recording method of Kaas and his collaborators. What did this method reveal?

6. Large zones of deprived cortex do not completely reorganize. Explain.

Multiple-Choice Study Questions

1. Prior to the study of Kaas and his collaborators, it had been shown that
    a. induction of strabismus in adult mammals can change the organization of their visual cortices.
    b. monocular deprivation in adult mammals can change the organization of their visual cortices.
    c. unilateral defocusing of the retinal image in adult mammals can change the organization of their visual cortices.
    d. all of the above
    e. none of the above

2. How long did Kaas and his collaborators wait after they produced the binocular lesions before they assessed their consequences?
    a. about 1 week
    b. between 2 and 6 months
    c. 3 days
    d. about 3 years
    e. between 2 and 6 hours

3. In their experiment, Kaas and his collaborators
    a. destroyed a small part of one retina of infant monkeys.
    b. removed one eye from infant monkeys.
    c. destroyed a small portion of one retina of adult cats and removed the other eye.
    d. removed one eye from infant cats and destroyed a small portion of the other retina.
    e. both a and b

4. In two adult cats, monocular retinal lesions that covered between 10° and 15° of the visual field were made, and the other eye was removed. In these two cases,
    a. there was no reorganization of their visual cortex.
    b. there was only partial reorganization of their visual cortex.
    c. neurons in the center of the deprived zone remained unresponsive to visual stimuli.
    d. both a and c
    e. both b and c

5. Eliminating the neural input to a small portion of the hand area of adult somatosensory cortex causes
    a. the neurons in this area of cortex to acquire receptive fields in another part of the hand.
    b. the corresponding area of the contralateral somatosensory cortex to take over the disrupted receptive fields.
    c. the corresponding area of the contralateral hand to have larger receptive fields.
    d. all of the above
    e. both b and c

6. The area of blindness in the visual field that is produced by a retinal lesion is called
    a. a meridian.
    b. a quadrant.
    c. an arbor.
    d. a receptive field.
    e. none of the above

The answers to the preceding questions are on page 290.

Food-For-Thought Questions

1. What do you think is the evolutionary advantage of retinotopic organization?

2. Compare the plasticity of the visual system with the plasticity of the somatosensory system. What do you think are the advantages of sensory system plasticity?

3. From a biopsychological perspective, the most exciting next step in this line of research would be the demonstration that retinotopic reorganization influences perception. If you were a biopsychologist, how would you go about demonstrating such an effect?

ARTICLE 3

Open Microsurgical Autograft of Adrenal Medulla to the Right Caudate Nucleus in Two Patients with Intractable Parkinson's Disease

I. Madrazo, R. Drucker-Colín, V. Díaz, J. Martínez-Mata, C. Torres, and J.J. Becerril
Reprinted from The New England Journal of Medicine, 1987, Volume 316, 831-834.

Parkinson's disease is a permanent motor disorder associated with the degeneration of dopamine-releasing neurons in the substantia nigra. It is characterized by (1) tremor, which is most severe when the patient is not moving, (2) bradykinesia (slowness of movement), and (3) muscular rigidity. As a result, in the advanced statges of Parkinson's disease, patients have great difficulty performing even the most routine motor tasks such as walking and talking. Although the drug levodopa (L-DOPA) temporarily alleviates the symptoms of some patients, it often produces debilitating side effects, which limit its usefulness. Because there is no known cure for Parkinson's disease, the following preliminary report of Madrazo and his colleagues generated considerable excitement.

The substantia nigra neurons that degenerate in cases of Parkinson's disease normally release dopamine into synapses in the caudate nucleus, an important subcortical motor structure. Madrazo and his colleagues treated two cases of Parkinson's disease by transferring dopamine-releasing cells from each patient's own adrenal medulla onto the surface of his right caudate nucleus. Backlund and his colleagues had previously used a similar autotransplant procedure in patients with Parkinson's disease, but their treatment had proven unsuccessful.

Before the autotransplants were performed, Madrazo and his colleagues took spectrophotographic and electromyographic measures of the patients' tremor. For example, in the spectrophotographic test, the patients held a tiny light in a dark room and made a series of prescribed arm movements while the paths of the light were recorded on photographic film. Following the autotransplantation of the adrenal medulla cells, both patients displayed vast improvement. Spectrophotographic and electromyographic activity was much improved, and there seemed to be substantial improvement in their ability to walk, talk, and perform other basic motor activities.

This is a preliminary study that leaves many important questions unanswered. Nevertheless, as a result of it, autotransplantation operations are currently being performed in many hospitals around the world. As I write these words, it is not clear whether the tremendous promise of this preliminary study is going to be fulfilled, but the value of autotransplantation in the treatment of Parkinson's disease may have become more apparent by the time these words reach you. Why don't you ask your instructor if she or he knows anything about the current status of this treatment: Has it proven to be a smashing success, a dismal failure, or something in between?

OPEN MICROSURGICAL AUTOGRAFT OF ADRENAL MEDULLA TO THE RIGHT CAUDATE NUCLEUS IN TWO PATIENTS WITH INTRACTABLE PARKINSON'S DISEASE

Ignacio Madrazo, M.D., D.Sc., René Drucker-Colín, M.D., Ph.D., Víctor Díaz, M.D., Juan Martínez-Mata, M.D., César Torres, M.D., and Juan José Becerril, M.D.

Abstract Recent experimental studies and one clinical case have suggested that grafting tissue from the adrenal medulla into the brain may ameliorate the signs of Parkinson's disease. We describe the treatment of two young patients (35 and 39 years old) with intractable and incapacitating Parkinson's disease, in whom fragments of the adrenal medulla were autotransplanted to the right caudate nucleus. Clinical improvement was noted in both patients at 15 and 6 days (respectively) after implantation and has continued in both. Rigidity and akinesia had virtually disappeared in the first patient at 10 months after surgery, and his tremor was greatly reduced. A similar degree of improvement was present in the second patient at three months.

We conclude that autografting of the adrenal medulla to the right caudate nucleus was associated with a marked improvement in the signs of Parkinson's disease in two patients, but our results are preliminary and further work is necessary to see whether this procedure will be applicable over the long term in other types of patients with Parkinson's disease. (N Engl J Med 1987; 316: 831-4.)

RECENT studies of animal models of Parkinson's disease have shown that grafting tissue from the adrenal medulla into the cerebral ventricles results in a reduction of the rotational behavior induced by apomorphine, similar to that resulting from substantia nigra grafts.[1-5] Since adrenal grafts do not reinnervate the host brain, this effect is presumably due to the release of dopamine.[3,5] However, it has been shown that chromaffin cells cultured in the presence of nerve growth factor[6] or placed as a graft in the anterior chamber of the eye are capable of developing nerve processes.[7] Since chromaffin cells are a good source of dopamine,[8] such cells could theoretically be transplanted into the brains of patients with Parkinson's disease. Recently, Backlund et al.[9] have reported on a successful transplantation of tissue from the adrenal medulla to the brain in such patients, using a cannula placed in the parenchyma of the caudate nucleus.

We describe two young patients with severe and rapidly progressing Parkinson's disease, in whom we used an open microsurgical technique to graft tissue of the adrenal medulla into the lateral ventricle in direct contact with the head of the caudate nucleus. Experimental studies had suggested that placing a graft in the periventricular and intraventricular regions, as opposed to placing it within the parenchyma, would lead to better growth and survival of the transplants[10-12] and would probably be the most efficacious procedure for intracerebral grafting of the adrenal medulla.[13] Our results suggest that grafting chromaffin cells in direct contact with both the cerebrospinal fluid and the caudate nucleus produced excellent amelioration of most of the clinical signs of Parkinson's disease in our two patients.

From the Departments of Neurosurgery and Urology, Hospital de Especialidades Centro Médico "La Raza," Mexico City, and the Department of Neurosciences, Instituto de Fisiología Celular, Universidad Nacional Autónoma de México, Mexico City. Address reprint requests to Dr. Drucker-Colín at the Departamento de Neurociencias, Instituto de Fisiología Celular, Universidad Nacional Autónoma de México, Apartado postal 70-600, 04510 México, D.F., Mexico.

Methods

Patients

Patient 1 was a 35-year-old man admitted to our neurology department in August 1984. He had no history of drug abuse, orthostatic hypotension, or familial parkinsonism. Parkinson's disease had started one year previously, with rapid progression of signs on the right side followed seven months later by signs on both sides. The patient could not tolerate anticholinergics, antihistamines, or amantadine. He received 750 mg of levodopa for short periods, which resulted in improvement, but he was unable to continue taking the drug because of severe gastrointestinal side effects. In less than three years, he was confined to a wheelchair and unable to perform even the most basic activities. A neurologic examination demonstrated no dementia, bilateral rigidity more severe on the right, bradykinesia, resting tremor, or speech impairment. In February 1986, autotransplantation was proposed to the patient and his family and, with their approval, to the hospital's ethics committee. Written consent was obtained from all parties involved. Before surgery, the patient's status was recorded on videotape, and spectrophotography and computed tomography of the skull and adrenal glands were performed. Surgery was performed on March 23, 1986.

Patient 2 was a 39-year-old man who had no history of orthostatic hypotension or drug abuse (including 1-methyl-4-phenyl-1,2,3,6-tetrahydropyridine [MPTP]) and whose 49-year-old sister had Parkinson's disease. Symptoms of the disorder had started when the patient was 33 years of age, with severe left-sided tremor and rigidity, which became bilateral six to seven months later and predominated in the upper extremities. A year later he was severely disabled and incapable of writing, eating, or performing other everyday activities on his own. His motor impairment made him completely dependent and confined him to living at home with assistance. His mental functions as assessed by psychometric evaluation were normal. Treatment with levodopa, 250 mg every four hours, was begun seven months after the initial symptoms. The response was satisfactory at first, but soon intolerance and unresponsiveness to the drug developed. Biperiden, trihyxyphenidyl, and bromocriptine were then given, without benefit.

The patient was admitted to our department for the first time in June 1986, with the following neurologic findings: incapacitating resting tremor of 4 or 5 cycles per second, predominating in the upper extremities, which became worse during stress or the performance of simple mental tasks; bilateral akinesia; bilateral rigidity with severe, predominantly right-sided, cogwheel sign; and severely impaired speech, writing, and gait. In September 1986, autotransplantation was offered to the patient and his family. With their approval, the proposal was submitted to the hospital's ethics committee, after which written consent was obtained from all parties involved. Before surgery, spectrophotography, electromyography, and computed tomography of the skull

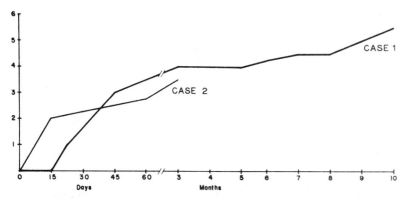

Figure 1. Performance Scores of Two Patients after Autotransplantation for Parkinson's Disease.

Scores are based on subjective observations by the patients' families and physicians. A score of 1 (on the y axis) indicates that the patient could not move out of a wheelchair, perform basic activities, or speak articulately; 2, that he needed help to perform basic activities or take a few steps, and pronounced a few words; 3, that he did not need help to perform basic activities or to take a few steps, and spoke well enough to communicate; 4, that he led a normal life at home, walked without help, and spoke fluently; 5, that he started work activities; and 6, that his life was normal.

and adrenal glands were performed. Transplantation was carried out on October 10, 1986.

Surgery

The same surgical technique was used in both patients — simultaneous right adrenalectomy and right frontal craniotomy. After extraction of the adrenal gland, the medulla was dissected under a surgical microscope, and several fragments of medullary adrenal gland weighing a total of 0.8 to 1 g were placed on a wet surface. Simultaneously the right caudate nucleus was approached with a surgical microscope, through the right lateral ventricle by a nontraumatic transcortical (F2 [second frontal circumvolution]) standard dissection. After the caudate nucleus had been identified, a bed approximately 3 by 3 by 3 mm was constructed on its head and the fragments of adrenal medulla were implanted within the bed. The grafted tissues were then anchored by three or four standard stainless-steel miniature staples (Johnson and Johnson). The graft was thus placed so that the grafted tissue was embedded within the caudate but still in contact with the cerebrospinal fluid.

RESULTS

Patient 1

The immediate postoperative course of Patient 1 was uneventful. Subjective and objective clinical improvement appeared 15 days after implantation (Fig. 1). From then on, functional recovery occurred on an almost daily basis. On the 17th postoperative day, the patient was able to walk with assistance, his speech was clearer, and his bradykinesia was

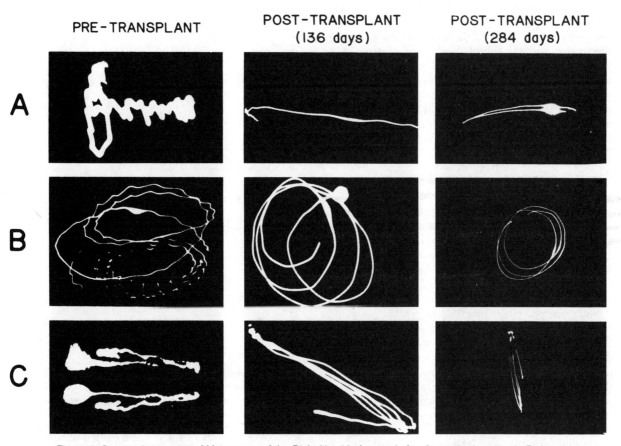

Figure 2. Spectrophotograms of Movements of the Right Hand before and after Autotransplantation in Patient 1.
The patient was asked to draw a straight line (A), to draw circles (B), and to perform the finger–nose test (C).

25

diminished considerably. Five months after surgery, he was able to walk and eat without help, carry out all other basic functions, and speak clearly and fluently. He did not require drugs. The level of tremor registered on spectrophotography at 136 days after surgery was dramatically reduced on the right side and slightly reduced on the left side (Fig. 2). However, rigidity and akinesia were practically absent on both sides. Moreover, his facial expressions have become almost normal. The biweekly evaluation six months after the operation continued to show improvement, and at present, 10 months after grafting, the patient is capable of coming to the outpatient clinic without any assistance and of playing soccer with his five-year-old son. He is considering working. Figure 2 shows through spectrophotographic recordings at 284 days after grafting the additional improvement in this patient, and Figure 1 shows the gradual improvements to date.

Patient 2

The immediate postoperative course of Patient 2 was also uneventful, and the amelioration of the signs of Parkinson's disease was almost immediate. Tremor, which was the most important clinical sign, was dramatically reduced six days after surgery. Figure 3 shows the results of electromyographic studies: before surgery, the tremor was severe — 4 or 5 cycles per second; 11 days after grafting, the frequency had diminished to about 1 cycle per second and the amplitude was very low. At 87 days after grafting, electromyograms showed complete absence of tremor. In addition, the rigidity was abolished. At this writing, more than three months after surgery, the patient can walk without help, he speaks clearly, and he has almost normal facial expressions. The most dramatic improvement has been the almost complete disappearance of tremor.

Discussion

The results in our two patients suggest that transplanting part of the adrenal medulla into the ventricle in contact with the caudate nucleus may in some cases markedly ameliorate most signs of Parkinson's disease. Whether this procedure would be of benefit in all cases of intractable Parkinson's disease is not known. We are also unable to explain the basis for the ipsilateral amelioration of the clinical signs, since one might expect only contralateral improvement.

Since our procedure involved surgical removal of a small piece of caudate nucleus, it could be argued that this in itself was responsible for the amelioration of the clinical signs. In fact, such surgical procedures have been tried before.[14] We do not believe, however, that the surgical removal of a fragment of the caudate was solely responsible for the benefits observed, since the clinical improvement was bilateral. It is more likely that the improvement resulted from the release of dopamine into the ventricles, from which it would then reach the contralateral caudate.

The central nervous system is an immunologically privileged organ (and therefore rejection is rare), which makes it an ideal site for tissue grafting (fetal or autologous).[15,16] The two main obstacles to successful function of a graft are the complex interconnection that must be established between the grafted cells and the surrounding neurons and the necessity for an adequate supply of nutrients to the grafted cells in their new environment. In the nigrostriatal system, experimental grafts have been implanted at three sites: the ventricular cavities, the cerebral parenchyma, and cavities in the cerebral parenchyma, where the grafted tissue is placed in contact with the cerebrospinal fluid.[17] Currently, it appears that the ventricular cavities may be the best sites for the grafts, because the cerebrospinal fluid immediately provides the tissue

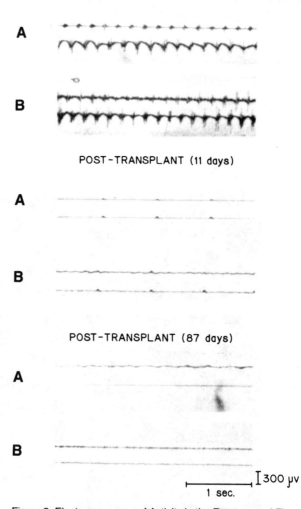

Figure 3. Electromyograms of Activity in the Extensor and Flexor Muscles before and after Autotransplantation in Patient 2.
Before surgery a high-intensity tremor (4 or 5 cycles per second) was observed in the activity of the extensor muscles (Tracing 1) and flexor muscles (Tracing 2). Panel A shows activity when the upper extremities were at rest, and Panel B when they were extended. The tremor had diminished by day 11 and disappeared by day 87.

with the required nutrients.[10,11] It also allows the chemical substances secreted by the grafted cells to be transported to other structures in the central nervous system and thus acts as a biologic infusion pump. The placement of grafts in a previously prepared cavity in the parenchyma, but still in contact with the cerebrospinal fluid, may be advantageous because it allows functional synapses to be established with other functionally or anatomically related structures.[18,19]

Chromaffin cells possess a high degree of phenotypic plasticity. When they are surrounded by adrenal cortex, they become rounded and absorb large amounts of epinephrine; but when they are removed from the gland and placed in culture, they change morphologically and biochemically, simulating catecholaminergic cells.[20,21]

The purpose of implanting chromaffin-cell grafts in the caudate nucleus is to restore nigrostriatal function. We have successfully demonstrated in two young patients that grafting of chromaffin cells into the lateral ventricle, with partial inclusion in the caudate nucleus, can result in amelioration of most signs of Parkinson's disease. Obviously, it remains to be seen whether this procedure will be of benefit in other types of patients, particularly those who are older.

REFERENCES

1. Freed WJ. Functional brain tissue transplantation: reversal of lesion-induced rotation by intraventricular substantia nigra and adrenal medulla grafts, with a note on intracranial retinal grafts. Biol Psychiatry 1983; 18:1205-67.
2. Freed WJ, Morihisa JM, Spoor HE, et al. Transplanted adrenal chromaffin cells in rat brain reduce lesion-induced rotational behaviour. Nature 1981; 292:351-2.
3. Freed WJ, Olson L, Ko G, et al. Intraventricular substantia nigra and adrenal medulla grafts: mechanisms of action and ³H spiroperidol autoradiography. In: Björklund A, Stenevi U, eds. Neural grafting in the mammalian CNS. Amsterdam: Elsevier, 1985:471-89.
4. Björklund A, Dunnett SB, Stenevi U, Lewis ME, Iversen SD. Reinnervation of the denervated striatum by substantia nigra transplants: functional consequences as revealed by pharmacological and sensoriomotor testing. Brain Res 1980; 199:307-33.
5. Dunnett SB, Björklund A, Stenevi U. Dopamine-rich transplants in experimental parkinsonism. Trends Neurosci 1983; 6:266-70.
6. Unsicker K, Krisch B, Otten U, Thoenen H. Nerve growth factor-induced fiber outgrowth from isolated rat adrenal chromaffin cells: impairment by glucocorticoids. Proc Natl Acad Sci USA 1978; 75:3498-502.
7. Olson L, Seiger A, Freedman R, Hoffer B. Chromaffine cells can innervate brain tissue: evidence from intraocular double grafts. Exp Neurol 1980; 70:414-26.
8. Snider SR, Carlsson A. The adrenal dopamine as an indicator of adrenomedullary hormone biosynthesis. Naunyn Schmiedebergs Arch Pharmacol 1972; 275:347-57.
9. Backlund E-O, Grandberg P-O, Hamberger B, et al. Transplantation of adrenal medullary tissue to striatum in parkinsonism: first clinical trials. J Neurosurg 1985; 62:169-73.
10. Perlow MJ, Kumakura K, Guidotti A. Prolonged survival of bovine adrenal chromaffin cells in rat cerebral ventricles. Proc Natl Acad Sci USA 1980; 77: 5278-81.
11. Rosenstein JM, Brightman MW. Intact cerebral ventricle as a site for tissue transplantation. Nature 1978; 276:83-5.
12. Wuerthele SM, Freed WJ, Olson L, et al. Effect of dopamine agonists and antagonists on the electrical activity of substantia nigra neurons transplanted into the lateral ventricle of the rat. Exp Brain Res 1981; 44:1-10.
13. Freed WJ, Cannon-Spoor HE, Krauthamer E. Intrastriatal adrenal medulla grafts in rats: long-term survival and behavioral effects. J Neurosurg 1986; 65:664-70.
14. Cooper IS. Surgical treatment of parkinsonism. Ann Rev Med 1965; 16:309-30.
15. Mason DW, Charlton HM, Jones A, Parry DM, Simmonds SJ. Immunology of allograft rejection in mammals. In: Björklund A, Stenevi U, eds. Neural grafting in mammalian CNS. Amsterdam: Elsevier, 1985:91-8.
16. Barker CF, Billingham RE. Immunologically privileged sites. Adv Immunol 1977; 25:1-54.
17. Lund RD, Hauschka SD. Transplanted neural tissue develops connections with host rat brain. Science 1976; 193:582-4.
18. Beebe BK, Møllgård K, Björklund A, Stenevi U. Ultrastructural evidence of synaptogenesis in the adult rat dentate gyrus from brain stem implants. Brain Res 1979; 167:391-5.
19. Bjorklund A, Johansson B, Stenevi U, Svendgaard N-A. Re-establishment of functional connections by regenerating central adrenergic and cholinergic axons. Nature 1975; 253:446-8.
20. Wurtman RJ, Pohorecky LA, Baliga BS. Adrenocortical control of the biosynthesis of epinephrine and proteins in the adrenal medulla. Pharmacol Rev 1972; 24:411-26.
21. Unsicker K, Rieffert B, Ziegler W. Effects of cell culture conditions, nerve growth factor, dexamethasone and cyclic AMP on adrenal chromaffin cells in vitro. In: Eränkö O, Soinila S, Päivärinta H, eds. Histochemistry and cell biology of autonomic neurons, SIF cells, and paraneurons. New York: Raven Press, 1980:51-9.

©Copyright, 1987, by the Massachusetts Medical Society
Printed in the U.S.A.

Glossary

adrenalectomy - removal of an adrenal gland

akinesia - a pathological poverty of movement; any disorder characterized by long periods of inactivity

apomorphine - a drug that activates dopamine receptors; when apomorphine is injected into an animal with a lesion to one substantia nigra, the dopaminergic substantia nigra neurons on the other side of the brain are activated, and this causes the animal to move in a circle away from the side of the lesion

autotransplanting - autografting; removing tissue from one part of an organism and implanting it in another part of the same organism

bradykinesia - a pathological slowness of movement

caudate nucleus - an important subcortical motor nucleus that is adjacent to the lateral ventricle; there are two caudate nuclei, one in the left hemisphere and one in the right hemisphere

chromaffin cells - dopamine-releasing cells of the adrenal medulla; they derive their name from their tendency to take up chromium salts

cogwheel sign - a common symptom of Parkinson's disease; when the arm of the patient is flexed by the physician, it moves in a series of jerks, just like a cogwheel

computed tomography - an X-ray technique that is used to view the living brain in three dimensions

craniotomy - any surgical procedure for opening the skull

dementia - severe intellectual deterioration

electromyography (EMG) - recording the gross electrical activity of muscles, usually from large electrodes on the surface of the skin

fetal - pertaining to the fetus

hypotension - low blood pressure

lateral ventricles - the left and right lateral ventricles are chambers in the cerebral hemispheres that are filled with cerebrospinal fluid

levodopa (L-DOPA) - a chemical precursor of dopamine, which is used in the treatment of Parkinson's disease; levodopa readily passes through the blood-brain barrier whereas dopamine does not

MPTP - a highly toxic drug that induces a disorder that is virtually identical to Parkinson's disease

parenchyma - the functional tissue of an organ, as opposed to its supportive or connective tissue

resting tremor - a tremor that is particularly pronounced when a patient is not moving; resting tremor is a symptom of Parkinson's disease

substantia nigra - a pair of brainstem nuclei that contain dopaminergic neurons whose axons project to the caudate nuclei; the dopaminergic neurons of the substantia nigra degenerate in cases of Parkinson's disease

Essay Study Questions

1. What are the symptoms of Parkinson's disease?

2. How do the authors think that an implant on one side of the brain can alleviate symptoms on both sides of the body?

3. A small piece of caudate nucleus was removed during the surgery; perhaps this removal is what had the therapeutic effect rather than the autotransplant. On what basis do the authors rule out this interpretation?

4. What evidence, if any, do Madrazo et al. provide to support their view that the adrenal autotransplantation procedure of Backlund et al. was not therapeutically effective because of the less-than-optimal site of their implants?

Multiple-Choice Study Questions

1. The adrenal graft performed by Madrazo et al. was held in place by
   a. surgical thread.
   b. brain screws.
   c. suction.
   d. staples.
   e. surgical glue.

2. Backlund and his colleagues were unsuccessful in their attempt to treat Parkinson's disease with adrenal autotransplants. Madrazo et al. suggest that this failure was due to the
   a. age of the patients.
   b. small number of patients.
   c. fact that the tissue was implanted inside the caudate rather than on its boundary with the lateral ventricle.
   d. use of MPTP.
   e. fact that there were not enough chromaffin cells in the transplant.

3. In both patients, the first signs of improvement were apparent within
   a. 15 days of surgery.
   b. 2 days of surgery.
   c. 10 weeks of surgery.
   d. 10 months of surgery.
   e. 5 months of surgery.

4. According to Madrazo et al., the best site for therapeutic adrenal autotransplantation is
   a. the lateral ventricle.
   b. the parenchyma of the caudate.
   c. a prepared site in the wall of the lateral ventricle.
   d. the substantia nigra.
   e. at the boundary between the caudate and substantia nigra.

5. Parkinson's disease is often associated with
   a. akinesia.
   b. bradykinesia.
   c. tremor at rest.
   d. all of the above
   e. both a and c

6. Madrazo and his colleagues measured the tremor of their two patients with
   a. electromyography.
   b. computed tomography.
   c. spectrophotography.
   d. all of the above
   e. both a and c

The answers to preceding questions are on page 290.

Food-For-Thought Questions

1. The procedure used by Madrazo and his colleagues is being adopted in hospitals throughout the world. Do you think that this one study warrants the widespread application of this procedure? Why do you think that most scientists are stressing the need for caution, whereas hospital administrators are racing to be among the first to adopt this procedure?

2. The authors provide no evidence for their hypothesis that the therapeutic effect of their autotransplantation procedure is attributable to the release of dopamine into the caudate. Design an experiment to test this hypothesis.

3. The authors suggest that dopamine released from the autotransplant travels throught the ventricular system to act on the caudate nucleus on the other side of the brain. If this were true, would it not make more sense to implant the dopaminergic tissue in the ventricular system half way between the two caudate nuclei? Discuss.

ARTICLE 4

Knowledge Without Awareness: An Autonomic Index of Facial Recognition by Prosopagnosics

D. Tranel and A.R. Damasio
Reprinted from Science, 1985, Volume 228, 1453-1454

Patients with bilateral damage to their occipital and temporal lobes sometimes display a neuropsychological disorder called prosopagnosia. Although their visual abilities appear to be largely intact, patients with prosopagnosia are unable to recognize faces—often, even their own.

In Tranel and Damasio's interesting experiment, two female prosopagnosic patients were each shown sets of 50 photographs of faces, 8 of which were randomly interspersed photographs of people who were familiar to them. Each set of 50 photographs was presented twice. First, each subject sat quietly as the effect of each photograph on their electrical skin conductance was recorded. Normally, increases in the ability of the skin to conduct electricity occur in response to emotion-inducing stimuli; these increases are called skin conductance responses (SCRs). During the second presentation of the photographs, each subject was asked to rate the familiarity of each face and to identify it if she could.

The two patients differed in their ability to recognize the familiar faces. Subject 1 did not recognize the faces of her family members or of famous personalities who had been familiar to her before the onset of her disorder. (She would not have recognized the faces of people whom she had met after the onset of her disorder either, but this was not formally tested.) Subject 1 was able to recognize the faces of people whom she had met before the onset of her disorder (e.g., family members and friends), but she could not recognize the faces of people whom she had gotten to know only after the onset of her disorder (e.g., her physicians and psychologists). Surprisingly, both patients displayed more and larger SCRs to the familiar than to the unfamiliar faces, even when they did not consciously recognize them or find them familiar.

The fact that the familiar faces, but not the unfamiliar faces, reliably induced large SCRs in the two prosopagnosic subjects proved that some part of their brain recognized the familiar faces although they were not consciously aware of it. This finding suggests that prosopagnosia is not, as is commonly assumed, a disorder in which no facial recognition occurs; it is a disorder in which facial recognition occurs but is unable to gain access to the circuits of the brain that mediate conscious awareness of the recognition.

Knowledge Without Awareness: An Autonomic Index of Facial Recognition by Prosopagnosics

Daniel Tranel and Antonio R. Damasio

Abstract. Prosopagnosia, the inability to recognize visually the faces of familiar persons who continue to be normally recognized through other sensory channels, is caused by bilateral cerebral lesions involving the visual system. Two patients with prosopagnosia generated frequent and large electrodermal skin conductance responses to faces of persons they had previously known but were now unable to recognize. They did not generate such responses to unfamiliar faces. The results suggest that an early step of the physiological process of recognition is still taking place in these patients, without their awareness but with an autonomic index.

Patients with prosopagnosia are unable to recognize visually the faces of persons they previously knew or ought to have learned without difficulty. They fail to experience any familiarity with those faces, and, even after they recognize the faces through other cues, such as voices, their physiognomies remain meaningless. Prosopagnosia is due to a complete failure to evoke memories pertinent to specific faces or to a defective evocation that fails to reach awareness. The condition is caused by bilateral damage to mesial occipitotemporal cortices or their connections.

Investigators of prosopagnosia have generally relied on the verbal report of the patient's experience as the sole index of recognition, an approach that does not address potential covert processes of which there may be no subjective awareness. In this study we used the electrodermal skin conductance response (SCR) as a dependent measure and found that two prosopagnosic patients generated significantly larger SCR's and responded more frequently to familiar faces than to unfamiliar ones (*1*). These results indicate that, despite their inability to experience familiarity with the visual stimulus and to provide verbal evidence of recognition, prosopagnosics still carry out some steps of the recognition process for which there is an autonomic index.

The subjects were two female patients with stable prosopagnosia caused by bilateral occipitotemporal damage, as determined from computerized tomography (CT) and nuclear magnetic resonance (NMR) imaging (*2*). We conducted several experiments. In each the patient was shown 50 black-and-white photographs of faces, depicting a full frontal pose on a white background (*3*). Forty-two of the faces were of persons entirely unfamiliar to the patient ("nontarget" faces) and eight were of persons with whom the patient was well acquainted ("target" faces). Both subjects were shown two sets of target faces selected from a period preceding the prosopagnosia (these target faces were randomly interspersed among the nontargets). In one of the sets, "family" faces, the target faces included those of the patient herself, family members, and close friends; in the other set, "famous" faces, the targets were famous politicians and actors. Subject 2 was exposed to a third set of target stimuli, "anterograde" faces, in which the targets were persons with whom the patient had had extensive contact since the onset of her illness but not before (physicians, psychologists, and so forth).

The subjects were given two presentations of each of the two sets of stimuli (or three sets, in the case of subject 2). During the first presentation skin conductance was recorded with Ag-AgCl electrodes from the thenar and hypothenar eminences of the nonpreferred hand on a Beckman type RM Dynograph recorder. Slides were presented for 2 seconds at intervals of 20 to 25 seconds. During the first viewing, no response was required of the subject; during the second, she was asked to verbally rate the familiarity of each face (*4*). Skin conductance was not recorded during the second presentation.

The results are presented in Table 1. As expected on the basis of her pervasive syndrome, subject 1 showed a complete failure to recognize any of the targets in the family and famous faces sets. Yet not only did she produce more frequent and consistent SCR's to the target stimuli, she also generated larger SCR's to the target faces than to the nontargets. The amplitude data were compared by the Mann-Whitney U test, a nonparametric test that avoids statistical assumptions not fulfilled by the data sets generated in this study. The average SCR amplitude for the target faces was significantly larger than that observed for the nontargets for both family faces ($U = 241$, $z = 4.01$, $P < 0.001$) and famous faces ($U = 265.5$, $z = 1.80$, $P < 0.05$) (*5*).

Subject 2 also evidenced more frequent and significantly larger SCR's to the target stimuli in the family faces ($U = 362$, $z = 4.63$, $P < 0.001$) and famous faces ($U = 204$, $z = 3.19$, $P < 0.001$) sets (Table 1), but, consistent with her lack of retrograde prosopagnosia, she also recognized accurately the familiar faces in these two sets. In the anterograde faces set, however, in which she was not able to recognize the target faces, she again produced more consistent and significantly larger SCR's to the target faces ($U = 283$, $z = 3.95$, $P < 0.001$). Thus this subject also showed a highly accurate autonomic index of recognition of familiar faces, despite a complete inability to experience familiarity with these faces and to recognize them formally.

The dissociation between the absence of an experience of recognition and the positive electrodermal identification may mean that in these subjects an early step of the physiological process of recognition is still taking place, but that the results of its operation are not made available to consciousness. Dissociations between overt recognition and unconscious discrimination of stimuli have been reported (*6*). Healthy subjects can show accurate autonomic discrimination of certain target stimuli, even when they are presented in a degraded or camou-

Table 1. Skin conductance response and verbal rating data for the two prosopagnosic subjects. For each category of faces (family, famous, and anterograde), two presentations of 8 target and 42 nontarget faces were made. The SCR data are based on the first presentation, while the verbal rating data are based on the second. Values in parentheses are standard deviations.

Sub-ject	First presentation				Second presentation; average verbal rating (4)	
	Stimuli responded to (%)		Average SCR amplitude (µS)		Target	Non-target
	Target	Non-target	Target	Nontarget		
Family faces						
1	71	12	0.934 (0.723)	0.048 (0.134)	6.0 (0.0)	6.0 (0.0)
2	100	36	1.660 (1.110)	0.146 (0.317)	1.0 (0.0)	5.1 (1.1)
Famous faces						
1	63	12	0.731 (0.652)	0.012 (0.034)	6.0 (0.0)	6.0 (0.0)
2	63	19	1.080 (1.420)	0.022 (0.052)	2.6 (1.9)	5.0 (1.2)
Anterograde faces						
2	75	17	0.345 (0.274)	0.022 (0.060)	4.4 (1.8)	4.6 (1.7)

flaged manner that precludes overt discrimination and identification (7). There is some parallel between such findings in healthy individuals and the observations described above, even if the mechanisms that lead to failure of recognition are different. Our results are also compatible with those of a recent study of prosopagnosia in a single patient, who showed discriminatory electrodermal responses to correct but not incorrect face-name matches (8).

We will attempt to interpret this "covert" recognition phenomenon in terms of a model of facial learning and recognition (9). The model includes step 1, perception; step 2, use of a template system, in which dynamic intramodal records of the elaboration of past visual perceptions of a given face can be aroused by the perception of that face (10); step 3, activation, in which multiple multimodal memories pertinent to the face are evoked; and step 4, a conscious readout of concomitant evocations that permits an experience of familiarity and either a verbal account of that experience or the performance of nonverbal matching tasks (11).

Prosopagnosia cannot be explained as being due to an impairment of the basic perceptual step (numerous indices of visual perception are normal, and patients are able to match unfamiliar faces and describe separate visual details of the faces). Nor can it be explained by an impairment of associated memories because they can be easily evoked through other channels. The defect may be explained, however, by an impairment of the activation step, which would either not take place or take place inefficiently. That, in turn, might be due to a dysfunction of the template system, which could be (i) intact but inaccessible to ongoing percepts, (ii) destroyed, or (iii) intact but prevented from activating multimodal memory stores. From the evidence above, it appears that facial templates are intact: the electrodermal "recognition" can be interpreted as being an index of successful matches between percepts, that is, correctly perceived target faces, and templates of those faces. Furthermore, the data on subject 2 suggest that, with respect to newly encountered faces, the process of template formation can proceed automatically in the absence of normal recognition processes.

The prosopagnosia of the two subjects can be viewed as a complete or partial blocking of the activation that normally would be triggered by template matching. From the anatomic specifications of the model (11), it appears that the blocking occurs either in white matter connections of the occipitotemporal region (linking both visual cortices to anterior temporal cortices, and the latter to multimodal sensory cortices) or in anterior temporal cortices.

According to the model, the findings presuppose the intactness, at least unilaterally, of the primary visual cortex and the inferior and mesial visual association cortex. Anatomic analyses of images of both patients obtained by CT and NMR verify these predictions (2). It is of great interest that the lesions that block activation of associated memories do not block the autonomic response. The anatomic substrates of the autonomic response remain to be elucidated.

DANIEL TRANEL
ANTONIO R. DAMASIO*
Department of Neurology,
Division of Behavioral Neurology,
University of Iowa College of
Medicine, Iowa City 52242

References and Notes

1. We previously showed, using the SCR as a dependent measure in neurologically intact subjects, that familiar faces have a notable "signal value" (D. Tranel, D. C. Fowles, A. R. Damasio, *Psychophysiology*, in press).
2. Subject 1 has both anterograde and retrograde prosopagnosia (she cannot recognize any faces that were familiar before the condition developed, including her own, nor has she learned any new faces during the 7 years of her prosopagnosia). She is a 62-year-old, right-handed woman who suffered bilateral strokes involving the occipitotemporal region. The lesion involves the white matter of both occipitotemporal regions but spares the most mesial and rostral regions of the inferior visual association cortex. As determined by a comprehensive neuropsychological assessment, her intellect, language, and visual perception are normal. Subject 2 has anterograde prosopagnosia only (she has not learned any new faces since the onset of her prosopagnosia 3 years ago). She is a 20-year-old, right-handed woman who had herpes simplex encephalitis leading to bilateral lesions of the occipitotemporal region. The lesions are located more anteriorly than in subject 1, so they also spare the mesial and rostral aspects of the inferior visual association cortex. Her language abilities are intact and her visual perception is compatible with normal recognition of faces learned before the onset of her illness.
3. The slides were constructed so that all the faces were of similar size. No slide contained features below the neckline, and no clothing was seen. The nontarget faces were selected so as to be similar to the targets in terms of age range and sex ratio. Brightness, contrast, and resolution were comparable in target and nontarget faces.
4. A six-point rating scale, ranging from "certain familiarity" (1) to "certain unfamiliarity" (6), was used. Ratings of 3 or less indicated some degree of familiarity with, or recognition of, the stimulus; conversely, ratings of 4 or greater indicated that the subject did not recognize the stimulus.
5. All P values are corrected for ties (data points with equal values), as recommended by S. Siegel [*Nonparametric Statistics for the Behavioral Sciences* (McGraw-Hill, New York, 1956), pp. 123–126].
6. J. K. Adams, *Psychol. Bull.* **54**, 383 (1957); M. F. Reiser and J. D. Block, *Psychosom. Med.* **27**, 274 (1965); C. Rousey and P. S. Holzman, *J. Pers. Soc. Psychol.* **6**, 464 (1967).
7. R. S. Lazarus and R. A. McCleary, *Psychol. Rev.* **58**, 113 (1951); R. S. Corteen and B. Wood, *J. Exp. Psychol.* **94**, 308 (1972).
8. The magnitude of the effect was not as large as the one reported here. The paradigm included both visual and verbal information, unlike our paradigms [R. M. Bauer, *Neuropsychologia* **22**, 457 (1984)].
9. A. R. Damasio, H. Damasio, G. W. Van Hoesen, *Neurology* **32**, 331 (1982).
10. The template system, which involves a dynamic process sensitive to the acquisition of new sensory information, permits the categorization of normal percepts according to physical structure. These dynamic templates do not contain information about the identity of a particular face. The latter information is stored not in a single site but rather in various sensory association cortices, including the visual, in both nonverbal and verbal forms. The templates serve as an interface between the repeated perception of a face and the multiple stored traces of information pertinent to the face; that is, the perception of a previously known face matches the respective template system, which in turn activates the pertinent multimodal memory stores.
11. The anatomic substrate of step 1 comprises bilateral visual system structures up to and including the primary visual cortex; the anatomical basis of step 2 is focused on bilateral mesial and inferior visual association cortices; the anatomic basis of step 3 includes bilateral anterior temporal structures, both mesial and lateral, and bilateral association cortices of different sensory modalities (9). Step 4 depends on the same association cortices.
12. We thank Dr. Hanna Damasio for providing the detailed anatomic analysis of CT and NMR images of the two subjects, Dr. Nelson Butters for referring subject 2 to our center, and Dr. Don Fowles for technical advice. Supported by National Institute of Neurological and Communicative Disorders and Stroke program project grant NS 19632-02.

* To whom correspondence should be addressed.

24 January 1985; accepted 4 April 1985

Glossary

autonomic - pertaining to the autonomic nervous system, that part of the peripheral nervous system that controls the organs of the body; skin conductance responses are mediated by the autonomic nervous system

bilateral - on both sides of the body

computerized tomography (CT) - a computerized X-ray technique for visualizing the living human brain in three dimensions

electrodermal - pertaining to the electrical properties of the skin

hypothenar eminence - the fatty portion of the palm of the hand that is touched by your little finger when you make a fist; a common site for attaching an electrodermal electrode

mesial - near the midline

nuclear magnetic resonance (NMR) - a technique for visualizing the living human brain in three dimensions; its powers of resolution are better than those of computerized tomography (CT)

occipitotemporal - involving both the occipital and temporal lobes of the brain

physiognomies - the facial features that identify an individual

prosopagnosia - a pathological difficulty in the recognition of faces that is not attributable to visual, intellectual, or motivational deficits; prosopagnosia is often associated with damage to the mesial portions of the occipital and temporal lobes

skin conductance responses (SCRs) - increases in the ability of the skin to conduct electricity that occur in response to arousing stimuli (e.g., in response to familiar faces)

thenar eminence - the fatty portion of the palm of the hand at the base of the thumb; a common site for attaching an electrodermal electrode

Essay Study Questions

1. Compare retrograde and anterograde prosopagnosia. Subject 1 experienced both disorders; which of the two disorders was experienced by Subject 2?

2. What autonomic indices, other than SCRs, might Tranel and Damasio have used in their study? Would their study have been more convincing if they had used more autonomic indices of recognition?

3. What evidence suggests that prosopagnosia is not the result of a simple visual impairment?

4. What evidence indicates that prosopagnosia is not the result of an inability to form memories?

5. To what do Tranel and Damasio attribute prosopagnosia?

6. Why were "nontarget faces" included in this study?

7. Why were tests of anterograde prosopagnosia not administered to Subject 1?

Multiple-Choice Study Questions

1. Prosopagnosia is often associated with bilateral damage to the
    a. hippocampus.
    b. primary visual cortex.
    c. temporal portions of occipital and parietal cortex.
    d. mesial portions of occipital and temporal cortex.
    e. temporal portions of frontal and parietal cortex.

2. The skin conductance response is
    a. an overt response.
    b. a covert response.
    c. an autonomic response.
    d. both a and c
    e. both b and c

3. The thenar eminence is in the
    a. temporal cortex.
    b. parietal cortex.
    c. frontal cortex.
    d. occipital cortex.
    e. none of the above

4. Prosopagnosics can
    a. match unfamiliar faces.
    b. describe individual details of faces.
    c. remember and identify people by nonvisual means.
    d. all of the above
    e. none of the above

5. The location of the brain damage in the two prosopagnosic subjects was determined by
    a. autopsy.
    b. Nissl staining.
    c. conventional X-ray technology.
    d. CT and NMR.
    e. SCR.

6. Subject 2 displayed an SCR to every
    a. family face.
    b. famous face.
    c. control face.
    d. familiar psychologist's face.
    e. familiar physician's face.

The answers to the preceding questions are on page 290.

Food-For-Thought Questions

1. Can this study of prosopagnosia be considered to be an empirical test of Freud's concept of the unconscious?

2. Tranel and Damasio point out that there is evidence that healthy subjects can show autonomic recognition of target stimuli that they do not consciously recognize. Have you ever experienced a situation in which you have experienced an emotional reaction to a stimulus (e.g., an odor, a face, or a place) that at first you did not consciously recognize? Describe it. What does your experience tell you about your own brain?

ARTICLE 5

A Face-Responsive Potential Recorded from the Human Scalp

D.A. Jeffreys

Reprinted from Experimental Brain Research, 1989, Volume 78, 193-202.

There are neurons in the superior temporal cortex of rhesus monkeys that seem to respond to faces but not to other complex visual stimuli. This suggests that evolutionary pressures may have granted facial stimuli a preferred status in the primate visual system. The fact that some human patients with combined temporal and occipital cortex damage cannot recognize faces even though their other perceptual abilities appear to be normal suggests that humans may also have face-specific neurons; however, this hypothesis has been difficult to test because research that involves invasive (i.e., penetrating the body) neurophysiological recording of the human brain is not usually possible for ethical reasons. Accordingly, Jeffreys recently took a noninvasive approach to the study of the hypothesized special status of faces in the human visual system. Jeffreys studied the changes in scalp EEG activity that were evoked in 9 human subjects by the tachistoscopic presentation of various visual stimuli, including faces. He found that in 8 of the 9 subjects stimuli that were perceived as faces produced a large positive wave, which was particularly obvious at electrode sites near the vertex (top of the head). This vertex positive potential (VPP) occurred between 150 and 200 milliseconds after the onset of each facial stimulus. Photographs, line drawings, caricatures, front views, side views, human faces, and nonhuman faces all elicited VPPs.

Jeffreys studied the features of VPPs in three subjects who had particularly large ones. The magnitude and polarity of VPPs were largely independent of the size of the face or its position in the visual field. However, presenting a face (2) against a patterned background, (3) to only one eye, (4) as a photographic negative, (5) upside down, or (6) with some of its facial elements missing increased the latency of VPPs and sometimes reduced their amplitude. These response properties are similar to those reported for single face-responsive neurons in the superior temporal cortex of the rhesus monkey.

In one important series of tests, Jeffrey's recorded VPPs between each of a series of active electrodes distributed in a line around the top of the head from ear to ear and connected reference electrodes over the mastoid projection of the temporal bones. He found that at electrode positions above the superior temporal lobe the VPP was positive, but at sites below the superior temporal lobe, it reversed its polarity (i.e., it was negative). This finding is consistent with the hypothesis that VPPs are generated by the activity of face-responsive neurons in the superior temporal lobes. There are few better examples of the productive and interesting interplay between biopsychological research on human and nonhuman subjects.

A face-responsive potential recorded from the human scalp

D.A. Jeffreys

Communication and Neuroscience Department, Keele University, Keele, Staffs. ST5 5BG, U.K.

Summary. Evoked potentials were recorded to the separate tachistoscopic presentation of a variety of faces and other simple and complex visual stimuli. A positive potential of 150–200 ms peak latency which responds preferentially, but not exclusively, to faces was identified in 8 out of 9 subjects. This potential, best recorded from midline central and parietal electrodes, was evoked by all face stimuli, including photographs, outline drawings, and fragmentary figures. Changes in stimulus size and other parameters which do not affect the clarity of the face, generally had little effect on the peak amplitude. Stimulus changes such as face inversion, reversing the contrast polarity of photographic images, and selectively removing particular facial features, produced a marked increase in latency but often only slight attenuation of this peak. These response properties correspond well with those reported for face-related single cells in the temporal cortex of the rhesus monkey. The scalp distribution of this face-responsive peak also appears consistent with bilateral sources in the temporal cortex.

Key words: Evoked potentials – Face responses – Temporal cortex – Human

Introduction

There have been many recent studies of the response properties of constituent neurones in temporal cortical areas of the rhesus monkey brain (for review see Maunsell and Newsome 1987). These studies have shown that, in contrast to the simple localized receptive fields of cells in occipital areas such as V1 and V2, the visual areas in temporal cortex contain cells with very large receptive fields and often very complex response properties. Cells have been described, for example, which respond best to complex 3-dimensional objects, to hands and, in particular, to faces (Bruce et al. 1981; Desimone et al. 1984; Perrett et al. 1982, 1984, 1985, 1987; Rolls 1984; Rolls and Baylis 1986; Rolls et al. 1985, 1987; Baylis et al. 1985).

According to Perrett et al. (1987), the face-related cells are found mainly in superior temporal sulcal cortex; they respond more strongly to faces or facial features than to other complex visual stimuli, and generally do not respond to geometric patterns (or to non-visual or emotive stimuli); they are relatively insensitive to size and location, or other parameters that do not affect the ability of humans to perceive the face; but their responses are attenuated and/or delayed by changes that do affect the perception of the face.

Assuming a broadly similar organization of the visual cortex in man and rhesus monkey, then we would expect to find face-sensitive neural mechanisms in localized regions of the temporal cortex in humans as well as in monkeys. There is of course no direct evidence of this, although it is known that localized occipito-temporal lesions can cause a selective inability to recognize familiar faces, i.e., prosopagnosia (see Damasio 1985 for review); there is also limited evidence that face-sensitive potentials can be recorded from the human scalp (Srebro 1985b; Bötzel and Grüsser 1987).

This paper describes an investigation of the visual evoked potentials (VEPs) recorded to a variety of face and non-face control stimuli. It forms part of a wider study of the functional organization of the human visual cortex, based on the identification of constituent VEP components of distinct surface topography and/or different response properties, which presumably reflect the underlying activity of geographically separate and/or physiologically distinct generators.

With this approach, which involves the recording of VEPs to many different stimuli in each individual subject (thus limiting the number of subjects), several distinct components have already been identified in the VEP recorded from posterior scalp locations in response to stationary patterned stimuli. These include 2 relatively long latency components, of apparent non-visuotopic cortical origin, which respond selectively to distinct, complex stimulus parameters (Jeffreys 1989; Jeffreys et al. 1986, 1987), in addition to 2 earlier components originating from visuotopic (occipital) regions and sensitive to local contours (Jeffreys and Axford 1972; Jeffreys 1977).

The results of the experiments described in this paper suggest that there is also a positive peak, of 150–200 ms latency, best recorded from midline central and parietal electrodes, which responds preferentially to faces. The response properties of this potential resemble in several respects those described above for the face-related cells of the monkey temporal cortex; and its distinctive surface topography appears to be consistent with a temporal cortical origin.

A preliminary account of some of these experiments was presented at a meeting of The Physiological Society (Jeffreys and Musselwhite 1987).

Methods

The basic stimulus was the sudden appearance of a stationary image into a previously diffuse field, presented by alternately switching 2 back-illuminated fields of a tachistoscope (Musselwhite and Jeffreys 1982). One (Stimulus) field contained a translucent photocopy or ink drawing of the appropriate face or non-face representation, and the other (Pre-stimulus) field was unpatterned; a 3rd unpatterned (Background) field was continuously illuminated. Field switching times were below 1 ms.

Several types of stimuli were used. These included fragmentary figures (Street 1931; Leeper 1935), photographs, outline drawings and adapted Mooney Faces (Mooney 1957; Parkin and Williamson 1987). Details of these faces and their dimensions within the 9 by 9 degrees square stimulus field are given in each figure. The stimulus duration, indicated in each figure by the horizontal line above the top time scale, was either 200 or 250 ms, and mean cycle period 1.1–1.2 s. The unpatterned areas of pre-stimulus and stimulus fields were set to the same luminance level, 136 cd/m^2; maximum contrast of the dark areas of the stimulus was 0.5, and viewing distance 64 cm.

In most experiments, VEPs were simultaneously recorded from 8 scalp locations, usually a coronal row of 5 electrodes at standard EEG placements, T3, C3, Cz, C4, and T4, and 3 additional midline electrodes at Fz, Pz, and Oz; with respect to a linked-earlobes reference. Additional recording locations and alternative references were used in more detailed topographic studies for 2 subjects.

The amplified and filtered (0.2–50 Hz) EEG signals were averaged over a 512 ms period (256 samples per channel at 2 ms sampling interval), starting 100 ms (occasionally 50 ms) before the stimulus onset, using a CED/CA Alpha computer. The VEPs were generally the average of the responses evoked by either 16 or 24 presentations of each stimulus, computed in 2 or 3 separate batches of 8 presentations to reduce subjective fatigue and the incidence of eyeblinks and EEG alpha activity. The responses to the first 2 or 3 repetitive presentations of the single stimulus figure in each run was excluded from the averaged response. The computer averaging programme allowed for the rejection of any interim responses with obvious EEG alpha or eyeblink contamination. For every stimulus, at least 2 non-consecutive runs were carried out and stored separately on floppy discs to monitor the repeatability of the averaged VEPs: the responses shown in most of the following figures were computed from the combined data for the 2 separate runs.

The stored VEP data were plotted off-line on a HP 7220A digital plotter. The plotting programmes allowed for optional addition or subtraction of any of the stored VEPs and for further digital filtering (CED 2nd-order Butterworth low-pass filter, plus reverse filtering to correct for phase shifts; corner frequency one tenth of sampling frequency). For clarity, this additional filtering was used for the illustrated response waveforms in this paper. The only quantitative analysis of the responses recorded in these experiments were simple measures of peak latency and amplitude, namely the mean post-stimulus sampling delays yielding maximal peak amplitude values (measured with respect to a −100 to +50 ms baseline) for the 2 repeated runs for each stimulus.

A total of 9 subjects were used, 5 of whom were inexperienced and unfamiliar with the purpose of the experiments. Only 2 experienced subjects (D.A.J. and M.J.M.) were available and/or suitable for the many experimental sessions required for some of the detailed studies of the VEP properties. One inexperienced subject (R.E.J.) took part in 2 sessions, the remainder only in single sessions. Subjects were asked to fixate on a small central fixation cross, but were given no other task except to report what they had seen at the end of each run.

Results

In the initial experiments, which were designed to look for scalp potentials sensitive to global rather than local stimulus parameters, the responses evoked by several fragmentary figures, originally developed by Street (1931) for Gestalt completion tests, were compared with those evoked by various control stimuli with no overall form (geometric patterns and arrays of discrete, irregularly shaped elements).

The results, illustrated in the left hand column of Fig. 1, showed that the more easily recognizable of the fragmentary figures evoked distinctive VEPs at midline central locations, whose predominant feature is a positive peak with a latency of 160–180 ms. This "vertex positive potential" (VPP) is often followed by a smaller and more variable "vertex negative potential" (VNP), of similar scalp topography, but this was not studied in these experiments. Of the fragmentary figures, the "man's head" (Fig. 1a) was always the most effective, and the "man on horseback" (Fig. 1f) the least effective stimulus for evoking the VPP. No such peak

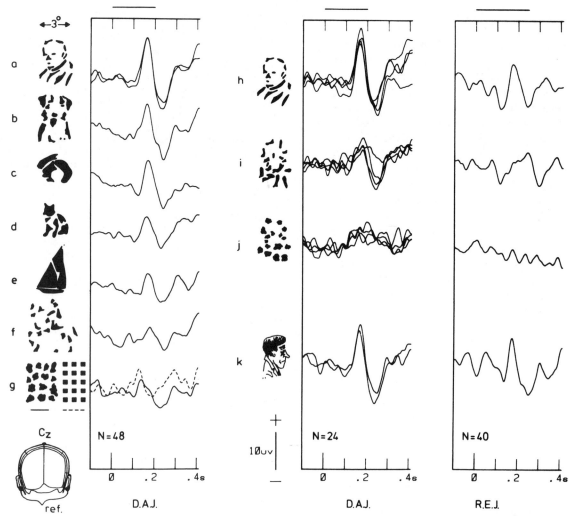

Fig. 1a–k. Left hand column: VEPs recorded from vertex (Cz) electrode to the illustrated fragmentary figures (**a–f**) and to arrays of irregular elements and squares (**g**). Subject D.A.J. Middle and right hand columns: VEPs recorded for subjects D.A.J. and R.E.J., respectively, to a fragmentary figure (**h**); to control stimuli of elements of corresponding location but different shape (**i**, **j**); and to an outline drawing of a profile face (**k**). The responses obtained in (2 or 4) non-sequential repeated runs have been superimposed in **a, h–k**. The measured (mean) peak latency (ms)/amplitude (uv) values were, for D.A.J.: **a** 164/10.6; **b** 163/6.8; **c** 160/5.6; **d** 168/4.5; **e** 172/3.3; **f** 181/2.8; **h** 168/11.6; **i** 180/5.2; **k** 170/9.1; and for R.E.J., **h** 174/4.5; **i** 211/4.7; **k** 176/6.3

was evoked by any of the control stimuli (e.g., Fig. 1g).

In further experiments, VEPs were recorded to the "man's head" fragmentary figure and to control stimuli containing constituent elements of corresponding location but different shape. The results in Fig. 1h–j show that such changes, which degrade the overall form of the figure, also markedly attenuate the VPP. These results also demonstrate, firstly, the consistent form of the VPP evoked by the same fragmentary figure at 4 different times during the course of a 2-day experiment (2 each day) for the experienced subject, D.A.J. (Fig. 1h); and, secondly, the essentially similar results, although poorer signal-to-noise ratio improvement, for the VEPs recorded to the fragmentary figure and control stimuli (Fig. 1h–j) for the inexperienced subject (R.E.J.).

In subsequent experiments, VEPs were recorded to a wider variety of both meaningful and non-meaningful stimuli. The results suggest: 1) that although the VPP can be evoked by some non-face images (e.g., Fig. 1e), a consistently larger amplitude response is produced by a face than by any other non-face stimulus so far tried; 2) that many different forms of face representation, ranging from realistic photographs to caricatures, are equally effective in evoking the VPP; and 3) that similar large VPPs are evoked by frontal, profile (Fig. 1k) and intermediate "half-profile" views

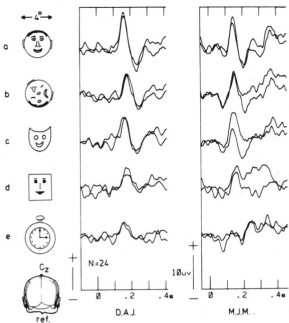

Fig. 2 a–e. Vertex-recorded VEPs to the illustrated outline figures for subjects, D.A.J. (left column) and M.J.M. (right column). Responses to 2 repeat runs have been superimposed in each case. Peak latency (ms)/amplitude (uv) values were, for D.A.J.: **a** 159/14.0; **b** 181/8.3; **c** 166/9.8; **d** 173/7.7; and **e** 155/5.8; and for M.J.M.: **a** 145/8.3; **b** 151/6.7; **c** 138/6.3; **d** 162/5.9; and **e** 195/4.3

of both animal and human faces; but that the back of the head is a less potent stimulus.

Figure 2 illustrates the responses obtained to a selection of line drawn stimuli for 2 (experienced) subjects. In both cases, the "cartoon" face evokes the largest and most clearly defined VPP, and even the distorted face (Fig. 2b) and simple "mask" (Fig. 2c) are more potent stimuli than the watch (Fig. 2e). Other relatively ineffective stimuli were: drawings and/or pictures of hands, feet, trees, flowers, and of various buildings and common household objects; single geometric shapes and letters; and displayed words and names of people and things.

Figure 3 illustrates the VEPs recorded in a further 6 (including 5 inexperienced) subjects, to the same line drawn face and watch stimuli as in Fig. 2. These results show that a positive peak of similar latency, but variable amplitude, is evoked by the face, but not the watch, in all but 1 (F.E.) subject. No results are shown for the remaining (experienced) subject because she was not tested with the stimuli shown in Fig. 3; but, a very small VPP was tentatively identified for other face stimuli.

In further studies of the influence of various spatial and temporal stimulus parameters, carried

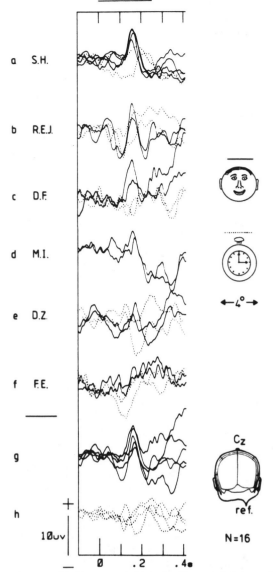

Fig. 3. a–f VEPs recorded to the illustrated face and watch drawings (continuous and dotted traces, respectively) in 1 experienced male subject (M.I.), and 5 inexperienced subjects; 2 of whom were male (S.H.; D.Z.) and 3 female. The mean response waveforms to the face and watch for the first 5 subjects have been superimposed in **g**, **h**, respectively

out on 3 subjects with relatively large VPP responses, the magnitude of the VPP was found to be largely independent of changes in either size or retinal site of stimulation which do not affect the clarity of the face stimulus. In Fig. 4, for example, responses of similar amplitude are evoked for faces ranging in size from 8.7 to 2.5 degrees (face length), although further size reductions, which make the face less distinct, result in an attenuation and increased delay of the VPP. In other experiments (not illustrated), a VPP of similar form and

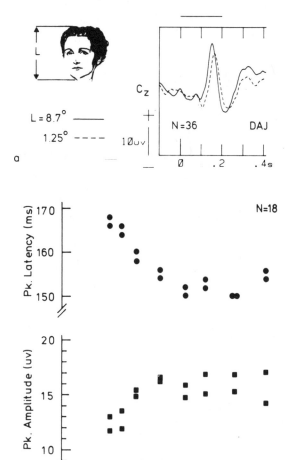

Fig. 4. a A comparison of the VEP waveforms evoked by large (continuous traces) and small (dashed traces) images of the same face; respective face length, L=8.7 and 1.25 degrees. **b** Peak latency and amplitude measures for the VEPs to each of 8 different face sizes. Data for 2 repeat runs for each stimulus size shown separately in **b**, but combined in **a**. Subject D.A.J., and linked mastoids reference

amplitude was found to be evoked by a circular outline face of 4 degrees diameter (see Fig. 2a) when fixated centrally, and at each of 4 different perimetric points: top, bottom, left and right, although the VPP latency was some 16–20 ms shorter (and, according to subjective reports, the face was seen more distinctly) for central, compared to eccentric fixation. The polarity, form and scalp distribution of the VPP remained unchanged by these variations in the retinal site of stimulation.

Several other changes in experimental conditions were also found to increase the latency and usually, but not always, to reduce the amplitude of the VPP. These included: 1) Presenting a face against a patterned instead of an unpatterned background. 2) Presenting a face monocularly instead of binocularly; this produced a VPP of corresponding amplitude but of 6–10 ms longer latency for subject D.A.J. 3) Replacing a normal picture of a face by its photographic negative. Such a contrast reversal invariably produced a latency increase (20–30 ms for subject, D.A.J.), and, for some, but not all, face stimuli, an attenuation of the VPP.

The VPP latency was also found be influenced by changes in face orientation. For example, as illustrated in Fig. 5a, b, the inversion or rotation of a well-defined face results in a delayed (by between 18 and 25 ms) and slightly attenuated VPP.

In further studies of the influence of stimulus orientation, VEPs were recorded the normal and inverted presentation of 2 adapted "Mooney Faces" (Mooney 1957; Parkin and Williamson 1987), which are only recognized as faces when upright. The results are illustrated for 2 subjects in Fig. 5c, d. These results show that whereas a relatively large VPP is evoked by each upright Mooney figure, no comparable peak is evoked when the same stimuli are inverted, and no longer seen as faces. This contrasts with the relatively large VPP evoked by the inverted stimulus in Fig. 5a, which is still seen as a face. As can be seen in Fig. 5, essentially similar results to those obtained for the experienced subject (D.A.J.), were also obtained for the inexperienced subject (R.E.J.), who was given no prior information about the nature of the stimuli, and was first presented with the 2 inverted Mooney faces.

Fig. 6 illustrates the results of selectively removing different constituent features of a face stimulus on the recorded VEPs in 2 subjects. These results show that the complete face evokes a large VPP (dotted traces) for each subject, although the latency of this peak is over 20 ms shorter for one (M.J.M.) than for the other. In both cases, the main result of removing selected internal and/or external facial features is to increase the VPP latency; there are no obviously systematic changes in peak amplitude. Moreover, the induced delay varies according to which particular facial feature is removed or presented. Removing only the nose and mouth, for example, has very little effect, whereas presenting the nose and mouth alone, produces VPP delays of more than 50 ms for both subjects.

The already described properties of the VPP were found to be largely independent of the temporal parameters of stimulation. For my stimuli, a presentation time of 20 ms or less is sufficient to

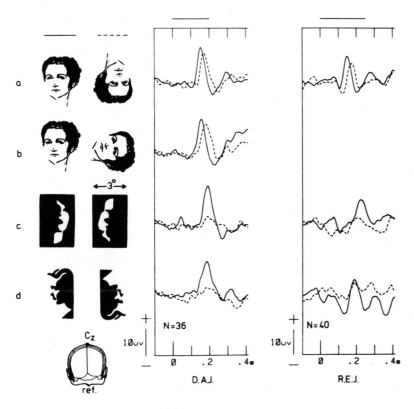

Fig. 5a–d. Vertex-recorded VEPs for an experienced (D.A.J.), and an inexperienced (R.E.J.) subject, recorded to: **a** the upright and inverted presentation of a well-defined face stimulus; **b** upright and rotated presentation of the same face; **c, d** upright and inverted presentation of adapted Mooney Face stimuli (Parkin and Williamson 1987) which are seen as faces only when upright. Measured mean peak latency (ms)/amplitude (uv) values for the continuous and dashed traces, respectively, were, for D.A.J.: **a** 152/12.9 and 170/10.8; **b** 154/12.7 and 179/10.9; **c** 193/14.4 and 205/2.8; **d** 188/12.6 and 210/3.0; and for R.E.J.: **a** 150/10.2 and 174/7.3; **c** 222/9.5 and 212/2.7; **d** 190/5.8 and 236/5.4

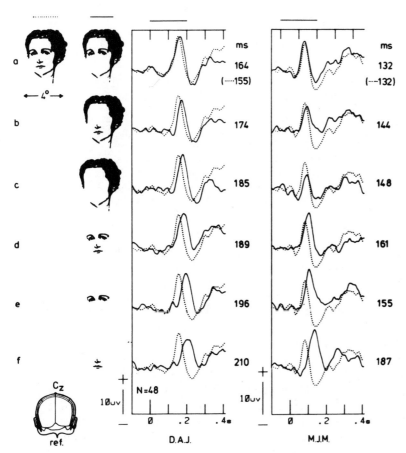

Fig. 6a–f. Vertex-recorded VEPs in 2 subjects (D.A.J. and M.J.M.) to a complete face (dotted traces) and to the same face but with the indicated missing features (continuous traces). Peak latency values shown on the right. Peak amplitude (uv) values for the dotted trace and for dashed traces in **a–f**, respectively, were, for D.A.J.: 12.7; 12.4, 10.8; 13.1; 12.6; 12.2; and 10.4; and for M.J.M.: 7.3; 8.0; 6.3; 3.8; 9.7; 10.6; and 10.0

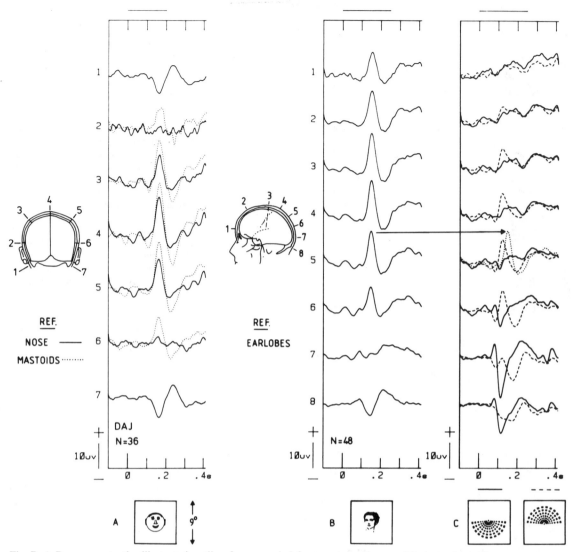

Fig. 7. A Responses to the illustrated outline face recorded from a coronal row of 7 electrodes with respect to a nose reference (continuous traces), and from the central 5 electrodes of the same row with respect to a linked mastoids reference (dotted traces). Electrode locations 1 to 7: left mastoid; T3; C3; Cz; C4; T4; right mastoid; respectively. **B** VEPs to an outline face stimulus recorded from a midline row of 8 electrodes. Electrode locations 1–8, Fp; Fz; Cz; midway between Cz and Pz; Pz; midway between Pz and Oz; Oz; inion; respectively. **C** VEPs to the lower (continuous traces) and upper (dashed traces) half-field presentation of the illustrated pattern, recorded from the same midline electrodes as in **B**. The dotted trace corresponds to the face-evoked response at electrode 5 in **B**. Subject D.A.J

evoke a VPP of near maximal amplitude, and further increases in stimulus duration have little effect on the recorded peak amplitude, waveform or latency (provided the interval between sequential stimulus presentations is at least twice as long as the stimulus duration).

In experiments with Dr. M.J. Musselwhite, using computer-generated faces, which will be reported separately, we were able to compare the VEPs to several different stimuli when presented either in random sequence in the same experimental run, or, as in this study, presented singly in separate runs. The results suggest that such changes in the mode of presentation do not affect the properties of the face-evoked VPP.

Studies of the VEP topography using various recording and/or common reference locations were carried out in 2 subjects. The main results, which were similar for both subjects and for the various face stimuli used, are illustrated in Fig. 7.

Figure 7A shows responses obtained with an outline face stimulus and a coronal row of 7 equally spaced electrodes extending from the left mastoid to the right mastoid. The VPP recorded from the 5 central electrodes with respect to a linked-mastoids reference (dotted traces) is seen to have a fairly symmetrical monophasic distribution with a maximum at the vertex. However, when the face-evoked responses are recorded from all 7 electrodes with respect to a nose reference, the VPP is seen to reverse polarity at the level of the T3 and T4 electrodes, from positive above to negative below.

Figure 7B, C show the responses evoked by a face, and by the lower and upper half-field presentation of a pattern of radially expanding discrete elements, respectively, recorded from a midline array of 8 electrodes. These results show, that the VPP is evoked only by the face stimulus, and that its scalp distribution differs from that of any of the major pattern-evoked peaks in Fig. 7C.

The face-evoked VPP also differs from the early pattern-evoked VEP components in that its polarity and scalp topography is independent of changes in retinal site of stimulation. This contrasts with the variable polarity and topography of the early pattern VEP peaks. In Fig. 7C, for example, the 2 initial peaks of the lower-field pattern VEPs (continuous traces), are respectively positive and negative, with localized occipital distributions; whereas the contemporary peaks of the upper-field pattern VEPs (dashed traces) are reversed in polarity and more widely and anteriorly distributed.

For the early pattern-evoked peaks, the response evoked by a full-field pattern is approximately the sum of the separate, and opposite polarity, contributions from the upper and lower half-fields (Jeffreys and Axford 1972; Jeffreys 1977). The apparent absence of such early components in the VEP recorded at posterior electrodes in Fig. 7B can be shown to be due to the cancellation of the opposite polarity contributions from upper and lower field portions of the centrally-fixated face stimulus. These early peaks can be clearly identified in the responses recorded at occipital electrodes to a face stimulus with a preponderance of contours in the lower visual field. Conversely, a face stimulus with many discrete contours in the upper visual field can evoke a pattern-related positive peak at parietal and central electrodes. Moreover, because of the partial temporal overlap of the respective waveforms of this pattern-evoked potential and the VPP (see Fig. 7C, electrode 5), these distinct potentials could merge to form an apparently single broad positive peak. The VPP, however, can be readily distinguished from the earlier positive pattern-evoked peak (and from any eyeblink potential artifact) by its distinctive longitudinal scalp distribution, when the VEPs are recorded from a midline array as in Fig. 7B.

Discussion

This paper has described a scalp potential with the following properties: it responds to images of faces and, to a lesser degree, some other familiar objects, but not to less meaningful control stimuli; it is evoked by a variety of face images, including distorted and incomplete representations; its amplitude is largely independent of the stimulus size, location, or other parameters which do not affect the clarity of the face image; but changes which do affect the perception of the stimulus figure, such as the inversion (Fig. 5a), or removal of some constituent features (Fig. 6), of a face stimulus, results in a delayed and often, but not always, attenuated response.

This "vertex positive potential" was identified in 8 of a total of 9 subjects, but it had a very variable amplitude and was in some cases too small for detailed study (see Fig. 3). In this small sample, there was no obvious connection between the VPP size and either the subject's gender or his/her familiarity with the aims of the experiment, although the largest VEPs were recorded for the author. The properties of the VPP were, however, very consistent for the 3 subjects used for the bulk of the experiments described in this paper.

The different surface topography of the VPP compared to that of any response peak evoked in the same subject by the more conventional pattern stimuli (Fig. 7) indicates that it has a distinct generative site. Its scalp distribution also provides clues to possible sites of origin of the VPP, although the exact location within the brain of the generators of this, or any other, VEP component cannot be determined unequivocally from surface VEP recordings alone (Nunez 1981).

For example, the results in Fig. 7A, show that, with a nose reference, the VPP (and VNP) reverses polarity at the T3 and T4 electrodes. For a truly "inactive" reference, such a potential distribution accords with that produced by *tangentially oriented* dipole generators i.e., with generative regions of cortex oriented *perpendicular* to the overlying scalp, directly beneath the site of the polarity reversal. Since the T3 and T4 scalp placements have been shown to be located close to the superior temporal sulci (Homan et al. 1987), then this would suggest bilateral sites of origin in cortex within the temporal sulci. However, since the form of the response distribution is dependent on the choice of reference electrode, the positions of the polarity reversals, i.e., electrodes T3 and T4 for the VPP [but, interestingly, at slightly higher coronal levels overlying the auditory cortex in the Sylvian fissures for the late peaks or of the auditory evoked potential (Vaughan and Ritter 1970)], do not necessarily indicate the underlying source locations (Nunez 1981).

The coronal topography of VPP is nevertheless consistent with a possible site of origin in either the superior temporal sulci, or ventral temporal cortex. The invariant polarity and scalp topography of the VPP for face stimuli presented in the different visual half-fields, also conforms with generative sites in non-visuotopic cortical areas.

Relation to other VEP studies

The VPP differs in polarity, latency and/or response properties from other scalp potentials, such as the classical "vertex potential" and the "cognitive" P300 component which also have maximal amplitudes at or near the vertex (for review, see Regan 1989). The non-specific "vertex potential", unlike the VPP, can be evoked by an unpatterned flash stimulus, and is characterized by an initial negative peak (N1) at approximately 140 ms followed by a positive peak (P2) at 200 ms (Hillyard and Picton 1979). The P300 potential, is generally evoked by task-relevant, unexpected stimuli, and so is unlikely to be recorded to the monotonous stimulus presentation used here to evoke the VPP.

There are very few reports of experiments involving a straightforward comparison of the responses evoked by face and non-face stimuli, and none is directly comparable to the present study. In one study, Small (1983) recorded VEPs to known and unknown faces and to geometric shapes, and found a right-greater-than-left amplitude asymmetry of the P300 peak for faces but not shapes. Although the illustrated VEPs in Small's fig. 3 show clear differences between the response waveforms recorded to the face and figure stimuli, they are difficult to relate to the present results because VEPs were recorded only from lateralized scalp locations, with respect to a midfrontal reference.

Srebro (1985b) computed the Laplacian derivative of VEPs to brief presentations of computer generated faces and simple geometric figures corrupted by varying amounts of noise. He reported basically similar results for the faces and the figures: in both cases, there was a difference between the Laplacian response associated with the report that the face or figure was seen and that associated with the report that the face or figure was not seen. This difference in the Laplacians had a simple waveform with peak activity at about 206 ms after stimulus onset, and was localized to both temporal lobes, with some degree of right hemispheric lateralization. Direct comparison is again complicated by the different recording techniques, but Srebro's subtracted Laplacian response clearly differs from the VPP in its equal sensitivity to simple shapes (which do not evoke a VPP) as well as to faces. Moreover, if the equivalent current generators of the VPP are in temporal cortex and oriented tangentially to the scalp surface, then they are relatively inaccessible to Laplacian recording techniques (Srebro 1985b).

Bötzel and Grüsser (1987) briefly described the VEPs to the transition between any 2 of 3 distinct pictorial categories: either line drawings of a face, a tree, and a chair; or photographs of faces, vases, and shoes. They found a triphasic, negative/positive/negative sequence of "face-specific" peaks, with respective latencies of 140–160 ms, 210–240 ms, and 300 ms, for the photographs. These peaks were most pronounced at the vertex electrode (Cz); more marked in female than male subjects; and recorded whether or not the experimental procedure included a recognition task. Bötzel and Grüsser suggested deep limbic structures as their likely sites of origin.

In experiments comparing the VEPs recorded to the transition from different pre-stimulus fields to the same face stimulus, I have found that a VPP of identical form and topography but longer latency is evoked with structured (various non-face pictures), compared to diffuse prestimulus fields. The actual latency increase varies according to the type of non-face pattern, but can be as much as 40 ms. It therefore seems likely that Bötzel and Grüsser's P210–240 and N300 peaks, best recorded at midline Cz and Pz electrodes, correspond to the the VPP and succeeding VNP peaks evoked by face stimuli in these experiments.

Conclusions

The properties of the VPP, in particular its sensitivity to faces; its relatively invariant response with respect to parameters such as size and position; and its delayed response for changes such as face inversion which hinder the perception of face; suggest that its generators play some role in the perception of complex and meaningful stimuli. The VPP also appears more likely to reflect a stage of essentially visual processing, probably in temporal cortex, than any general psychological response to the appearance of a face, for the following reasons:

1) The properties of the VPP correspond to those of face-related cells in the monkey's temporal cortex (Perrett et al. 1987). 2) The VPP scalp topography is consistent with a temporal cortical origin. 3) The VPP latency is considerably shorter

than most "cognitive" EP potentials (Regan 1989); and for optimal stimulus conditions (as in Fig. 7), it is only some 30–40 ms later than that of the (local) contour-specific "C2" component, produced in a visuotopic region of occipital cortex (Jeffreys and Axford 1972; Jeffreys 1977, 1989). [By comparison with onset latencies of 80–160 ms for the face-related neurones (Perrett et al. 1982, 1984), the VPP had typical peak values of 150–200 ms (onset latencies 120–170 ms) for the various faces used here.] 4) The VPP has a brief, and highly consistent (Fig. 1h), waveform, similar to those of the earlier stimulus-specific peaks (e.g., Fig. 7C).

More detailed and quantitative studies of the VPP, and of its relationship to other VEP components, are now needed, using more subjects and a greater variety of stimuli, to confirm (or not) and extend the findings and conclusions of these preliminary experiments. The results so far suggest that such studies can provide valuable information about the analysis of faces and other meaningful visual stimuli in the human brain.

Acknowledgements. I wish to thank Grant Rockley for his technical assistance, Steven Duffy for his help in preparing the figures, and all the subjects for their patient cooperation.

References

Baylis GC, Rolls ET, Leonard CM (1985) Selectivity between faces in the responses of a population of neurons in the cortex in the superior temporal sulcus of the monkey. Brain Res 342:91–102

Bötzel K, Grüsser O-J (1987) Potentials evoked by face and nonface stimuli in the human electroencephalogram. Perception 16:A21

Bruce C, Desimone R, Gross CG (1981) Visual properties of neurons in a polysensory area in superior temporal sulcus of the Macaque. J Neurophysiol 46:369–384

Damasio AR (1985) Prosopagnosia. Trends Neurosci 8:132–135

Desimone R, Albright TD, Gross CG, Bruce C (1984) Stimulus-selective properties of inferior temporal neurons in the macaque. J Neurosci 4:2051–2062

Hillyard SA, Picton TW (1979) Event-related brain potentials and selective information processing in man. In: Desmedt JE (ed) Cognitive components in cerebral event-related potentials and selective attention. Karger, Basel, pp 1–52

Homan RW, Herman J, Purdy P (1987) Cerebral location of international 10–20 system electrode placement. Electroenceph Clin Neurophysiol 66:376–382

Jeffreys DA (1977) The physiological significance of pattern visual evoked potentials. In: Desmedt J (ed) Visual evoked potentials in man: new developments. Clarendon Press, Oxford, pp 134–167

Jeffreys DA (1989) Evoked potential studies of contour processing in human visual cortex. In: Kulikowski JJ, Dickinson CM (eds) Contour and colour. Pergamon, London, pp 525–541 (in press)

Jeffreys DA, Axford JG (1972) Source locations of pattern-specific components of human visual evoked potentials. I. Component of striate cortical origin. II. Component of extrastriate cortical origin. Exp Brain Res 16:1–40

Jeffreys DA, Musselwhite MJ (1987) A face-responsive visual evoked potential in man. J Physiol 390:26P

Jeffreys DA, Murphy F, Musselwhite MJ (1986) Human pattern-evoked potentials recorded with textured backgrounds. J Physiol 382:97P

Jeffreys DA, Musselwhite MJ, Murphy F (1987) A study of the late components of the pattern-onset VEP. Electroenceph Clin Neurophysiol 67:51P

Leeper R (1935) A study of a neglected portion of the field of learning – the development of sensory organization. J Gen Psychol 46:41–75

Maunsell JHR, Newsome WT (1987) Visual processing in monkey extrastriate cortex. Ann Rev Neurosci 10:363–401

Musselwhite MJ, Jeffreys DA (1982) Pattern-evoked potentials and Bloch's law. Vision Res 22:897–903

Mooney CM (1957) Age in the development of closure ability in children. Can J Psychol 11:219–226

Nunez PL (1981) Electric fields of the brain. Oxford University Press, New York

Parkin AJ, Williamson P (1987) Cerebral lateralization at different stages of facial processing. Cortex 23:99–110

Perrett DI, Rolls ET, Caan W (1982) Visual neurones responsive to faces in the monkey temporal cortex. Exp Brain Res 47:329–342

Perrett DI, Smith PAJ, Potter DD, Mistlin AJ, Head AS, Milner AD, Jeeves MA (1984) Neurones responsive to faces in the temporal cortex: studies of functional organization, sensitivity to identity and relation to perception. Hum Neurobiol 3:197–208

Perrett DI, Smith PAJ, Potter DD, Mistlin AJ, Head AS, Milner AD, Jeeves MA (1985) Visual cells in the temporal cortex sensitive to face view and gaze direction. Proc R Soc Lond B223:293–317

Perrett DI, Mistlin AJ, Chitty AJ (1987) Visual neurones responsive to faces. Trends Neurosci 10:358–364

Regan D (1989) Human brain electrophysiology: evoked potentials and evoked magnetic fields in science and medicine. Elsevier, Amsterdam

Rolls ET (1984) Neurons in the cortex of the temporal lobe and in the amygdala of the monkey with responses selective to faces. Hum Neurobiol 3:209–222

Rolls ET, Baylis GC, Leonard CM (1985) Role of low and high spatial frequencies in the face-selective responses of neurons in the cortex in the superior temporal sulcus in the monkey. Vision Res 25:1021–1035

Rolls ET, Baylis GC (1986) Size and contrast have only small effects on the responses to faces of neurons in the cortex of the superior temporal sulcus of the monkey. Exp Brain Res 65:38–48

Rolls ET, Baylis GC, Hasselmo ME (1987) The responses of neurons in the cortex in the superior temporal sulcus of the monkey to band-pass spatial frequency filtered faces. Vision Res 27:311–326

Small M (1983) Asymmetrical evoked potentials in response to face stimuli. Cortex 19:441–450

Srebro R (1985a) Localization of visually evoked cortical activity in humans. J Physiol 360:233–246

Srebro R (1985b) Localization of cortical activity associated with visual recognition in humans. J Physiol 360:247–260

Street RF (1931) A Gestalt completion test. Teachers College, Columbia University, New York

Vaughan HG, Ritter W (1970) The sources of auditory evoked responses recorded from the human scalp. Electroenceph Clin Neurophysiol 28:360–367

Received May 17, 1988; received in final form June 5, 1989 / Accepted June 9, 1989

Glossary

binocularly - involving two eyes

coronal - refers to any vertical plane through the brain that cuts straight across it; for example, a vertical plane that cuts through both ear lobes is a coronal plane

EEG signals - electroencephalographic signals; in human subjects, they are typically recorded through large scalp electrodes; EEG signals are complex signals that reflect the total electrical activity of the brain

fixate - look at a point without moving one's eyes

Gestalt - an approach to the study of vision that emphasizes that the perceived whole of a stimulus has properties that are more than the sum of the properties of its parts

mastoid - a portion of the temporal bone of the skull; the mastoid is just in front of and a bit above the ear canals; mastoid electrodes are those placed over the mastoid; because there is no neural tissue directly under the mastoid, it is a good location for the placement of reference electrodes because of its relative electrical silence

monocularly - involving one eye

monophasic - refers to any signal that is composed of a single, unidirectional change from baseline followed by a return

Mooney Faces - drawings of faces that are used in certain kinds of perception research; their major feature is that they are not perceived as faces if they are presented upside down

occipital lobes - the lobes of the cerebral hemispheres that are situated at the back of the head; the cortex of the occipital lobes mediates vision

polarity - refers to the relation between the electrical charge at two sites; in EEG recording, it refers to the relation, positive or negative, between the voltage at the active electrode and the voltage at the reference electrode

posterior - toward the back of the brain

prosopagnosia - a brain-damage-produced deficit in the recognition of faces

receptive field - the receptive field of a visual system neuron is the area in the visual field within which stimuli can influence that neuron's firing

reference electrodes - an EEG trace is a record of the difference in voltage between two electrodes as it changes over time; it is common to record the difference in voltage between an electrode that is placed directly over the cortex and another electrode, called a reference electrode, which is placed at a sight that is electrically silent, so that any change in the voltage difference between the two sites can be attributed to changes at the active electrode; common sites for reference electrodes are the tip of the nose, an ear lobe, or over the mastoid portion of the temporal bone

sulcal cortex - the cortex in a cerebral sulcus (i.e., the cortex in a small cerebral fissure)

superior - in neuroanatomy, "superior" means toward the top of the brain; hence, the superior temporal sulcus is a sulcus near the top of the temporal lobe

tachistoscope - a research device for the precise presentation of visual stimuli

temporal lobes - the lobes of the cerebral hemispheres that are adjacent to the temples; there are many neurons in the cortex of the superior temporal sulcus that respond specifically to faces that are presented in their receptive fields

topography - refers to the distribution of a signal over the cortex

vertex - the very top of the head

vertex positive potential (VPP) - a positive EEG potential that is often recorded from the vertex between 160 and 180 milliseconds after the presentation of a recognizable face

visual evoked potential (VEP) - any physiological signal evoked (i.e., elicited) by the presentation of a visual stimulus

Essay Study Questions

1. What is an average evoked response? Why did Jeffreys study average evoked responses rather than evoked responses? Jeffreys averaged the evoked responses recorded from each subject; why didn't Jeffreys average the evoked responses across subjects to get a group average?

2. What is a VPP? What observations suggested that VPPs are correlated with the perception of faces?

3. What variables influence the amplitude and latency of VPPs?

4. What evidence is there that VPPs are generated in humans by the activity of neurons in the superior temporal lobes?

5. What are the major differences between VPPs and the small, short-latency components of visual evoked responses?

Multiple-Choice Study Questions

1. Face-responsive neurons have been discovered in the cortex of the superior sulcus in the
    a. temporal lobes of rhesus monkeys.
    b. temporal lobes of people.
    c. parietal lobes of rhesus monkeys.
    d. occipital lobes of rhesus monkeys.
    e. frontal lobes of rhesus monkeys.

2. The small, short-latency components of visual evoked responses recorded from posterior scalp locations are
    a. sensitive to changes in contour.
    b. sensitive to changes in position.
    c. particularly sensitive to faces.
    d. all of the above
    e. both a and b

3. The traces plotted by Jeffreys in the various figures of his article are, strictly speaking,
    a. individual visual evoked responses.
    b. bursts of unit activity.
    c. average visual evoked responses.
    d. Gestalt evoked responses.
    e. fragmentary evoked responses.

4. The test with the Mooney figure was particularly informative because the
    a. Mooney figure is a face, and yet it produced no VPP.
    b. Mooney figure is not a face, and yet it produced a VPP.
    c. up-side-down Mooney figure was not seen as a face, and it did not produce a VPP.
    d. upside-down Mooney figure was not seen as a face, and yet it produced a VPP.
    e. Mooney figure produced a VPP both when it was in its normal orientation and when it was upside-down.

5. The magnitude of the VPP was greatest
    a. over the superior temporal lobe.
    b. at the vertex.
    c. over the mastoid.
    d. at the tip of the nose.
    e. none of the above

6. The earlobes, the nose, and the mastoids are good sites for
    a. safety pins.
    b. electrodes.
    c. VPPs.
    d. face-specific neurons.
    e. reference electrodes.

7. Paradoxically, when the vertex positive potential was recorded from near the ear, it was
    a. negative.
    b. slower.
    c. faster.
    d. bigger.
    e. not sensitive to faces.

The answers to the preceding questions are on page 290.

Food-For-Thought Questions

1. Progress in biopsychology often depends on the productive interplay between research on human and nonhuman subjects. Discuss this point with respect to Jeffreys' experiments. How might the relation between vertex positive potentials and superior-temporal-lobe face-responsive neurons be studied in monkeys?

2. The uninitiated tend to equate research quality with random sampling, large samples, and the lavish application of statistical tests. Discuss with respect to Jeffreys' article.

ARTICLE 6

Vision Guides the Adjustment of Auditory Localization in Young Barn Owls

E.I. Knudsen and P.F. Knudsen
Reprinted from Science, 1985, Volume 230, 545-548.

In order for an animal to function effectively, it is important that its auditory map of the external world be aligned with its visual map: objects must sound to be where they look to be. This is particularly true for animals such as barn owls, whose nocturnal predatory habits depend on their ability to localize their prey under very dim illumination. The ability of the barn owl's auditory system to accurately localize sounds is impressive because auditory space is not mapped directly onto the cochlea of the inner ear in the same way that visual space is mapped onto the retina. Consequently, the auditory system must compute the location of a sound from differences between the two ears in the sound's loudness and timing—sounds from the right are louder in the right ear, and they arrive there first. Because the precise interpretation of binaural sound-location cues depends on the size and shape of each animal's head and ears, their interpretation must be continually adjusted as the animal grows. The following experiment of Knudsen and Knudsen demonstrates that information provided by the visual system guides the adjustment of auditory localization in young barn owls.

Knudsen and Knudsen placed an ear plug in one ear of each of a group of baby barn owls; he left the plugs in place for several weeks. During this period, the owls learned to accurately orient their heads toward brief sounds and lights in a dark test chamber. Eye-glass frames were mounted on each of the young barn owls 1 week before its ear plug was removed. When the plugs were removed, all of the young barn owls oriented accurately toward brief lights, but they made systematic errors in their orientation toward brief sounds. As soon as this stage of the testing was completed, the eye-glass frames were fitted with either (1) opaque occluders, which eliminated all visual input, or (2) prisms that displaced visual input by 10°, or (3) the eye-glass frames were left blank as a control procedure. Twenty-eight days later, the control barn owls displayed no systematic errors in auditory localization, whereas the errors of the barn owls that had been wearing the visual occluders were unchanged. In contrast, the owls that had been wearing the prisms displayed errors in auditory localization that accurately matched the visual localization error that had been introduced by their prisms.

These results demonstrate that in developing barn owls—and presumably in developing humans—the visual system provides the spatial information that is critical for the fine-tuning of auditory localization. When there is a slight difference between where a young barn owl sees objects to be and where it hears them to be, the auditory system learns to reinterpret the binaural cues from which it infers the location of sounds in such a way that perceived auditory location is in accord with perceived visual position.

SCIENCE

Vision Guides the Adjustment of Auditory Localization in Young Barn Owls

Abstract. *Barn owls raised with one ear plugged make systematic errors in auditory localization when the earplug is removed. Young owls correct their localization errors within a few weeks. However, such animals did not correct their auditory localization errors when deprived of vision. Moreover, when prisms were mounted in front of their eyes, they adjusted their auditory localization to match the visual error induced by the prisms, as long as the visual and auditory errors were within the same quadrant of directions. The results demonstrate that, during development, the visual system provides the spatial reference for fine-tuning auditory localization.*

ERIC I. KNUDSEN
PHYLLIS F. KNUDSEN
*Department of Neurobiology,
Stanford University School of
Medicine, Stanford, California 94305*

Sensory space does not project directly onto the sensory surface of the ear in the way that it does onto the eye or the body surface. As a consequence, the auditory system must derive spatial information indirectly from a variety of acoustic cues. These spatial cues depend on the size and shape of the head and ears and hence change as the head and ears grow. How then, might an animal maintain accurate sound localization during maturation? Young barn owls adjust their interpretation of auditory spatial cues on the basis of experience (1). Owls raised with one ear occluded learn, within 4 to 6 weeks, to localize sounds accurately using the altered auditory cues imposed by the earplug. When the earplug is removed, young owls make large localization errors, which they correct over a period of weeks.

We have investigated the signal that guides the adjustment of auditory localization in maturing barn owls. Of all senses, vision provides the most detailed spatial information to the brain. Even in owls, in which hearing is highly developed and visual acuity is relatively poor, the spatial resolving power of vision is still superior to that of audition (2). Moreover, in humans, vision strongly influences the perception of sound source location (3). Therefore, we hypothesized that vision plays an important role in adjusting errors in auditory localization in barn owls.

We monaurally occluded nine baby barn owls aged 26 to 44 days (4). The animals remained monaurally occluded for 41 to 97 days (Table 1). During this time they were trained to orient their heads toward auditory (noise burst) and visual (light-emitting diode) stimuli, which were presented at random locations in a darkened, sound-attenuating chamber (5). One week before an earplug was removed, we attached spectacle frames to each owl (6). The mean of more than 100 responses to visual stimuli with the spectacle frames empty defined the head-centered spatial origin for each owl. During the experiments, the accuracy of auditory and visual localization was computed on the basis of this reference value.

The experiments began on the day the earplug was removed. Immediately after earplug removal, the responses of the owl to auditory and to visual stimuli were measured. Because auditory and visual stimuli were presented independently and in a darkened chamber, visual capture did not influence these responses (3). Owls that had been raised with the right ear plugged oriented to the right and above the auditory stimulus, whereas owls that had been raised with the left ear plugged oriented to the left and below the auditory stimulus (7). Two animals were allowed normal vision and served as controls; five animals were fitted with Fresnel prisms (8), which deviated vision by 10°, and two animals were fitted with opaque occluders which prevented vision totally (Table 1). After 28 days, the opaque occluders were replaced with prisms. Most of the owls were exposed to several different prism orientations, each orientation being maintained for periods of 22 to 68 days. The final experiment in every case was to remove the prisms and follow the recovery of accurate auditory localization.

Owls 1 and 2, which were permitted normal vision, adjusted their auditory errors rapidly (Fig. 1, A and B). The errors of both birds diminished at average rates of 0.7° per day, and after 28 days these birds were localizing sounds with mean errors of less than 3°, our criterion for normal localization accuracy (1). In contrast, owls 3 and 4, which had their eyes covered with opaque occluders after the earplugs were removed, maintained constant auditory errors for periods of 28 days (Fig. 1C). Since without vision these animals did not adjust their auditory errors, vision must be essential to trigger the adjustment process, to guide the adjustment, or both.

The role of vision was clarified by the first experiment done on owl 5. As shown in Fig. 1D, the bird had an initial auditory error of right 6.4° and up 8.4°. Prisms were mounted on this owl to

Table 1. Auditory and visual histories of all owls studied. Abbreviations: L, left; R, right; U, up; D, down.

Owl	Auditory history			Visual history (sequence of visual experiments)
	Ear occluded	Age (days) at which earplug was		
		Inserted	Removed	
1	L	27	113	Control
2	R	35	76	Control
3	L	31	75	Occluders; prisms: LU, RU, off
4	R	36	85	Occluders; prisms: R, RU, RD, off
5	R	43	92	Prisms: RU, U, off, L, D, off
6	R	28	94	Prisms: LD, LU, U, off
7	R	44	141	Prisms: R, off
8	R	26	75	Prisms: LU, L, off
9	L	28	78	Prisms: LU, LD, off

deviate its visual world in the same direction (the actual prism setting was right 5.4° and up 8.4°). After 29 days, the owl's auditory error was essentially unchanged at right 5.4° and up 7.0°. Thus, when vision is allowed, but there is no mismatch between visual and auditory space, no adjustment of sound localization occurs.

We investigated the extent to which vision controls the adjustment process by establishing various mismatches between visual and auditory space. It became apparent immediately that a visual-auditory mismatch could induce the selective adjustment of either the horizontal or the vertical component of an auditory localization error (Fig. 2).

Moreover, in 12 of 12 experiments, the owls adjusted their auditory localization to match the visual error induced by the prisms, as long as the visual error was in the same quadrant of directions as the initial auditory error. For example, owls with initial auditory errors to the right and up adjusted their auditory errors to match prism-induced visual errors of straight-right (two experiments), straight-up (two experiments), various directions to the right and up (two experiments), or back to 0°, 0° (six experiments).

In contrast, whenever either the horizontal or the vertical component of the prism-induced visual error was opposite in direction to the initial auditory error, the owl either adjusted only partially (four experiments), or failed to make any systematic adjustment at all (four experiments). For example, owl 9 experienced auditory and visual errors that were in opposite vertical directions (Fig. 2B): its initial auditory error was left 7.8° and down 8.5° and the visual error induced by the prisms was left 5.2° and up 8.4°. Although this owl adjusted the vertical component of its auditory error by 9.3° in just 17 days, wearing the prisms an additional 28 days did not cause a significant upward error. The auditory error never went more than 2.1° into the upward direction. A similar result is shown for owl 4 (Fig. 2C). However, in this older bird, the adjustment was slower and even less complete. Owl 8 had auditory and visual errors that were in opposite horizontal directions (Fig. 2D): its initial auditory error was right 8.3° and up 11.5°, and its prism-induced visual error was left 5.0° and up 8.7°. After 38 days, the auditory error had changed to left 1.9° and up 6.8°, but during the subsequent 36 days no further adjustment was observed.

The experimental series performed on owl 6 (Fig. 2E) also demonstrated the

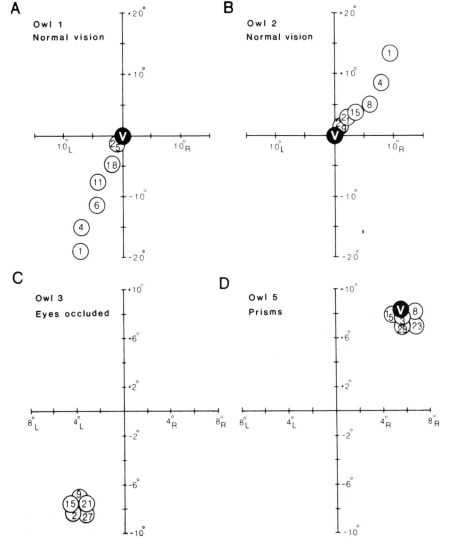

Fig. 1. Auditory localization errors following earplug removal in four barn owls that had been raised with one ear plugged. The origin of each coordinate system was defined by the mean of more than 100 responses by the owl to the visual stimulus measured without opaque occluders or prisms in the spectacle frames. Each open circle represents the mean of 15 to 25 responses of the owl to the auditory stimulus; the number indicates the number of days after earplug removal on which the data were gathered. Standard deviations of the auditory responses ranged from 0.8° to 3.8°, but were typically between 1.0° and 2.5°. The owls were tested every 2 to 3 days, but because the data overlapped extensively, only a few representative points are shown. Circled V's indicate the visual error attributable to the prisms.

limited effectiveness of a visual-auditory mismatch in guiding the adjustment of sound localization when the visual and auditory errors were in opposite directions. In experiment 1, the prisms were oriented so that the visual error (to the left and down) was opposite in both dimensions to the auditory error (to the right and up). Although the discrepancy between auditory and visual localization was large in this case, it was well within the range of adjustment of auditory localization (for example, Fig. 1, A and B). Nevertheless, the owl's auditory error remained essentially unchanged for 26 days. In experiment 2, the prisms were oriented so that the visual and auditory errors had the same vertical components, but the visual error was still leftward and the auditory error rightward. After 23 days, the auditory error was unchanged. However, in experiment 3, the prisms were oriented so that the visual error was 10° straight up, and the owl adjusted its auditory error in just 13 days to right 1.8° and up 8.8°. Finally, the prisms were removed, and the owl corrected the residual vertical component of its auditory error in 9 days. This animal made horizontal and vertical adjustments, but only when the visual and auditory errors were in the same quadrant of directions.

These results are consistent with the hypothesis that a mismatch between visual and auditory space establishes a corrective force that guides a reinterpretation of auditory cues. When vision is prevented, or when visual and auditory space are matched but incorrect, no corrective force is generated and no adjustment in auditory localization occurs. Apparently, spatial information provided by other senses is inadequate to cause such an adjustment. However, the influence of vision had definite limitations in these experiments. Vision altered the magnitude, but not the sign, of the auditory error. This implies that there is, in addition, a nonvisual spatial referent that

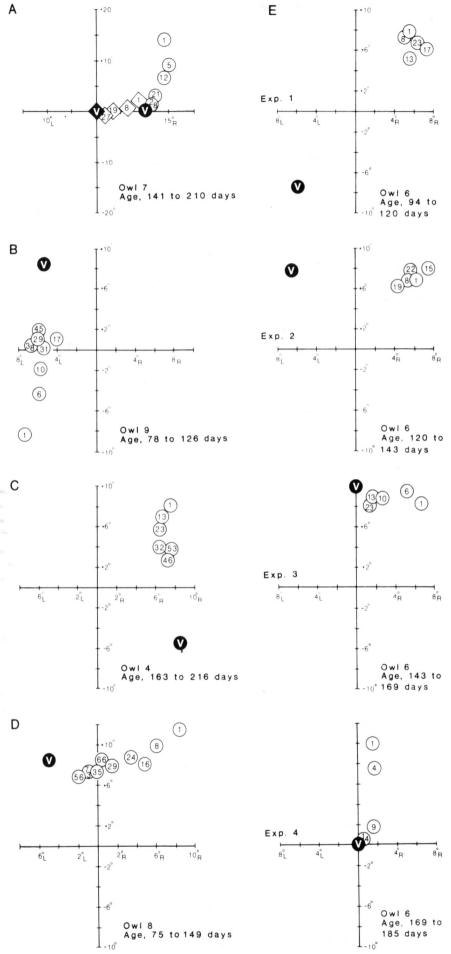

Fig. 2. Adjustments of auditory localization errors by birds wearing prisms. The origin of each coordinate system was based on more than 100 visual responses by the bird before prisms were installed. Numbered circles represent mean auditory errors. The numbers indicate the number of days since the prisms were installed. The circled V's represent the owl's mean visual error caused by the prisms. As in Fig. 1, only a few representative points are shown. In (A), circles represent data from the first experiment on owl 7; the diamonds represent data from the second experiment (prisms off). In (E), data from four sequential experiments on owl 6 are presented in the order in which they were performed.

1 NOVEMBER 1985

confines the adjustment process to one quadrant of directions.

The influence of vision on auditory localization apparently decreases with age. Before an owl is 50 to 60 days old, it can adjust its auditory error in any direction, presumably guided by vision (1). In the experiments presented here, in which the owls were between 75 and 220 days old, younger birds made larger and more rapid adjustments than did older birds (for example, compare Fig. 2B with 2C), but in no case could an owl be induced to change the sign of its auditory error. Finally, adult owls (more than 7 months old) maintain auditory localization errors indefinitely, even when they experience normal vision (1). Thus, the ability of vision to generate a corrective force, or the ability of the auditory system to respond to it, or both, diminish with age. This corrective force exerted by the visual system on the auditory system provides a mechanism for fine-tuning the associations that underlie auditory localization during development.

References and Notes

1. E. I. Knudsen, S. D. Esterly, P. F. Knudsen, *J. Neurosci.* **4**, 1001 (1984); E. I. Knudsen, P. F. Knudsen, S. D. Esterly, *ibid.*, p. 1012.
2. J. D. Pettigrew, *Proc. R. Soc. London Ser. B* **204**, 435 (1979); E. I. Knudsen, G. C. Blasdel, M. Konishi, *J. Comp. Physiol.* **133**, 1 (1979).
3. J. R. Lackner, *Neuropsychologia* **11**, 29 (1973); B. R. Shelton and C. L. Searle, *Percept. Psychophys.* **28**, 589 (1980); D. H. Warren, R. B. Welch, T. J. McCarthy, *ibid.* **30**, 557 (1982).
4. Long-term monaural occlusion was accomplished by suturing a dense, foam-rubber plug (E.A.R. Corporation) into the external meatus while the animal was anesthetized with halothane and nitrous oxide.
5. The method used in this study has been described in detail (1). Briefly, we measured auditory localization by comparing the accuracy with which an owl oriented its head to auditory and to visual stimuli. The stimuli were presented from a remotely controlled, movable speaker and light source, which were positioned at a new, random location before every trial. Auditory stimuli consisted of repeated noise bursts presented at 10- to 50-dB sound-pressure level (re 20μPa). Visual stimuli consisted of a continuous glow from a light-emitting diode. A trial consisted of a presentation of either the auditory or the visual stimulus; the final head orientation (as indicated by an infrared beam reflected from a mirror mounted on the owl's head) relative to the true location of the stimulus was recorded. The stimulus continued until the animal settled on a particular direction. A test session included 15 to 25 responses to auditory stimuli and an equivalent number to visual stimuli. When an owl responded to a stimulus with a quick movement of the head followed by a steady fixation, it was given a food reward. Because the reward was not contingent on the location to which the owl oriented, the reward did not bias the localization response.
6. Owls were anesthetized with halothane and nitrous oxide, and a metal plate was secured to the skull with screws and dental cement. The spectacle frames were bolted to this plate. Each frame was 10 mm in diameter and permitted approximately a 70° field of view as determined by retinoscopy. The frames did not physically obstruct or interfere with the external ears.
7. The auditory localization error was defined as the difference between the owl's mean response to the acoustic stimulus, based on 15 to 25 responses measured that day, and the reference value, which was based on more than 100 visual responses recorded before prisms or occluders were installed. The vertical component of the barn owl's auditory error results from the fact that, due to an asymmetry in its external ears, interaural intensity differences indicate the vertical location of the sound source.
8. The prisms were 20-diopter Fresnel (Optical Sciences Group). They were oriented in the spectacle frames through the use of an optical bench. Their orientation on the animal was determined by comparing the visual responses of the bird before and after mounting the spectacles.
9. We thank J. Middlebrooks, S. du Lac, and S. Esterly for reviewing the manuscript. Supported by grants from the March of Dimes (1-863), Sloan Foundation, the National Institutes of Health (RO1 NS 16099-05), and a neuroscience development award from the McKnight Foundation.

29 April 1985; accepted 24 July 1985

Glossary

auditory localization - identifying the location of a sound source; this is done on the basis of binaural (involving two ears) cues

monaurally - pertaining to one ear

occlude - to block off; visual occluders eliminate vision, and auditory occluders attenuate sound entering the ear canals

prism - a wedge of transparent glass, which refracts (bends) light

Essay Study Questions

1. Owl 5 was fitted with prisms, but they did not correct its auditory localization error. Why?

2. Why does monaural occlusion of the right ear produce auditory localization errors to the left of the sound when the occluder is in place, and then produce errors to the right of the sound when the occluder is removed?

3. Knudsen and Knudsen found that the ability of a young barn owl's visual system to educate its auditory system is limited—only certain errors can be corrected. Describe the evidence that supports this conclusion.

4. What evidence suggested that the ability of a barn owl's visual system to educate its auditory system declines with age?

Multiple-Choice Study Questions

1. Which sensory system has greater powers of spatial resolution than the auditory system?
    a. olfactory system
    b. gustatory system
    c. visual system
    d. all of the above
    e. none of the above

2. Occlusion of the left ear produced an auditory localization error
    a. to the right of the sound when the plug was removed.
    b. to the left of the sound when the plug was removed.
    c. below the sound when the plug was removed.
    d. both a and c
    e. both b and c

3. Twenty-eight days after the plugs were removed, the spatial localization errors of the control owls in Knudsen and Knudsen's experiment were normal, that is, they averaged less than
    a. 3°.
    b. 3%.
    c. 3 centimeters.
    d. 3 meters.
    e. 3 inches.

4. Knudsen and Knudsen found that a visual-auditory mismatch can produce a
    a. vertical correction in vision.
    b. horizontal correction in audition.
    c. vertical correction in audition.
    d. all of the above
    e. both b and c

5. Accurate adjustments of the auditory localization response to visual-auditory mismatches occurred only
    a. in barn owls over 60 days old.
    b. in female barn owls.
    c. when the visual and auditory errors were in the same quadrant of directions.
    d. when the visual and auditory localization errors were in opposite directions.
    e. both a and d

6. Adult barn owls (more than 7 months old) correct auditory localization errors when
    a. their vision is occluded.
    b. both of their ears are occluded.
    c. when they experience normal vision.
    d. they are wearing prisms.
    e. none of the above

The answers to the preceding questions are on page 290.

Food-For-Thought Questions

1. Although Knudsen and Knudsen do not mention it in their paper, one ear of barn owls is higher than the other. From the data provided in the paper, can you figure out which ear is typically higher? What do you think is the survival value of nonhorizontal ears?

2. In some cases the study of species that have special nonhuman abilities has much to tell us about humans. Discuss with respect to the study of barn owl audition.

ARTICLE 7

Studies of Instrumental Behavior With Sexual Reinforcement in Male Rats (*Rattus norvegicus*): II. Effects of Preoptic Area Lesions, Castration, and Testosterone

B.J. Everitt and P. Stacey

Reprinted from Journal of Comparative Psychology, 1987, volume 101, 407-419.

    The castration of male animals, including male humans, leads to a decline in their copulatory activity, that can be reversed by replacement injections of the testicular hormone, testosterone. Testosterone is thought to influence copulatory behavior by acting on receptors located in two adjacent areas of the brain, the medial preoptic area and the anterior hypothalamus. In support of this theory is the finding that bilateral lesions that damage these areas have been shown to disrupt the copulatory behavior of the male members of a variety of laboratory species. Furthermore, the copulatory behavior of castrated male animals can be restored by tiny implants of testosterone directly into the medial-preoptic anterior-hypothalamic area.
    It is widely assumed that both castration and medial-preoptic anterior-hypothalamic lesions eliminate copulatory behavior by abolishing the motivation of males to engage in sexual activity; however, there are other possible explanations. To clarify this issue, Everitt and Stacey studied the effects of medial-preoptic anterior-hypothalamic lesions (Experiment 1) and castration (Experiment 2) on the rate at which male rats press a lever to gain access to receptive females on a second-order reinforcement schedule. Lever presses were occasionally reinforced by a "gift from the sky"; reinforced lever presses sprang the lock on an overhead trap door and a receptive female rat dropped into the chamber next to the lever-pressing male.
    In contrast to the widely accepted view, Everitt and Stacey found that male rats with lesions of the medial preoptic area and anterior hypothalamus displayed no decline in sexual motivation. Although mounts, intromissions, and ejaculations were almost totally abolished by the lesions, the lesioned rats still lever-pressed for receptive females at their usual high rate. Castrated male rats also failed to fit the expected stereotype. Although castration obolished their copulatory behavior, it produced only a partial decline in their lever pressing for receptive females. Replacement injections of testosterone produced a rapid recovery in their lever pressing, while the recovery of their copulatory behavior lagged far behind.
    The results of these innovative experiments suggest that the disruptive effects of castration on male copulatory behavior are only in part a reflection of motivational deficits, and that motivational deficits play no part in the abolition of male copulatory behavior by medial-preoptic anterior-hypothalamic lesions.

Studies of Instrumental Behavior With Sexual Reinforcement in Male Rats (*Rattus norvegicus*): II. Effects of Preoptic Area Lesions, Castration, and Testosterone

Barry J. Everitt and Pamela Stacey
Department of Anatomy, University of Cambridge, Cambridge, England

> We studied effects of lesions to the medial preoptic area (POA), castration, and testosterone replacement on instrumental and unconditioned sexual behavior in male rats. We achieved instrumental measures of sexual motivation by training males to work for an estrous female, presented in an operant chamber under a second-order schedule of reinforcement. POA lesions abolished mounts, intromissions, and ejaculation but did not disrupt instrumental responses, investigation of the female, or abortive mounting attempts. Castration abolished attempts to copulate and also caused a marked decrease in instrumental responses. Testosterone resulted in the prompt reinstatement of instrumental responses and more gradual recovery of unconditioned sexual behavior. We discuss these results in terms of the motivational and performance effects of these neuroendocrine manipulations.

Sexual behavior in male mammals depends on testicular androgens (Beach & Holz-Tucker, 1949; Bermant & Davidson, 1974). Castration is followed by a decline in sexual activity over a time course varying from weeks (in the rat: Beach, 1942; Davidson, 1966b; Södersten, Hansen, Eneroth, Wilson, & Gustafsson, 1979) to months (in the monkey: Phoenix, Slob, & Goy, 1973) and even years (in man: Bancroft, 1978). These effects of castration are reversed by treatment with testosterone or with combinations of its metabolites, and the speed of reversal varies with the prior period of hormone deprivation (Beach & Holz-Tucker, 1949; Bermant & Davidson, 1974; Grunt & Young, 1953).

These well-established data are complemented by those concerning the putative sites of action of sex steroids. Bilateral lesions of the medial preoptic-anterior hypothalamic areas (mPOA/AHA) of males of a variety of species prevent the display of sexual behavior, despite the presence of circulating testosterone (Giantonio, Lund, & Gerall, 1970; Ginton & Merari, 1977; Hansen, Köhler, Goldstein, & Steinbusch, 1982; Hart, 1974; Hart, Haugen, & Peterson, 1973; Heimer & Larsson, 1966–1967; Kelley & Pfaff, 1978; Lisk, 1967; Slimp, Hart, & Goy, 1978). Conversely, the sexual activity of castrated males can be restored by implanting testosterone (or its metabolites) directly within the mPOA/AHA; the steroid is much less effective in other hypothalamic or extrahypothalamic sites (Christensen & Clemens, 1974; Davidson, 1966a; Lisk, 1967; Pfaff, 1980).

The work was supported by the Medical Research Council (Program Grant PG7307226 and Project Grant G8501695N).

We thank Trevor Robbins, Joe Herbert, Barry Keverne, Brad Powers, Pauline Yahr, Stefan Hansen, and Tony Dickinson for many helpful discussions and criticism of the manuscript. We also gratefully acknowledge the constructive criticisms of Ben Sachs and another, anonymous referee.

Correspondence concerning this article should be addressed to Barry J. Everitt, Department of Anatomy, University of Cambridge, Downing Street, Cambridge, England, CB2 3DY.

A general feature of the interpretation of these data on the effects on sexual behavior of hormonal and hypothalamic manipulations is that they have a direct impact on sexual motivational or arousal processes (for discussion, see Bindra, 1976; Grossman, 1967; Konorski, 1967; Pfaff, 1982; Toates, 1980, 1986). Data from the more recent studies of unconditioned sexual behavior, however, suggest that the situation is not as simple as it at first seems. Indeed, it is far from clear which aspects of sexual behavior are regulated by testosterone through its actions either on the hypothalamus or elsewhere in the central nervous system.

First, the castrated male rat and the mPOA/AHA-lesioned male rat are different. The former is generally quiescent in the presence of a female in heat, although occasional episodes of anogenital investigation and weak mounting attempts are displayed (Beach, 1942; Hansen & Drake af Hagelsrum, 1984; Södersten et al., 1979). On the other hand, the mPOA/AHA-lesioned male vigorously investigates a female and makes frequent, abortive mounting attempts (Hansen & Drake af Hagelsrum, 1984; Hansen et al., 1982; Hart, 1986; Heimer & Larsson, 1966–1967; Malsbury & Pfaff, 1974). These mounting attempts are often referred to as "climbings," distinguished from mounts normally made by an intact male because they do not involve pelvic thrusting or intromission (Hansen & Drake af Hagelsrum, 1984).

Furthermore, it has become apparent that mPOA/AHA lesions and, to a lesser and qualitatively different extent, castration induce the development of other behavior in male rats paired with estrous females. Such responses include excessive grooming, scratching, and if a drinking apparatus is made available, excessive drinking (Hansen & Drake af Hagelsrum, 1984). Such displacement behavior, as it has been called (Tinbergen, 1952), may reflect an underlying process of behavioral thwarting or frustration. In other words, although such males are motivated to respond sexually, they cannot do so because either the neural or the endocrine manipulation impairs the capacity to do so.

The ability to separate motivational from so-called performance factors (the ability to execute more reflexive consummatory responses, such as mounting and intromitting) would clearly aid the unraveling of the complexities of the consequences of manipulating neural and endocrine systems involved in controlling the display of sexual behavior. One approach is to train animals to emit arbitrary, instrumental responses, which do not depend on the ability to perform copulatory reflexes and associated movement patterns, to gain access to a sexual incentive stimulus, such as an estrous female. The rather limited literature on attempts to do so reveals only limited success in obtaining reliable measures of instrumental behavior for sexual reinforcement with which the effects of neuroendocrine manipulations could be studied (see Toates, 1986).

Male rats will learn to run a maze or a simple runway to gain access to a receptive female (Beach & Jordan, 1956; Denniston, 1954), and that performance is better on the task when the male is allowed to ejaculate rather than only intromit (Hogan & Roper, 1978). Sexual preference tests of various sorts reveal an apparently similar picture (Hetta & Meyerson, 1978; Merkx, 1983, 1984). The willingness with which males run the maze and their willingness to cross an electrified grid to gain access to a receptive female are apparently reduced by castration (Hogan & Roper, 1978; Warner, 1927). In similar experiments, researchers have also investigated the effects of hormones in females (Drewett, 1973; Meyerson & Lindstrom, 1973). However, the contribution of feedback about the impaired ability to copulate following castration to the impaired instrumental behavior in these experiments remains difficult to assess (see also Toates, 1986).

Male rats can be trained to press a lever to gain access to a female (Beck, 1971, 1974, 1978; Beck & Chmielewska, 1976; Hogan & Roper, 1978; Jowaisas, Taylor, Dewsbury, & Malagodi, 1971; Larsson, 1956; Sachs, Macaione, & Fegy, 1974; Schwartz, 1955), but the rates of responding are frequently extremely low (Hogan & Roper, 1978, but see Jowaisas et al., 1971) and therefore difficult to make use of quantitatively following neuroendocrine manipulations. In another article in this issue (Everitt, Fray, Kostarczyk, Taylor, & Stacey, 1987), we describe a novel method of measuring instrumental behavior with sexual reinforcement in male rats. The procedure used involves presenting an estrous female under a second-order schedule of reinforcement. Instrumental responses are maintained at a high rate during the schedule by the presentation of brief, visual stimuli that gain reinforcing properties through prior pairing with the estrous female (Everitt et al., 1987). The ability to generate high rates of instrumental behavior has provided a method of examining the nature of the effects of manipulating the neuroendocrine mechanisms controlling the expression of sexual behavior. We report here the effects of lesions to the preoptic-anterior hypothalamic areas, castration, and testosterone replacement on instrumental and unconditioned sexual behavior in the male rat.

Experiment 1

In Experiment 1, we reexamined the nature of the effects on sexual behavior of lesions to the preoptic area in male rats by assessing their impact on instrumental behavior with sexual reinforcement, in addition to copulatory responses, measured directly when males and females interacted sexually immediately after the instrumental period. In this way, we attempted to assess the possibly separate contribution of disrupted motivational and performance (consummatory) factors to the overall and profound deficit in sexual behavior that is known to follow damage to this part of the brain (see Kelley & Pfaff, 1978).

Method

Subjects and apparatus. The subjects were 24 male rats of the Lister hooded strain weighing 250 g at the start of the experiment. They were housed in pairs and maintained on a reversed light–dark cycle (12 hr light and 12 hr dark, lights off at 8:00 a.m.). All males received their food (20 g daily) late in the afternoon at the conclusion of the day's testing. They gained weight during the course of the experiment so that their body weights did not differ significantly from those of males of similar age fed ad libitum. Details of the apparatus used for studying instrumental and interactional sexual behavior may be found in the accompanying article (Everitt et al., 1987).

Preliminary training. All males were given twice-weekly observation sessions with a female made receptive by sequential treatment with estradiol (5 µg, s.c. 4 hr before testing) and progesterone (500 µg, s.c. 4 hr before testing) until their copulatory performance had stabilized. Eight pairs of males, matched as nearly as possible for their sexual performance (on mean intromission latency, inter-intromission interval, and ejaculation latency measures; see Dewsbury, 1979; Sachs, 1978), were then given six sessions in an operant chamber, during which they were allowed to ejaculate twice, so that they became accustomed to the female's entry from the trapdoor box in the roof (Everitt et al., 1987). The remaining four pairs of males were assigned to work for food reinforcement.

Second-order schedules. We then instituted a second-order FRx(FRy:S) schedule of presentation of the reinforcer exactly as described in Experiment 3 of the accompanying methodological article (Everitt et al., 1987). When each male was responding in a stable manner at FR10(FR10:S) on the assigned lever, we changed the schedule to the type FIz(FR10:S) with the fixed interval increasing from 5 to 15 min over five sessions (FR = fixed ratio and FI = fixed interval). When responding had stabilized, we again matched males as closely as possible for both their conditioned and unconditioned responses, and we took four groups, two responding for food and two for an estrous female, for surgery.

Sexual behavior. On entry of the female into the operant chamber at the end of the period of the male's instrumental responding, sexual interaction was allowed to proceed until the first intromission following ejaculation. The following measures were recorded both manually and with the aid of an event recorder connecting through a laboratory interface to a CUBE microcomputer (Control Universal, Cambridge, England): (a) latency to the first mount, (b) latency to the first intromission, (c) latency to ejaculate (from the first adequate mount or intromission, whichever occurred first), (d) postejaculatory refractory period (PEI, the time from ejaculation to the next intromission), (e) number of mounts without intromission to ejaculation, (f) number of mounts with intromission to ejaculation, (g) number of inadequate mounts or climbs (mounts not involving pelvic thrusting or forelimb palpation of the female's flanks), (h) number of episodes of anogenital investigation by the male, and (i) number of episodes of vigorous scratching by the male's hind limb of his flank and head.

Preoptic area lesions. Males were anesthetized with Avertin (tribromoethanol in tertiary amyl alcohol, 0.1 ml/100 g body weight, intraperitoneal) and their heads were shaved and they were then placed in a stereotaxic frame (David Kopf, California). Lesions were induced by the infusion, through 30-g stainless steel cannulae attached to 5 μl syringes (Precision Sampling, Baton Rouge, Louisiana), of the excitotoxin N-methyl aspartic acid (0.12 M in Sorensen's phosphate buffer, pH 7.0–7.4 adjusted with KOH). Infusions of 1 μl were made on each side sequentially over a period of 1 min, with the cannula left in place for a further 2–3 min. Sham operations consisted of identical infusions of the phosphate buffer vehicle (pH 7.0–7.4). Four males in the food group received preoptic lesions, and 4 were sham operated. Ten males in the sex group received preoptic lesions, and 6 were sham operated. Subsequently, 2 of the lesioned males in the sex group died (1 while under anesthesia), and 1 was subsequently found to have a small unilateral lesion (see below) and was excluded from the behavioral analysis. Thus, the behavioral experiments were conducted on 8 males responding for food (4 lesions and 4 controls) and 13 males responding for an estrous female (7 lesions and 6 controls).

Lesion assessment. At the conclusion of the experiment, males were perfused transcardially under barbiturate anesthesia with 50 ml 0.9%(w/v) NaCl at room temperature, followed by 150 ml 10%(w/v) formal saline. Brains were removed and postfixed for 2 hr and then transferred to a 10% solution of formal sucrose and left for 48 hr or until the brain sank. After blocking, sections were cut at 60 μm on a freezing microtome, and every other one was mounted on egg-albumen-coated glass slides. Following oven drying, sections were stained with cresyl violet and coverslipped under permamount.

Lesions were reconstructed in two ways: First, areas of marked neuronal loss, including but not restricted to areas of gliosis, were drawn onto standard frontal sections of the rat brain. These are represented in the Results section. Second, areas of neuronal loss on each section were drawn, with the aid of a drawing attachment on the microscope (Leitz Dialux 20), onto separate sheets of paper. Each lesion profile was then digitized by using a Grafpad graphics tablet (British Micro, Watford, England) attached to the user port of a British Broadcasting Corporation microcomputer (Acorn, Cambridge, England). The cross-sectional areas of the lesions were then computed and integrated over their anterior–posterior extent to give lesion volumes (uncorrected for shrinkage).

Statistical analysis. Behavioral results were subjected to two-way analysis of variance (ANOVA), with treatment as the between-groups variable and days as the within-groups variable, or analysis of covariance (ANCOVA), with preoperative performance as the covariate (GENSTAT, Alvey, Galwey, & Lane, 1982; Numerical Algorithms Group, 1977). In the case of significant interactions, planned comparisons were made at each level of the interaction by using the appropriate error term and the standard errors of the differences (*SED*s) between means (Dunnett's *t*, see Weiner, 1962). The *SED*s are shown on individual figures in the Results.

Results

Lesion assessment. Although generally restricted to the preoptic-anterior hypothalamic areas, the lesions induced by infusions of N-methyl aspartic acid were rather variable in extent (see Figures 1, 2, and 3). Anteriorly, lesions occasionally extended as far forward as the junction between the preoptic area and the caudal, medial septal nuclei. Posteriorly, lesions sometimes extended back to the level of the suprachiasmatic nuclei but not far enough to reach the ventromedial nuclei. Dorsally, the anterior commissure appeared to contain naturally the spread of neurotoxin, whereas ventrally, the suprachiasmatic and supraoptic nuclei were spared. In two instances, the lesion extended into the lateral preoptic/hypothalamic areas and abutted the substantia innominata.

Analysis of lesion volume by using the digitizer confirmed the variability of damage induced by N-methyl aspartate. The

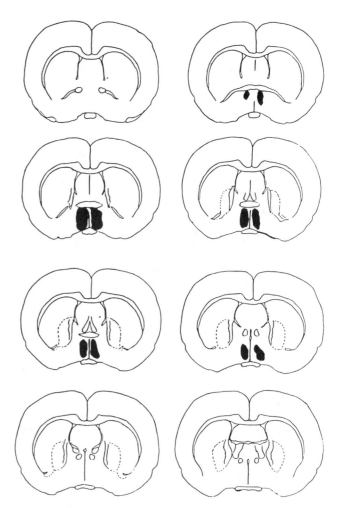

Figure 1. Schematic representation of n-methyl-D-aspartic-acid-induced lesions of the preoptic-anterior hypothalamic area reconstructed from camera lucida drawings of individual sections stained with cresyl violet. The lesioned area always includes the medial preoptic and anterior hypothalamic areas (Figure 1) but may be variable in its dorsal and lateral extents (Figures 2 and 3). (Sections are redrawn from the atlas of Paxinos & Watson, 1982.)

largest lesion was 7.55 mm³, and the smallest, bilateral lesion was 1.70 mm³. One animal was rejected from the lesioned group following this analysis. The lesion was restricted to only one side and was 1.3 mm³. (This animal showed no postsurgical change in sexual behavior.)

Unconditioned sexual behavior. On the first postoperative test, conducted 8 days after surgery, only two lesioned males mounted, one intromitted, and none showed the behavioral ejaculatory pattern. Figure 4 shows the proportions of males displaying these behaviors over the four postoperative tests, which occurred at weekly intervals, and the results of analysis (chi-square), which revealed significant differences between shams and lesions in each case.

However, observation of the preoptic-area-lesioned males revealed that the gross changes in their behavior (represented in Figure 4) were not obviously due to a lack of interest in or arousal by the female when she entered the chamber from the trapdoor box. Thus, all males made repeated abortive mount-

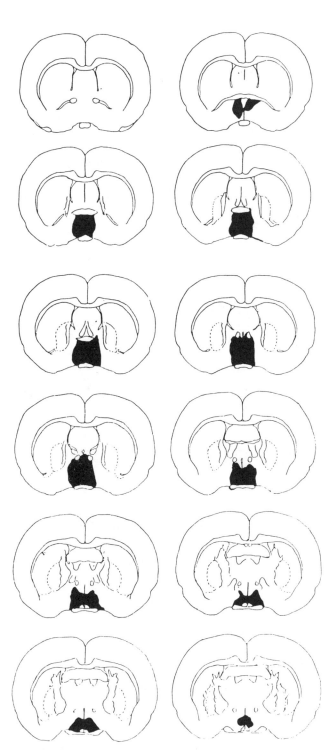

Figure 2. Schematic representation of n-methyl-D-aspartic-acid-induced lesions of the preoptic-anterior hypothalamic area reconstructed from camera lucida drawings of individual sections stained with cresyl violet. The lesioned area always includes the medial preoptic and anterior hypothalamic areas (Figure 1) but may be variable in its dorsal and lateral extents (Figures 2 and 3). (Sections are redrawn from the atlas of Paxinos & Watson, 1982.)

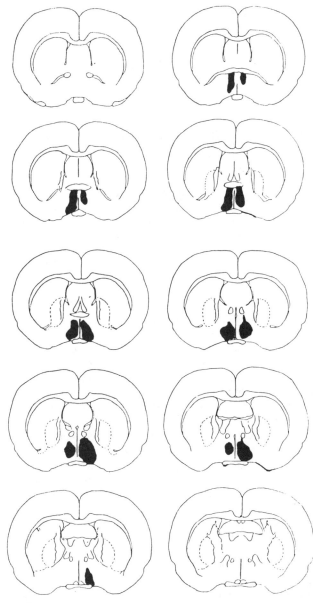

Figure 3. Schematic representation of n-methyl-D-aspartic-acid-induced lesions of the preoptic-anterior hypothalamic area reconstructed from camera lucida drawings of individual sections stained with cresyl violet. The lesioned area always includes the medial preoptic and anterior hypothalamic areas (Figure 1) but may be variable in its dorsal and lateral extents (Figures 2 and 3). (Sections are redrawn from the atlas of Paxinos & Watson, 1982.)

ing attempts, in which they climbed the females but failed to thrust or palpate with their forelimbs (Figure 5). An ANCOVA (with preoperative score as the covariate) revealed a significant Lesion × Test Day interaction, $F(3, 33) = 4.995$, $p < .01$, and specific planned comparisons revealed that shams and lesions differed on each postoperative test day ($p < .01$ in each case).

The lesioned males also showed a marked increase in the frequency of anogenital investigations of the female—ANCOVA with preoperative score as the covariate, main effect of lesion, $F(1, 11) = 27.67$, $p < .01$. Planned comparisons revealed that

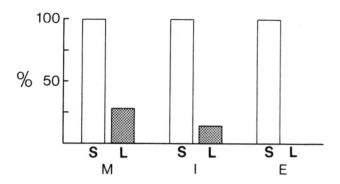

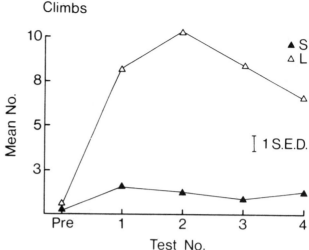

Figure 4. The effects of preoptic area (POA) lesions on the percentage of males mounting (M), intromitting (I), and ejaculating (E): $X^2 (1, N = 13) = 5.11, p < .05$ (percentage mounting); $X^2 (1, N = 13) = 7.29, p < .025$ (percentage intromitting); and $X^2 (1, N = 13) = 9.29, p < .01$ (percentage ejaculating). (S = mean percentage following sham surgery; L = mean percentage following POA lesions.)

lesions and shams differed significantly on each postoperative test day ($p < .01$ in each case, see Figure 5).

Analysis of the frequencies of vigorous hindlimb scratching also revealed a significant Lesion × Test Day interaction, $F(3, 33) = 3.36, p < .05$, and planned comparisons indicated that lesioned rats displayed this behavior more frequently than did shams on each postoperative test day ($p < .01$) but that it occurred less frequently on Test 1 than on all other postoperative tests ($p < .05$, Figure 5).

Instrumental behavior. Analysis of the response rate (responses/15 min) revealed a significant Operation × Test Day interaction, $F(3, 33) = 3.31, p < .05$ (ANCOVA with preoperative score as covariate). Planned comparisons between groups on each test day and between test days showed that both lesioned and sham-operated rats responded at a lower rate than they did preoperatively on Test Day 1 ($t = 2.0$ and 2.9, respectively, $p < .05$) and that lesioned rats responded significantly less than did controls on Postoperative Test Day 4 (Figure 6). Thus, preoptic lesions were not associated with a decline in responding for the female on the first three tests; that is, they did not differ from sham-operated controls, even though their sexual interaction with females once earned was markedly impaired. Only on Test 4 did preoptic males work at a slightly lower rate to earn the receptive female.

In those males working for food reinforcement under the second-order schedule, analysis revealed a significant main effect of test day, $F(4, 24) = 8.81, p < .01$, which was due to the significantly lower rates of responding of both groups on the first postoperative test ($t = 2.7$ for lesions and 4.5 for shams, $p < .025$; see Figure 6). There was no significant effect of the lesion.

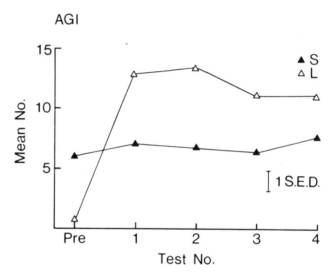

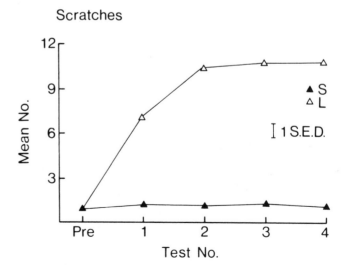

Figure 5. The effects of preoptic area (POA) lesions on the mean numbers of abortive mounts (Climbs), anogenital investigation (AGI), and vigorous hindlimb scratching (Scratches). (S = sham-operated controls; L = POA lesion; S.E.D. = standard error of the difference of the means; Pre = preoperative.)

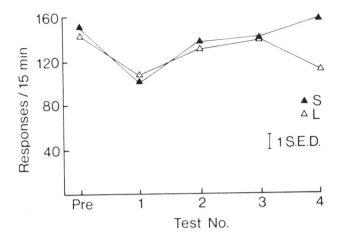

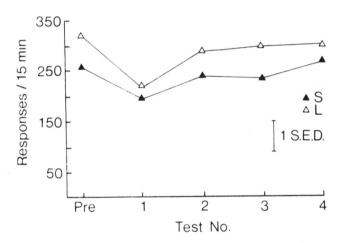

Figure 6. The effects of preoptic area (POA) lesions on instrumental responses by animals working for an estrous female (top panel) or food (bottom panel) presented under a second-order schedule of reinforcement. (S = sham-operated controls; L = POA lesion; S.E.D. = standard error of the difference of the means; Pre = preoperative.)

although it impairs sexual performance, damage to the preoptic area does not apparently diminish sexual interest in the female. Such a view is directly supported by the observation that preoptic area lesions did not reduce the willingness of males to work for a receptive female on the first three weekly tests following surgery. Thus, the results indicate a dissociation between instrumental and consummatory measures of sexual behavior and strongly suggest that the primary impact of preoptic area lesions is on the latter, that is, on performance aspects of sexual behavior, rather than on the former, motivational or arousal processes. The small decrease in responding seen on the fourth test may indicate an extinction-like process, because the males had experienced three tests in which attempts to copulate were unsuccessful and, therefore, not reinforcing (see Everitt et al., 1987). Responding for food on the second-order schedule was also unaffected by the lesion.

Experiment 2

Sex steroids bind to motor neurons in the spinal cord (Breedlove & Arnold, 1980) and affect sensory mechanisms associated with copulation (Hart, 1973; Pfaff, 1981; Sachs, 1983), in addition to actions within the brain, particularly the hypothalamus. These hormones are widely accepted as affecting the display of sexual behavior in large part by exerting motivational or arousal effects within the central nervous system (see Grossman, 1967, Hart, Wallach, Melesed'Hospital, 1983; Konorski, 1967; Pfaff, 1982; Toates, 1986). However, as with the effects of preoptic area lesions, the nature of the behavioral consequences of androgen deprivation in males has recently been questioned (Hansen et al., 1982), and the extraordinarily long period between castration and the loss of sexual interest by the males of many species (see the introduction) remains difficult to explain in terms of actions of steroids on a "central motive state" (Bindra, 1976). In Experiment 2, we studied the effects of castration and testosterone replacement both on responding under the second-order schedule for a receptive female and on consummatory behavior in an attempt to understand better the nature of the behavioral processes altered by these manipulations.

Method

Subjects and apparatus. The subjects used were 32 male rats of the Lister hooded strain weighing 250 g at the start of the experiment. They were housed in pairs in a room under a reversed light–dark cycle (12 hr light and 12 hr dark, lights off at 8:00 a.m.). All males received their food (20 g daily) late in the afternoon at the conclusion of the day's testing and gained weight during the course of the experiment so that their body weights did not differ significantly from those of free-fed rats of similar age. The apparatus used was identical to that described in the accompanying methodological article (Everitt et al., 1987).

Preliminary training. All males were given twice-weekly observation sessions in a large arena with a female made sexually receptive with sequential estradiol and progesterone treatment (see above). Sixteen pairs of males, matched as far as possible for their sexual performance, were then given six sessions in the operant chambers so that they became accustomed to the female's entry from the trapdoor box in the roof (Everitt et al., 1987).

Summary

The results of Experiment 1 confirm those of a number of earlier studies in that they clearly demonstrate the marked impact of lesions of the medial preoptic area on sexual behavior in the male rat (see the introduction for references). In addition, the increased incidence of abortive mounts, anogenital investigation, and hindlimb scratches seen to follow these lesions also confirms the observations of Hansen and Drake af Hagelsrum (1984) and lends support to the assertion that

Second-order schedules. Training of the animals under the second-order schedule proceeded exactly as described in the Method section of Experiment 1. After 14 sessions, males were again matched, as far as possible, in terms of their stabilized instrumental and unconditioned sexual behavior and assigned to one of two groups—future castrated or sham-operated subjects.

Sexual behavior. Measures of sexual behavior were recorded in the way described in the Method section of Experiment 1. Observations of unconditioned sexual behavior were conducted (a) preoperatively; (b) on Tests 1, 2, 3, 5, and 9 postoperatively, prior to implantation of testosterone; and (c) on Tests 1, 3, and 5 after implantation of testosterone.

Surgical procedures. Males were anesthetized with Brietal (sodium methohexitone, 0.1 ml/100 g body weight), and the testes were exposed transcrotally. The testicular vessels were ligated, the testes were removed, and the wound was closed with chromic gut sutures. Sham-operated males received only the incision and closure steps. All males were left 40 days before being tested in the operant chambers. They received no contact with females during this period, being left undisturbed with the male with which they were paired. They received nine weekly tests under the second-order schedule following castration (or sham surgery) prior to implantation with testosterone.

Following the period of testing of the castrates (9 weeks from the first test, i.e., 14 weeks after castration), castrated males received testosterone implants (the free steroid within silastic tubing—Dow Corning 602-285, 0.157 cm inside diameter × 0.318 cm outside diameter × 30 mm long—implanted subcutaneously between the scapulae under light ether anesthesia); controls received only an incision and wound closure. The first postimplantation test was conducted 8–10 days later and thereafter on five occasions at weekly intervals.

Statistical procedures. Behavioral results were subjected to two-way ANOVAS (GENSTAT, Alvey et al., 1982; Numerical Algorithms Group, 1977) with treatment as the between-subjects variable and days as the within-subjects variable. In the case of significant interactions, planned comparisons were made at each level of the interaction by using the appropriate error term and the *SED*s (see above). *SED*s are shown on the data figures in the Results.

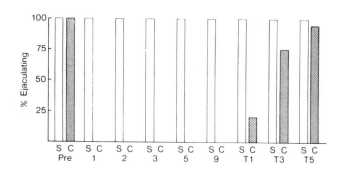

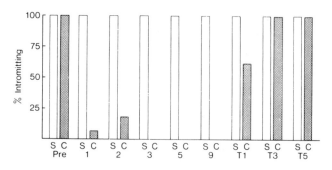

Figure 7. The effects of castration and testosterone replacement on the proportion of males ejaculating and intromitting. (S = sham-operated controls; C = castrates; Pre = preoperative score; 1–9 = weekly tests beginning 6 weeks after castration; T1, T3, and T5 = data from the first, third, and fifth weekly tests after testosterone implantation.)

Results

Unconditioned sexual behavior. Prior to castration, there was no significant difference between the two designated groups on any behavioral measure. On the first test, which occurred 6 weeks after castration or sham surgery (immediately following the first session of responding on the second-order schedule, see below), no animal displayed an ejaculatory pattern, and only one male intromitted (Figure 7). A similar picture was apparent on the second test, when three males showed one intromission pattern, and none ejaculated, but for the remainder of this phase of the experiment, no castrated animal displayed either behavior (Figure 7).

Most castrated males mounted on the first and second postsurgical tests, but this fell to very low levels by the third test and remained low thereafter (Figure 8). An ANOVA revealed a significant Castration × Test interaction, $F(8, 240) = 9.58$, $p < .01$, and planned comparisons on each test day showed castrates to differ significantly from controls in Weeks 3, 5, and 9. Mount latencies progressively lengthened following surgery—Castration × Test interaction, $F(8, 240) = 23.198$, $p < .01$—with castrates having significantly longer latencies on each test day (Figure 8). The frequencies of abortive mounts (climbs) made by castrates increased markedly on Postoperative Test 1 and declined progressively over the next four tests (Figure 8). An ANOVA showed a significant Castration × Test interaction, $F(8, 240) = 21.02$, $p < .01$, and planned comparisons on each test day showed castrates to be significantly different from controls on all but the last postsurgical test (Figure 8).

Castrated males anogenitally investigated females at a significantly higher rate than did controls on each postcastration test day—Castration × Days interaction, $F(8, 240) = 7.18$, $p < .01$, after square-root transformation, which did not alter the level of significance—although levels of this behavior fell during the course of these observations (Figure 9). The ancillary behavior of vigorous hindlimb scratching also increased markedly in castrates from the first postsurgical test—$F(8, 240) = 17.526$, $p < .01$, after square-root transformation, which did not alter the level of significance—and remained elevated for the remainder of the pretestosterone replacement tests (Figure 9).

Implantation of testosterone capsules subcutaneously reversed all the above changes in behavior, although at different rates. Details of the analyses are presented in Figures 7–9. On

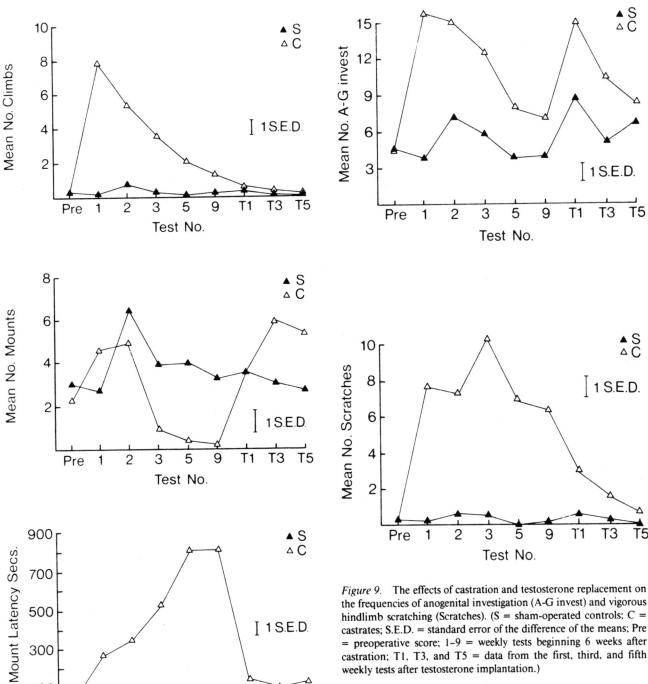

Figure 8. The effects of castration and testosterone replacement on the numbers of mounts, abortive mounting attempts (Climbs), and the latency to mount. (S = sham-operated controls; C = castrates; S.E.D. = standard error of the difference of the means; Pre = preoperative score; 1–9 = weekly tests beginning 6 weeks after castration; T1, T3, and T5 = data from the first, third, and fifth weekly tests after testosterone implantation.)

Figure 9. The effects of castration and testosterone replacement on the frequencies of anogenital investigation (A-G invest) and vigorous hindlimb scratching (Scratches). (S = sham-operated controls; C = castrates; S.E.D. = standard error of the difference of the means; Pre = preoperative score; 1–9 = weekly tests beginning 6 weeks after castration; T1, T3, and T5 = data from the first, third, and fifth weekly tests after testosterone implantation.)

the first postreplacement test, a quarter of the males showed ejaculatory patterns, whereas two thirds showed intromissions (Figure 7). There was a marked increase in the number of mounts and a reduction in mount latencies so that controls and castrates did not differ significantly from each other (Figure 8). The number of mounts made by testosterone-implanted castrates was significantly greater than the number for controls on Test Days 3 and 5 (Figure 8). The frequencies of climbs, which had already decreased to a low level by the last pretestosterone test, remained very low for the remainder of the experiment (Figure 8). Levels of anogenital investigation remained significantly higher than they did in controls on Postimplantation Day 1, but they decreased by Test 5 so that there was no significant difference between the two groups (Figure 9). Similarly, numbers of hindlimb scratchings, which had decreased by Postcastration Test 9 but remained significantly higher than for controls, decreased even more after implantation of testosterone so that there were no significant differences between the groups on Posttestosterone Test Day 3 (Figure 9).

Instrumental behavior. The response rates of the two groups of males under the second-order schedule are shown in Figure 10. An ANCOVA (with preoperative response rate as the covariate) revealed a significant Castration × Test Day interaction and no effect of preoperative response rate, $F(13, 390) = 3.59$, $p < .01$. Individual planned comparisons on each test day showed there to be no significant difference on the first postcastration test (conducted 6 weeks after surgery, with no intervening sexual interaction with females) but significant differences between the two groups on each remaining test day up to Test 9, the last before implantation of testosterone.

In the first session following implantation with testosterone (with no intervening sexual interaction with females), there was a marked increase in response rates of castrated males so that they did not differ significantly from controls on the same day or differ from their own precastration response rates but were significantly higher than their own rates on Castration Test Day 9 ($p < .01$, see Figure 10). On the remaining test days, castrated, testosterone-implanted males were indistinguishable from controls in their response rates under the second-order schedule (Figure 10).

Summary

The results of this experiment, taken overall, demonstrate some dissociation of instrumental and consummatory measures of sexual behavior and present evidence of a motivational effect of testosterone. These results are considered in detail in the Discussion but may be summarized as follows: Castration followed by 6 weeks of sexual inactivity resulted in a population of males who never ejaculated, rarely intromitted, and occasionally mounted. They were not disinterested in females, however, persisting in anogenital investigations and making occasional abortive mounting attempts for a further 9 weeks. During this time, they also showed increased levels of hindlimb scratching, confirming the observations of Hansen and Drake af Hagelsrum (1984), who argued that this is a behavior born out of the thwarting of sexual response tendencies. Testosterone completely reversed these changes in behavior.

Examination of response rates under the second-order schedule revealed a less straightforward picture. On the first postcastration test, when males had no feedback about their sexual competence, they worked for females at a rate lower than but not significantly different from that of controls. Thereafter, when they experienced their impaired copulatory performance when interacting directly with females, castrated males did respond at a lower rate, thereby earning fewer conditional stimuli paired with presentation of a female than did controls but, nonetheless, continued to respond at a rate 50% to 60% that of controls, for the duration of the castration period. By contrast, testosterone treatment resulted in a prompt increase in responding on the first postimplantation test, that is, before the males had any feedback concerning the improvement in their consummatory competence. This is evidence, therefore, of an action of testosterone on an arousal or motivational mechanism.

Discussion

Effects of Preoptic Area Lesions

There have been numerous reports of the effects of lesions to the preoptic area on sexual behavior in a variety of male animals (rat: Hansen & Drake af Hagelsrum, 1984; Hansen et al., 1982; Heimer & Larsson, 1966–1967; Ginton & Merari, 1977; hamster: Powers, Bergondy, & Newman, 1987; cat: Hart et al., 1973; dog: Hart, 1974; goat: Hart, 1986; and monkey: Slimp et al., 1978; see Kelley & Pfaff, 1978, for a review), and the results of the present experiment amply confirm the abolition of mounting, intromitting, and ejaculatory patterns of behavior. However, the early assumption that POA-lesioned animals are sexually uninterested or unaroused by a female in heat has been questioned by a series of observations, both casual and precise, of the sexual and associated behaviors that survive POA lesions (Hansen et al., 1982; Hansen & Drake af Hagelsrum, 1984; Hart, 1986; Ryan & Frankel, 1978; Slimp et al., 1978).

Thus, male rhesus monkeys continue to masturbate and yawn (a sexually related and testosterone-dependent behavior, Slimp et al., 1978); dogs and cats continue to urine mark when placed with females (Hart, 1974; Hart & Voith, 1978); goats persist in their flehmen, self-inurination, and penis-licking displays (Hart, 1986); and male rats continue to anogenitally investigate, clasp, and climb the female for many

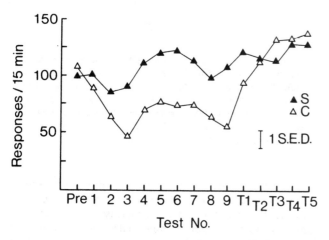

Figure 10. The effects of castration and testosterone replacement on instrumental responses of males working for an estrous female presented under a second-order schedule of reinforcement. (S = sham-operated controls; C = castrates; S.E.D. = standard error of the difference of the means; Pre = preoperative score; 1–9 = weekly tests beginning 6 weeks after castration; T1–T5 = data from the weekly tests after testosterone implantation.)

weeks following castration (Hansen & Drake af Hagelsrum, 1984; Hughes, Everitt, & Herbert, in press; present results). Hansen and Drake af Hagelsrum also described the marked increase in vigorous hindlimb scratching displayed by POA-lesioned males when paired with females in heat but not when alone—a result also confirmed in the present experiments—and demonstrated the development of adjunctive drinking by POA males in the same circumstances. The argument advanced to explain this pattern of results, that these ancillary behaviors arose out of the thwarting of a sexual response tendency (Hansen & Drake af Hagelsrum, 1984), indeed, is formally equivalent to the displacement behavior described by Tinbergen (1952).

Measurement of instrumental behavior under the second-order schedule of sexual reinforcement, behavior that is an expression of sexual interest that does not depend on consummatory competence, has provided further evidence that POA-lesioned male rats are not sexually unarousable or disinterested. Rather, with their having worked to gain access to a female and investigated and pursued her, it is the capacity to display the consummatory responses of mounting and intromission on proximal contact with a female that is impaired. Thus, instrumental and consummatory responses are dissociated by the lesion and focus attention more clearly on the nature of the deficit induced by it. This deficit appears, therefore, to be of a performance rather than a motivational/arousal nature (see Beach, 1956) and indicates that an important function of the preoptic area in regulating the expression of sexual behavior in the male rat may be to engage the neural mechanisms that subserve thrusting and other genitopelvic copulatory reflexes and that are believed to to located in the brainstem and spinal cord (see also Hansen, 1982; Hansen et al., 1982; Hansen & Drake af Hagelsrum, 1984). The fact that preoptic neurons take up testosterone and its aromatized metabolite, estradiol (Pfaff, 1980; Stumpf & Grant, 1975), suggests that this may be one mechanism by which steroid hormones influence the expression of species-specific copulatory responses.

This is not, however, the only possible explanation. The fact that the deficit is one in forming and integrating copulatory responses and is only revealed, therefore, in the presence of the female may also indicate that it is the classification of the female as a sexual stimulus that is impaired following POA lesions. This may both underlie the performance deficit and explain why arbitrary responses to an abstract stimulus, albeit that both are sexual in nature, are preserved. An intriguing parallel with the findings of Slimp et al. (1978) is thus revealed: They observed that preoptic-area-lesioned male rhesus monkeys failed to mount, intromit, and ejaculate with females but masturbated to ejaculation in their absence. Incentive motivational theory says much about responses to stimuli but little about how such stimuli are classified. This appears to be a fruitful area for future research on the neuroendocrine mechanisms underlying sexually motivated and other behaviors.

Effects of Castration and Testosterone Replacement

There is little consensus as to which psychological processes are affected by testosterone withdrawal and replacement. Clearly, castrated rats do not intromit or ejaculate and mount only rarely (Beach, 1942; Bermant & Davidson, 1974; Hansen & Drake af Hagelsrum, 1984). The present results—males were left 6 weeks before being allowed to interact with a receptive female—amply confirm this observation. Whereas this change in overt behavior has been taken to indicate that castrated males are not sexually aroused by estrous females (e.g., Davidson, 1966b; Singer, 1972), the persistence of copulatory and precopulatory behavior in long-term castrates has been taken to indicate changes of a different kind. Thus, Beach (1942) and Grunt (1954), observing castrated rats and guinea pigs, respectively, drew attention to the excitement displayed in the presence of receptive females, despite the absence of intromission patterns. Indeed, it is well recognized that castrated rats, cats, dogs, guinea pigs, and monkeys will investigate, pursue and attempt to mount (climb) females long after the ability to thrust and intromit has been lost (Beach, 1942, 1970; Hart, 1968; Hansen et al., 1982; Phoenix et al., 1973; Rosenblatt & Aronson, 1958). Castrated males will also display displacement behavior, such as scratching and drinking (Hansen & Drake af Hagelsrum, 1984), as was confirmed in the present experiment, in which long-term castrated males showed high levels of vigorous hindlimb scratching when placed with a receptive female, in addition to abortive mounting attempts.

There have been several attempts to assess the motivational effects of sex steroids by using various forms of instrumental technique. Warner (1927) reported that the willingness of male rats to cross an electrified grid to gain access to a female was reduced by castration, whereas females tolerated more intense levels of shock at estrus than at diestrus to gain access to a male. However, the effects of sex hormones on shock thresholds per se seem not to have been directly studied, so purely sensory changes induced by steroids could underlie the reported results. A more recent study in which this technique was used in female rats (Meyerson & Lindstrom, 1973) also pointed to a small effect of estradiol on sexual motivation. On the other hand, female rats in estrus will run as fast for and choose equally both potent and nonpotent males, and ovariectomized females will behave identically (Drewett, 1973). This led Drewett to conclude that estrus does not involve stimulation of any behavior more complex than the unlearned motor patterns associated with it.

Beach and Jordan (1956) reported that male rats would learn to run a straight alley for an estrous female reward and that running was decreased by castration, an effect reversed by treatment with testosterone. However, the males apparently also ran for an anestrous female. Furthermore, Beach (1956) described how the time spent traversing the alley increased "*when* [italics added] the consummatory responses began to weaken and disappear" (p. 12).

Use of a "plus maze" (Merkx, 1983, 1984) or other sexual preference procedure (Hetta & Meyerson, 1978) revealed that male rats show a preference for females in estrus, rather than anestrus, although castration resulted in a decrease in approach to both estrous and anestrous females (Merkx, 1984). Testosterone replacement resulted in a resurgence of interest in both types of females that later (after the opportunity to interact several times) shifted to favor slightly the estrous females (Merkx, 1984). In a similar mate-choice test, it was revealed that castration decreased and testosterone restored the preference of male rats for a receptive female (Edwards & Einhorn, 1986). However, lesions of the POA and dorsolateral midbrain tegmentum also caused very similar changes in preference. Thus, although providing interesting and useful data, this paradigm does not discriminate between the sexual behavioral deficits of the three preparations, even though they clearly differed in their sexual interactions with females. Using a novel variant of this form of procedure, that of place preference (in this case, preference for a place habitually associated with copulation with a receptive female over one associated with an unreceptive female, but with the females absent during the test phase), Miller and Baum (in press) also demonstrated a decrease in preference for the place where copulation had occurred following castration. This is an interesting observation because the male's response does not depend on cues emitted by the female during the test. It will be of value to observe POA-lesioned males in this kind of situation (see above).

An important consideration in trying to establish the nature of the effects of a steroid hormone on sexual behavior is to attempt to separate primary effects on an arousal/motivational mechanism (Beach, 1956) from those on performance (or the copulatory–ejaculatory mechanism of Beach, 1956). This is largely because, as indicated above, a primary change in the latter may underlie a decrease in the former, precisely because failure during repeated attempts to copulate will become nonreinforcing and thus lead secondarily to apparent changes in motivational variables. In the present experiments therefore, some emphasis was placed on the first session of instrumental responding under the second-order schedule following both castration and testosterone replacement, because it occurs before any overt sexual interaction with a receptive female and, therefore, before any feedback about consummatory competence can be obtained.

Six weeks following castration, males continued to respond under the second-order schedule at a rate not significantly different from that of controls. Only on the second test (7 weeks after castration and, therefore, after one episode of failed sexual interaction with the female) did responding drop to levels lower than those of controls. Even after 14 weeks, the males continued to work at half the control rate to earn females with whom they could not copulate but whom they investigated and occasionally attempted to mount. In the present experiments, we did not obtain a pure measure of an arousal (see Hart et al., 1973) or motivational (Beach, 1956) effect of testosterone withdrawal, because it is possible to argue that the decline in instrumental responses seen in the second session after castration was in part a response to the copulatory deficit experience in the first session. Perhaps a different result would have been obtained if 10 or 12 weeks had been allowed to elapse between castration and the first test, when the consequences of testosterone withdrawal may have been more complete. Such an experiment, although interesting, would be difficult to undertake because the long period without experience of the second-order schedule may be incompatible with retention of the response. Furthermore, it is difficult to envisage a testosterone-dependent neural mechanism subserving a sexual arousal process that survives testosterone deprivation for so long. Persistence of peripheral effects of testosterone on genital reflexes as a mechanism for maintaining sexual responsiveness also seems unlikely, because these are now known to be more rather than less sensitive to testosterone withdrawal (Hart et al., 1983; Meisel, O'Hanlon, & Sachs, 1984). Although the precise functions of these reflexes during coitus are well understood (O'Hanlon & Sachs, 1986; Sachs, 1982, 1983), the neuroendocrine mechanisms regulating and coordinating them and their importance in maintaining sexual behavior in the rat remain unclear.

Replacement of testosterone to the long-term castrate produced results that more clearly support an effect on a motivational process. In the first session following replacement, males showed a marked increase in their willingness to work for a female so that their response rates did not differ from those of controls or from their own precastration performance recorded 14 weeks earlier. Here, then, is direct evidence to suggest that testosterone exerts effects on the male's behavior that are independent of an improvement in consummatory competence—effects, therefore, on incentive motivation that govern appetitive responses, such as those measured under the second-order schedule.

Conclusions

Some authors have suggested that the value of the distinction between motivation and performance is doubtful: What can be conceptually distinguished may have no basis in the organization of the processes governing sexual motivation (e.g., Davidson, 1980; Sachs, 1978; Sachs & Barfield, 1976). One test that can be applied to this issue is double dissociation. If variation of one set of factors specifically affects one of the hypothetical processes, whereas variation of another set, by contrast, selectively influences the other, this would constitute evidence for the separable existence of the two hypothetical processes. In the experiments reported here, in which males performed instrumental responses to gain access to a receptive female presented under a second-order schedule of reinforcement, the effects of lesions to the POA and castration followed by testosterone replacement were reinvestigated and dissociated. Lesions to the POA do not decrease instrumental responses or sexual interest in the female but do prevent the display of mounts, intromissions, and ejaculation. POA-lesioned males also develop displacement behavior. Castration, on the other hand, does decrease instrumental behavior, in addition to copulatory behavior, and both effects are reversed by testosterone replacement. Instrumental responses emitted to earn a receptive female are affected especially rapidly by the hormone and before any feedback concerning improved copulatory performance with the female has been obtained. These data argue for a motivational effect of testosterone but not of manipulations of the preoptic area. The precise nature of the deficit in the case of the latter, particularly its relation to problems of stimulus encoding, will require investigation by other behavioral methods if it is to be elucidated.

The long-term nature of the behavioral effects of testosterone is not particularly conducive to experimental study by using the second-order schedule of reinforcement. We suggest, however, that investigation of the nature of the effects on sexual behavior of short-term, reversible manipulations of neuroendocrine systems is well suited to such an experimental paradigm. Our preliminary results on manipulations of the dopamine and β-endorphin systems (Everitt, 1987; Hughes, Herbert, & Everitt, 1987) encourage this view.

References

Alvey, N., Galwey, N., & Lane, P. (1982). *An introduction to Genstat.* London: Academic Press.

Bancroft, J. (1978). The relationship between hormones and sexual behaviour in humans. In J. B. Hutchison (Ed.), *Biological determinants of sexual behaviour.* Chichester, England: Wiley.

Beach, F. A. (1942). Copulatory behavior in prepuberally castrated male rats and its modification by estrogen administration. *Endocrinology, 31,* 679–683.

Beach, F. A. (1956). Characteristics of masculine "sex drive." In M. R. Jones (Ed.), *Nebraska Symposium on Motivation* (Vol. 4, pp. 1–31). Lincoln: University of Nebraska Press.

Beach, F. A. (1970). Coital behavior in dogs: VI. Long-term effects of castration upon mating in the male. *Journal of Comparative and Physiological Psychology Monographs, 70* (3, Part 2).

Beach, F., & Holz-Tucker, A. M. (1949). Effects of different concentrations of androgen upon sexual behavior in castrated male rats. *Journal of Comparative and Physiological Psychology, 42,* 433–453.

Beach, F. A., & Jordan, L. (1956). Effects of sexual reinforcement upon the performance of male rats in a straight runway. *Journal of Comparative and Physiological Psychology, 49,* 105–111.

Beck, J. (1971). Instrumental conditioned reflexes with sexual reinforcement in rats. *Acta Neurobiologiae Experientis, 31,* 153–156.

Beck, J. (1974). Contact with male or female conspecifics as a reward for instrumental responses in estrous and anestrous female rats. *Acta Neurobiologiae Experientis, 34,* 615–630.

Beck, J. (1978). A positive correlation between male and female response latencies in the mutually reinforced instrumental sexual responses in rats. *Acta Neurobiologiae Experientis, 38,* 153–156.

Beck, J., & Chmielewska, J. (1976). Contact with estrous female as a reward for instrumental response in a growing male rat from the 3rd up to the 14th week of life. *Acta Neurobiologiae Experientis, 36,* 535–543.

Bermant, G., & Davidson, J. M. (1974). *Biological bases of sexual behavior* (pp. 97–177). New York: Harper & Row.

Bindra, D. (1976). *A theory of intelligent behavior* (Chap. 9, pp. 179–205). New York: Wiley.

Breedlove, S. M., & Arnold, A. P. (1980). Hormone accumulation in a sexually dimorphic motor nucleus of the rat spinal cord. *Science, 210,* 564–566.

Christensen, L. W., & Clemens, L. G. (1974). Intra-hypothalamic implants of testosterone or estradiol and resumption of masculine sexual behavior in long-term castrated male rats. *Endocrinology, 95,* 984–990.

Davidson, J. M. (1966a). Activation of the male rat's sexual behavior by intracerebral implantation of androgen. *Endocrinology, 79,* 783–794.

Davidson, J. M. (1966b). Characteristics of sex behavior in male rats following castration. *Animal Behaviour, 14,* 266–272.

Davidson, J. M. (1980). The psychobiology of sexual experience. In J. M. Davidson & R. J. Davidson (Eds.), *The psychobiology of consciousness* (pp. 271–332). New York: Plenum Press.

Denniston, R. H. (1954). Quantification and comparison of sex drives under various conditions in terms of learned response. *Journal of Comparative and Physiological Psychology, 47,* 437–440.

Dewsbury, D. A. (1979). Factor analyses of measures of copulatory behavior in three species of muroid rodents. *Journal of Comparative and Physiological Psychology, 93,* 868–878.

Drewett, R. F. (1973). Sexual behaviour and sexual motivation in the female rat. *Nature, 242,* 476–477.

Edwards, D. A., & Einhorn, L. C. (1986). Preoptic and midbrain control of sexual motivation. *Physiology and Behavior, 37,* 329–335.

Everitt, B. J. (1987). [The effects of alpha flupenthixol and 6-hydroxydopamine lesions of the dorsal and ventral striatum on sexual behavior in male rats]. Unpublished observations.

Everitt, B. J., Fray, P., Kostarczyk, E., Taylor, S., & Stacey, P. (1987). Studies of instrumental behavior with sexual reinforcement in male rats (*Rattus norvegicus*): I. Control by brief visual stimuli paired with a receptive female. *Journal of Comparative Psychology, 101,* 395–406.

Giantonio, G. W., Lund, N. L., & Gerall, A. A. (1970). Effects of diencephalic and rhinencephalic lesions on the male rat's sexual behavior. *Journal of Comparative and Physiological Psychology, 73,* 38–46.

Ginton, A., & Merari, A. (1977). Long range effects of MPOA lesions on mating behavior in the male rat. *Brain Research, 120,* 158–163.

Grossman, S. P. (1967). *A textbook of physiological psychology.* New York: Wiley.

Grunt, J. A. (1954). Exogenous androgen and non-directed hyperexcitability in castrated male guinea pigs. *Proceedings of the Society for Experimental Biology and Medicine, 85,* 540–542.

Grunt, J. A., & Young, W. C. (1953). Consistency of sexual behavior patterns in individual male guinea pigs following castration and androgen therapy. *Journal of Comparative and Physiological Psychology, 46,* 138–144.

Hansen, S. (1982). Hypothalamic control of motivation: The medial preoptic area and masculine sexual behaviour. In P. Sodersten (Ed.), *Behavioural neuroscience in Scandinavia. Scandinavian Journal of Psychology* (Suppl. 1), 121–126.

Hansen, S., & Drake af Hagelsrum, L. J. K. (1984). Emergence of displacement activities in the male rat following thwarting of sexual behavior. *Behavioral Neuroscience, 98,* 868–883.

Hansen, S., Köhler, C., Goldstein, M., & Steinbusch, H. W. M. (1982). Effects of ibotenic acid-induced neuronal degeneration in the medial preoptic area and lateral hypothalamic area on sexual behavior in the male rat. *Brain Research, 239,* 213–232.

Hart, B. L. (1968). Role of prior experience in the effects of castration on sexual behavior of male dogs. *Journal of Comparative and Physiological Psychology, 66,* 719–725.

Hart, B. L. (1973). Effects of testosterone proprionate and dihydrotestosterone on penile morphology and sexual reflexes of spinal male rats. *Hormones and Behavior, 4,* 239–246.

Hart, B. L. (1974). Medial preoptic-anterior hypothalamic area and sociosexual behavior of male dogs: A comparative neuropsychological analysis. *Journal of Comparative and Physiological Psychology, 86,* 328–349.

Hart, B. L. (1986). Medial preoptic-anterior hypothalamic lesions and sociosexual behavior of male goats. *Physiology and Behavior, 36,* 301–305.

Hart, B. L., Haugen, C. M., & Peterson, D. M. (1973). Effects of medial preoptic-anterior hypothalamic lesions on mating behavior of male cats. *Brain Research, 54,* 177–191.

Hart, B. L., & Voith, V. L. (1978). Changes in urine spraying, feeding and sleep behavior following medial preoptic-anterior hypothalamic lesions in cats. *Brain Research, 145,* 406–409.

Hart, B. L., Wallach, S. J. R., & Melese-d'Hospital, P. Y. (1983). Differences in responsiveness to testosterone of penile reflexes and copulatory behavior of male rats. *Hormones and Behavior, 17,* 274–283.

Heimer, L., & Larsson, K. (1966–1967). Impairment of mating

behavior in male rats following lesions in the preoptic-anterior hypothalamic continuum. *Brain Research, 3*, 248-263.

Hetta, J., & Meyerson, B. J. (1978). Sexual motivation in the male rat. *Acta Physiologica Scandinavica*, Suppl. 453, 1-67.

Hogan, J. A., & Roper, T. J. (1978). A comparison of the properties of different reinforcers. In J. S. Rosenblatt, R. A. Hinde, C. Beer, & M.C. Busnel (Eds.), *Advances in the study of behaviour* (Vol. 12, pp. 1-64). New York: Academic Press.

Hughes, A. M., Everitt, B. J., & Herbert, J. (in press). The effects of β-endorphin infused into the medial preoptic area and bed nucleus of the stria terminalis on sexual and ingestive behaviour in male rats. *Neuroscience*.

Hughes, A., Herbert, J., & Everitt, B. J. (1987). [The effects of β-endorphin on instrumental and unconditioned sexual behavior in the male rat]. Unpublished observations.

Jowaisas, D., Taylor, J., Dewsbury, D. A., & Malagodi, E. F. (1971). Copulatory behavior of male rats under an imposed operant requirement. *Psychonomic Science, 25*, 287-290.

Kelley, D. B., & Pfaff, D. W. (1978). Generalizations from comparative studies on neuroanatomical and endocrine mechanisms of sexual behaviour. In J. B. Hutchison (Ed.), *Biological determinants of sexual behaviour* (pp. 225-254). Chichester, England: Wiley.

Konorski, J. (1967). *Integrative activity of the brain: An interdisciplinary approach*. Chicago: University of Chicago Press.

Larsson, K. (1956). Conditioning and sexual behavior in the male albino rat. *Acta Physiologica Gotoburgensia*. Stockholm, Sweden: Almquist & Wiksell.

Lisk, R. D. (1967). Neural localization for androgen activation of copulatory behavior in the male rat. *Endocrinology, 80*, 754-761.

Malsbury, C. W., & Pfaff, D. W. (1974). Neural and hormonal determinants of mating behavior in adult male rats. A review. In L. V. DiCara (Ed.), *Limbic and autonomic nervous systems research* (pp. 85-136). New York: Plenum Press.

Meisel, R. L., O'Hanlon, J. K., & Sachs, B. D. (1984). Differential maintenance of penile responses and copulatory behavior by gonadal hormones in castrated male rats. *Hormones and Behavior, 18*, 56-64.

Merkx, J. (1983). Sexual motivation of the male rat during the oestrous cycle of the female rat. *Behavioural Brain Research, 7*, 229-237.

Merkx, J. (1984). Effect of castration and subsequent substitution with testosterone, dihydrotestosterone and oestradiol on sexual preference behaviour in the male rat. *Behavioural Brain Research, 11*, 59-65.

Meyerson, B. J., & Lindstrom, L. H. (1973). Sexual motivation in the female rat. *Acta Physiologica Scandinavica*, Suppl. 389.

Miller, R. L., & Baum, M. J. (in press). Naloxone inhibits mating and conditioned place preference for an estrous female in male rats soon after castration. *Brain Research*.

Numerical Algorithms Group. (1977). *"GENSTAT," a general statistical program*. Oxford, England: Oxford University Press.

O'Hanlon, J. K., & Sachs, B. D. (1986). Fertility of mating in rats (*Rattus norvegicus*): Contributions of androgen-dependent morphology and actions of the penis. *Journal of Comparative Psychology, 100*, 178-187.

Paxinos, G., & Watson, C. (1982). *The rat brain in stereotaxic coordinates*. Sydney, Australia: Academic Press.

Pfaff, D. W. (1980). *Estrogens and brain function*. New York: Springer Publishing.

Pfaff, D. W. (1982). Neurobiological mechanisms of sexual motivation. In D. W. Pfaff (Ed.), *The physiological mechanisms of motivation* (pp. 287-317). New York: Springer-Verlag.

Phoenix, C. H., Slob, A. K., & Goy, R. W. (1973). Effects of castration and replacement therapy on sexual behavior of adult male rhesuses. *Journal of Comparative and Physiological Psychology, 84*, 472-481.

Powers, J. B., Newman, S. W., & Bergondy, M. L. (1987). MPOA and BNST lesions in male Syrian hamsters: Differential effects on copulatory and chemoinvestigatory behavior. *Behavioral Brain Research, 23* 181-195.

Rosenblatt, J. S., & Aronson, L. R. (1958). The decline in sexual behavior in male cats after castration with special reference to the role of prior sexual experience. *Behaviour, 12*, 285-338.

Ryan, E. L., & Frankel, A. I. (1978). Studies on the role of the medial preoptic area in sexual behavior and hormonal responses to sexual behavior in the mature male laboratory rat. *Biology of Reproduction, 19*, 971-983.

Sachs, B. D. (1978). Conceptual and neural mechanisms of masculine copulatory behavior. In T. E. McGill, D. A. Dewsbury, & B. D. Sachs (Eds.), *Sex and behavior* (pp. 267-295). New York: Plenum Press.

Sachs, B. D. (1982). Role of striated penile muscles in penile reflexes, copulation and induction of pregnancy in the rat. *Journal of Reproduction and Fertility, 66*, 433-443.

Sachs, B. D. (1983). Potency and fertility: Hormonal and mechanical causes and effects of penile actions in rats. In J. Balthazart, E. Prove, & R. Gilles (Eds.), *Hormones and behaviour in higher vertebrates* (pp. 86-110). West Berlin, Federal Republic of Germany: Springer-Verlag.

Sachs, B. D., & Barfield, R. J. (1976). Functional analysis of masculine copulatory behavior in the rat. In J. S. Rosenblatt, R. A. Hinde, E. Shaw, & C. Beer (Eds.), *Advances in the study of behavior* (Vol. 7, pp. 91-154). Orlando, FL: Academic Press.

Sachs, B. D., Macaione, R., & Fegy, L. (1974). Pacing of copulatory behavior in the male rat: Effects of receptive females and intermittent shocks. *Journal of Comparative and Physiological Psychology, 87*, 326-331.

Schwartz, M. (1955). Instrumental and consummatory measures of sexual capacity in the male rat. *Journal of Comparative and Physiological Psychology, 48*, 328-333.

Singer, J. J. (1972). Anogenital explorations and testosterone induced sexual arousal. *Behavioral Biology, 7*, 743-747.

Slimp, J. C., Hart, B. L., & Goy, R. W. (1978). Heterosexual, autosexual and social behavior of adult male rhesus monkeys with medial preoptic-anterior hypothalamic lesions. *Brain Research, 142*, 105-122.

Södersten, P., Hansen, S., Eneroth, P., Wilson, C., & Gustafsson, J.-A. (1979). Testosterone in the control of sexual behaviour. *Journal of Steroid Biochemistry, 12*, 337-346.

Stumpf, W. E., & Grant, L. D. (1975). *Anatomical neuroendocrinology*. Basel, Switzerland: Karger.

Tinbergen, N. (1952). "Derived" activities: Their causation, biological significance, origin and emancipation during evolution. *Quarterly Review of Biology, 27*, 1-32.

Toates, F. M. (1980). *Animal behaviour—A systems approach*. Chichester, England: Wiley.

Toates, F. M. (1986). *Motivational systems*. Cambridge, England: Cambridge University Press.

Warner, L. H. (1927). A study of sex behavior in the white rat by means of the obstruction method. *Comparative Psychology Monographs; 4* (22).

Weiner, B. J. (1962). *Statistical principles in experimental design*. New York: McGraw-Hill.

Received January 15, 1987
Revision received July 11, 1987
Accepted July 11, 1987 ∎

Glossary

ad libitum - this latin term means that one is at liberty to engage in the response whenever one wishes; thus an animal that is fed ad libitum is one that has continuous access to food

androgens - a class of gonadal hormones; the levels of androgens are generally higher in males than in females

anestrous female - a female mammal that has an estrous cycle but is not in a state of estrus (i.e., that is not fertile and in a state of sexual receptivity)

anterior hypothalamus - the anterior (i.e., the front) portion of the hypothalamus; the anterior hypothalamus is just behind the medial preoptic area

cannulae - fine tubes through which chemicals can be injected into specific sites in the brain

castration - removal of the gonads; removal of testes from males or ovaries from females

consummatory response - any response that consummates (i.e., completes) a sequence of motivated behavior

copulatory - referring specifically to the act of sexual intercourse

diestrus - one of the stages of the estrous cycle during which the female is not receptive

displacement behavior - when an animal that is motivated to engage in one behavior cannot for some reason engage in that behavior, it may excessively engage in other behaviors such as grooming or drinking; in such a context, these other behaviors are referred to as displacement behaviors; displacement behaviors are thought to be a sign of frustration

ejaculate - to eject sperm

estradiol - a gonadal hormone; the levels of estradiol are generally higher in females than in males

estrous females - in some species females will accept the sexual advances of males during only a particular phase of their hormonal cycle; during this phase they are fertile and are said to be estrous females

estrus - the period of the estrous cycle when the female is receptive; <u>estrus</u> is a noun, and <u>estrous</u> is an adjective

instrumental response - an arbitrary voluntary response, such as a lever press, that is performed for reinforcement

intromission - insertion of the penis into the vagina

medial - toward the midline

medial preoptic area - an area of the brain adjacent to the anterior hypothalamus; it is just above and just in front of the optic chiasm

mount - part of the copulatory sequence of many mammalian species; the male climbs on the female from behind and clasps her hindquarters

N-methyl aspartic acid - the neurotoxin that the authors injected into the medial preoptic area and anterior hypothalamus to induce lesions

postejaculatory interval (PEI) - the duration between an ejaculation and the next intromission

progesterone - a gonadal hormone; the levels of progesterone are generally higher in females than in males

reversed light-dark cycle - a daily cycle of laboratory lighting in which the subjects are maintained in darkness during the day and in light during the night; this makes it more convenient to test nocturnal laboratory animals, such as rats, during the dark part of their daily cycle, when they are more active

scapulae - the shoulder blades

second-order schedule - a schedule of reinforcement that is the product of two different schedules; for example, in the present experiments, every 10th lever press was reinforced by a light that had previously been paired with the presentation of a receptive female, and the first light to occur after each 15-minute interval was accompanied by a receptive female

sham operation - an operation performed as a control procedure in an experiment; it is performed just like the real operation except that the surgical manipulation under investigation is not performed; for example, a sham operation in a chemical lesion study often involves anesthetizing the subject and lowering a cannula into the target structure, but not making the lesion

silastic tubing - a type of multipurpose medical tubing; in this study a short length of tubing was filled with testosterone and implanted under the skin so that the testosterone was released gradually through the tubing walls into the body

steroid hormones - any hormone that is derived from cholesterol; all of the hormones released by the gonads (i.e., by the ovaries and testes) are steroid hormones

subcutaneously - under the skin

testicular - from the testes

testosterone - the most potent androgen; levels of testosterone are generally higher in males than in females

Essay Study Questions

1. Describe the effects of castration on male copulatory behavior. What are the effects of testosterone replacement injections?

2. What are the effects on male copulatory behavior of bilateral lesions that destroy both the medial preoptic area and the anterior hypothalamus?

3. Prior to the experiments of Everitt and Stacey, there was evidence to challenge the common view that testosterone maintains the sexual motivation of adult males by acting on receptors in the medial preoptic area and the anterior hypothalamus. Summarize it.

4. Describe the behavior toward receptive females of male rats with medial-preoptic anterior-hypothalamic lesions. What do these observations suggest?

5. What effects did castration have on copulatory behavior and lever pressing for estrous females?

6. What were the main conclusions of Everitt and Stacey?

Multiple-Choice Study Questions

1. Castration of male mammals produces a decline in their sexual behavior, which
    a. in some cases can be very gradual.
    b. can be restored by replacement injections of testosterone.
    c. can be restored by the implantation of testosterone into the medial-preoptic anterior-hypothalamic area.
    d. all of the above
    e. both a and b

2. Male rats do not normally intromit
    a. without ejaculating.
    b. on the first date.
    c. without mounting.
    d. without engaging in displacement behavior.
    e. after engaging in displacement behavior.

3. In Experiment 2, behavioral testing commenced about
    a. 2 hours after castration.
    b. 1 day after castration.
    c. 1 week after castration.
    d. 40 days after castration.
    e. 2 years after castration.

4. Castration reduced the lever pressing of males for receptive females to
    a. about 5% of control values.
    b. between 10% and 20% of control values.
    c. between 50% and 60% of control values.
    d. between 80% and 90% of control values.
    e. about 95% of control values.

5. The fact that the eventual disruption of lever pressing was not apparent on the first postcastration test day suggests that
    a. experience of the inability to copulate may contribute to the reduction in sexual motivation experienced by castrated male rats.
    b. it takes a few weeks for the testosterone to be cleared from the system of castrated rats.
    c. it takes a few weeks for male rats to recover from castration.
    d. testosterone acts on a general, rather than a specific, arousal mechanism.
    e. both b and c

6. On the basis of their data, Everitt and Stacey suggest that the medial preoptic area and anterior hypothalamus may include neurons whose function is to
    a. engage mechanisms that subserve thrusting and other genitopelvic copulatory reflexes.
    b. recognize the receptive female rat as an appropriate sexual stimulus.
    c. maintain a high level of sexual motivation.
    d. all of the above
    e. both a and b

The answers to the preceding questions are on page 290.

Food-For-Thought Questions

1. As things turned out, the "food condition" of Experiment 1 served little purpose. Why did Everitt and Stacey include it?

2. The experiments of Everitt and Stacey provide a particularly compelling illustration of the importance of behavioral analysis in neuroscientific research. Discuss.

3. Some theories of motivation are based on the premise that the primary motivating force in the life of an organism is the maintenance of homeostasis (i.e., the constancy of their internal bodily environment). Discuss with respect to the results of Everitt and Stacey.

ARTICLE 8

Escalation of Feline Predation Along a Gradient From Avoidance Through "Play" to Killing

S.M. Pellis, D.P. O'Brien, V.C. Pellis, P. Teitelbaum, D.L. Wolgin, and S. Kennedy
Reprinted from Behavioral Neuroscience, 1988, volume 102, 760-777.

Biopsychologists often focus their analyses on one or two easy-to-measure objective indices of the behavior that they are studying—on indices such as the number of lever presses, the number of errors in a maze, the amount of food consumed, or the latency to reach a goal box. The following article by Pellis and his colleagues is one of the best examples of the insights that can be gained through the careful analysis of entire behavioral sequences, as opposed to single behavioral indices. The focus of their experiments was the predatory behavior of the cat.

Pellis and his collaborators concluded that the reaction of each cat to its prey can be represented as a point along a continuum from avoidance to efficient killing. Support for this theory came from its ability to account for the effects of antianxiety drugs on cats' interactions with mice. An injection of an antianxiety drug shifted each cat's behavior along the continuum towards efficient killing: cats that usually avoided mice "played" with them after an injection, those that usually "played" with them killed them, and those that usually killed them killed them more quickly.

Bilateral lesions of the lateral hypothalamus initially eliminated all defensive and predatory behaviors. Although the predatory behaviors recovered progressively over the ensuing weeks, the defensive behaviors did not. The fact that the antianxiety drug diazepam facilitated attack throughout the recovery period even though the lesioned cats displayed no defensive behavior suggested that the facilitatory effect of diazepam on predatory attack is not totally attributable to its ability to suppress defensive tendencies.

In their final experiment, Pellis and his collaborators transcribed the behavior of both the cat and the mouse during 30 sequences of predatory attack. To do this, they used the Eshkol-Wachmann Movement Notation System, which is a method of translating complex body movements into a set of written symbols that characterize the movements of individual body segments during a behavioral sequence—it was originally designed to provide a written record of the movements of a dance, in the same way that sheet music provides a written record of a song. This fine-grained behavioral analysis of predatory attack sequences revealed that the playful-appearing movements of cats during predatory attack are far from playful; they appear to be designed to protect a hesitant cat from injurious bites. Accordingly, the proverbial cat playing with a mouse is not a "playing" cat at all; it is a cat vacillating between attack and defense.

Escalation of Feline Predation Along a Gradient From Avoidance Through "Play" to Killing

Sergio M. Pellis, Dennis P. O'Brien, Vivien C. Pellis, and Philip Teitelbaum
University of Illinois at Urbana-Champaign

David L. Wolgin and Susan Kennedy
Florida Atlantic University, Boca Raton

In this article, we show that feline predation involves a continuous gradient of activation between defense and attack and that predatory "play" results from an interaction of the two. Benzodiazepines (oxazepam, diazepam) escalated attack toward killing, so that cats that had avoided mice prior to the drug now played with them, cats that had originally played now killed, and cats that killed mice now did so with less preliminary contact. In such shifts, no sharp demarcation between play and predation was evident. Lateral hypothalamic lesions disrupted the escalation of attack. During recovery, attack was escalated once again along the gradient toward killing, but in the absence of both defense and play. A similar result was obtained in intact killers and nonkillers by the application of mild tail pinch. These results suggest that play with prey is a misnomer for predatory behavior that fails to escalate along the gradient between defense and attack. Movement notation analysis revealed that playful movements are adaptive in that they protect the cat from injury.

Although many definitions of play behavior have been proposed (see Appendix I in Fagen, 1981), none have been completely satisfactory. Activities are classified as play if they share some of the following characteristics: "... they are pleasurable, but lack obvious motivation or consummatory activities; behavior patterns are exaggerated, there is a degree of flexibility in the behavioral sequences, and... sequences are truncated; they... have no obvious functional significance; they are organism-dominated rather than stimulus-dominated, and are apparently conducted for their own sake with relative relaxation" (Einon, 1983, p. 212). Most commonly, play is reported for young animals (Fagen, 1981). However, adults from various taxa of mammalian carnivores are sometimes observed to interact with their prey in a "playful" manner (Eisenberg & Leyhausen, 1972). No case of such adult play is better known than that of the domestic cat (Leyhausen, 1979). Compared with instances of actual predation, in which the prey is killed, predatory play by the cat involves repeated contact in the absence of injury to the prey, a reduced range of the behavior patterns used, and different, seemingly less functionally organized, sequencing of predatory behavior patterns (Biben, 1979). Such findings support the possibility that predatory play is a behavioral category that is independent from predation.

An alternative view, based on the work of Adamec and his colleagues (Adamec, 1978; Adamec & Stark-Adamec, 1983; Adamec, Stark-Adamec, & Livingston, 1980a, 1980b, 1980c) is that predatory play and attack are on a continuum. That is, playful behavioral characteristics are by-products of the processes governing predation, not a reflection of a different motivational category of behavior. Adamec and Stark-Adamec (1983) have shown that predatory attack is often accompanied by defensive behavior. The degree of such behavior varies among cats. Efficient killers exhibit little defense; their attack consists of a relatively stereotyped sequence of approach from behind, pinning the prey with the forepaw, and biting the prey on the neck (Adamec & Stark-Adamec, 1983). Timid cats, however, exhibit a more defensive pattern of attack, consisting of repeated approach, tentative contact with the prey, and frequent withdrawal. The vacillation between approach and withdrawal conveys the impression that the timid cat is not "serious" in its attack, and thus appears to be playful (cf. Wolgin, 1982). When looked at in this way, many of the playful characteristics listed earlier may be interpreted as by-products of such vacillation. Two lines of evidence are consistent with this view. First, predatory play has been observed to escalate into killing, particularly in young cats (Leyhausen, 1979; Wolgin, 1982). Second, hunger can induce an otherwise playful cat to kill, whereas increased size of prey can induce an otherwise efficient killer to engage in a more playful pattern of attack (Adamec, 1975; Biben, 1979).

This work was supported in part by a grant from the Harry Frank Guggenheim Foundation to Sergio Pellis, by National Institutes of Health Grant R01 NS 11671 to Philip Teitelbaum, by National Research Service Award F32 NS 07305 to Dennis O'Brien, and by grants from the Division of Sponsored Research, Florida Atlantic University to David L. Wolgin.

We wish to thank Nancy O'Connell and Gerri Lennon for typing the manuscript, and Priscilla Kehoe, at Florida Atlantic, and Steve Barzci, at Illinois, for helping to test the cats.

We thank Hoffman-LaRoche, Nutley, New Jersey, and Wyeth Laboratories, Radnor, Pennsylvania, for their generous donations of diazepam and oxazepam, and two unknown reviewers for their helpful criticisms.

Dennis P. O'Brien is now at the Department of Veterinary Medicine, Veterinary Teaching Hospital, University of Missouri; Vivien C. Pellis and Philip Teitelbaum are now at the Department of Psychology, University of Florida; and Susan Kennedy is now at the Department of Medical Microbiology and Immunology, The Ohio State University College of Medicine.

Correspondence concerning this article should be addressed to Sergio M. Pellis, who is now at the Department of Psychology, University of Florida, Gainesville, Florida 32611.

In this article we provide additional support for the view that predatory play and attack are on a continuum. First, we show that benzodiazepines, which are used clinically to treat anxiety, produce an escalation of attack from play to killing. Second, we show that in cats recovering from lateral hypothalamic lesions, defensive components of attack are absent. When defense is absent, playful attack is correspondingly absent. In such cats, benzodiazepines facilitate attack by increasing the intensity of attack components. Third, in both normal and brain-damaged cats, activation by mild tail pinch short circuits the escalation of attack. Where escalation is rapid, play is absent. Finally, we provide a functional analysis of playful attack, in which the movements of both cat and prey are simultaneously correlated using the Eshkol-Wachman Movement Notation (Eshkol & Wachmann, 1958). Such analysis shows that many of the seemingly exaggerated and inappropriate movements by the cats during play are actually adaptive in avoiding the prey's teeth, thus confirming the interpretation that play results from the interaction of predatory and defensive patterns of behavior.

General Method

Animals

Seventeen adult domestic shorthair cats (10 female and 7 male) were used as subjects. They were individually housed at 20–22°C on a 12:12 hr light/dark cycle. Testing was conducted between 1100 and 1500 hr, during the light phase of the cycle. Purina Cat Chow and water were provided ad lib. Thirteen of the cats were born and raised in the laboratories of the University of Illinois. The remaining 4 were born and raised at the laboratories of Florida Atlantic University. Prey were adult albino laboratory mice (40–60 g) and adult Long-Evans hooded rats (250–350 g).

Drugs

Diazepam (Hoffman-LaRoche) was dissolved in a solution containing 40% propylene glycol, 10% ethanol, and 50% water to yield a concentration of 5 mg/ml. It was administered sc. Oxazepam (Hoffman-LaRoche) was suspended in sterile saline and Tween 80 (5% w/v). Its concentration was adjusted to yield a constant injection volume of 1 ml/kg, which was administered ip. Doses of the benzodiazepines are specified in the appropriate sections of the results.

Experimental Procedures

Testing was performed in one of two observation cages (104 × 76 × 74 cm at Illinois and 125 × 60 × 100 cm at Florida), each with three solid walls, a screen mesh top, and a Plexiglas front. All the cats were habituated to the enclosures before testing began. Each cat was weighed, injected with drug or vehicle, and placed in the observation area. Mice were then introduced into the enclosure (details are given later for each experiment). Most of the sessions were videotaped on a Sony cassette recorder (Model No. SLO-323). Some predatory sequences were filmed with a 16-mm Bolex movie camera at 24 fps. For subsequent analysis the videotapes were viewed on a monitor. The 16-mm films were viewed on a Lafayette motion analyzer, which allowed frame-by-frame inspection.

Part 1. Escalation of Predatory Behavior by the Administration of Benzodiazepines

Benzodiazepines have been shown to enhance predation in several species of mammalian carnivore (Apfelbach, 1978; Leaf, Wnek, Gay, Corcia, & Lamon, 1975; Leaf, Wnek, & Lamon, 1984; Wolgin & Servidio, 1979). Enhancement of predatory behavior by benzodiazepines might be due to facilitation of attack (directly or indirectly by increasing hunger; e.g., Della-Fera, Baile, & McLaughlin, 1979; Mereu, Fratta, Chessa, & Gessa, 1976) or to the reduction of defense (by decreasing fear of the prey; e.g., Sepinwall & Cook, 1978). Our preliminary observations showed that when administered benzodiazepines, cats that otherwise avoided mice were induced to attack playfully, whereas killing was enhanced and play was reduced in cats that killed. These otherwise paradoxical findings make sense if play represents an equilibrium point between approach and withdrawal, so that the actual effect of a benzodiazepine on an individual cat depends on how defensive it was prior to the drug. In this part of the article, we document that benzodiazepines shift the balance between attack and defense so that each cat is escalated farther along the gradient toward killing. In such shifts, no sharp demarcation between play and predation was evident.

Experiment 1: Effects of Oxazepam on Predatory Behavior

As a first approach, we evaluated the effect of various doses of oxazepam on the predatory behavior of cats that killed only after a relatively protracted playful interaction with the mouse.

Method

Four cats (3 males and 1 female), all of which killed mice, were used as subjects. The cats were tested with the vehicle and with each of three doses of oxazepam (0.3, 1.0, and 3.0 mg/kg) in counterbalanced order. Testing was conducted in the second observation enclosure described earlier (125 × 60 × 100 cm). Treatments were administered once or twice per week. Twenty minutes after injection of the drug, a live mouse was introduced into the enclosure for 30 min. Latencies to initial mouth contact and to kill mice were scored with a stopwatch. Two of the cats were given satiation tests in which a new mouse was given 30–60 s after the cat finished eating the previous mouse. If the cat killed a mouse but did not eat it, a new mouse was given at the end of 30 min. The tests were continued until the cat failed to kill a mouse within 30 min. The data were analyzed by means of repeated measures analysis of variance.

Results and Discussion

The mean latencies to kill following injection of the vehicle and each of three doses of oxazepam are shown in Figure 1. At doses of 1.0 and 3.0 mg/kg, oxazepam greatly reduced the latency to kill. Compared with the undrugged state, the cats were more vigorous and persistent in their attack, particularly early in the session when the undrugged cats were slower and more cautious in their movements. This resulted in earlier mouth contact with the prey (Table 1). Moreover, when

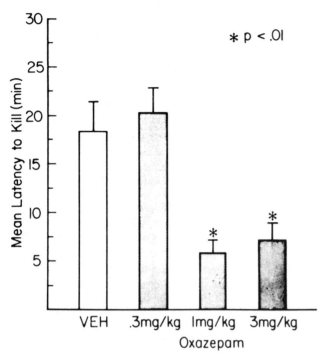

Figure 1. Mean latencies to the killing bite following injection of oxazepam (0.3, 1.0, and 3.0 mg/kg) or vehicle (VEH). (Vertical lines indicate standard errors.)

drugged, the cats rarely retracted their heads when near the mouse, whereas in the undrugged state, they frequently did so during the early moments of attack. Thus, under the influence of oxazepam, the cats were less hesitant in initiating an attack and less defensive in their interaction with the mice.

In the 2 cats given satiation tests, oxazepam increased the number of mice killed and eaten, particularly at the higher doses (Figure 2). In general, the latency to kill the second mouse was shorter than that for the first. The latencies for subsequent mice were generally longer, until eventually the mice were no longer killed (latency >30 min). Both with and without the drug, increasing levels of satiation were accompanied by a more playful pattern of attack, consisting of repeated swiping and aborted attempts to bite the mouse. In addition, the cats frequently paused for a while between bouts of attack. During such pauses, they withdrew from the mouse and sat down or paced around the cage while meowing. Cats given oxazepam, satiated after killing two or more mice, would often stop, lower their heads, and rub their snouts along the mouse's body, in a manner reminiscent of cats with trigeminal deafferentations (MacDonnell & Flynn, 1966a, 1966b) or lateral hypothalamic lesions (Wolgin & Teitelbaum, 1978). This suggests that satiation deactivates the sensory fields for biting.

Experiment 2: Escalation of Attack with Diazepam

To provide a more quantitative assessment of the effects of benzodiazepines on attack components, additional tests were conducted with cats exhibiting a wider range of responses toward prey, from cats that completely avoided mice to cats that walked up to and killed them with a single bite to the neck. These tests also afforded the opportunity to evaluate another benzodiazepine, diazepam.

Method

Ten cats (9 females and 1 male) were used as subjects. They were randomly assigned to receive either diazepam (4 mg/kg) or an equal volume of vehicle. Five to seven days later, each cat received the opposite treatment. The dose of diazepam was chosen on the basis of pilot work. Testing was conducted in the observation enclosure described earlier (104 × 76 × 74 cm). The frequency of predatory behavior patterns (see Table 2 for their description) was scored from the videotapes. Only those behavior patterns used to contact the prey directly were scored, so that the overall strength of the attack could be compared between cats. Behavior patterns leading to contact (e.g., stalking) or those occurring after contact had occurred (e.g., head shaking while holding the prey in the mouth after biting) were strongly dependent on the surroundings and on the behavior of the prey, respectively, and are analyzed elsewhere (Pellis & Officer, 1987).

A live mouse was introduced 30 min after injection. If the mouse was killed, it was removed, and 10 min later another mouse introduced, and so on up to 90 min postinjection, allowing a maximum of six mice to be presented. Any mouse that was seriously injured but not killed was scored as a kill, removed, and immediately anesthetized with an overdose of phenobarbital. If the first mouse was not killed, it was left in the cage for 30 min, then removed, and a second mouse introduced for 30 min. If the second mouse also was not killed, the testing session was ended.

Results

All cats increased their level of attack on prey. However, the effect of diazepam on predatory behavior varied with the predrug pattern of attack. In nonkillers, which previously avoided mice, diazepam induced a shift to a tentative pattern of attack, resulting in repeated, but relatively gentle, contact with the prey. In cats that had previously interacted with mice only tentatively, diazepam induced a stronger form of attack, in which the prey was killed. In cats that previously had killed, but only after repeated tentative contact with the prey, the drug induced killing with less preliminary contact.

Analysis of the behavior patterns used to contact prey revealed three categories (see Table 2): striking laterally with the palm of the forepaw at right angles to the ground (touch, swipe), striking vertically with the palm of the forepaw horizontal to the ground (tap, bat, pin), and orofacial contact

Table 1
Latencies (in min) to Initial Mouth Contact With the Mouse (Experienced Cats)

		Oxazepam	
Cat	Vehicle	1 mg/kg	3 mg/kg
41	5.00	2.00	3.25
42	22.50	0.50	0.50
43	5.75	2.00	0.75
44	4.00	2.25	0.50
M	9.30	1.75*	1.25*

* $p < .05$. Because of heterogeneity of variance, a logarithmic transformation of the data [$x' = \log(x + 10)$] was performed before statistical analysis.

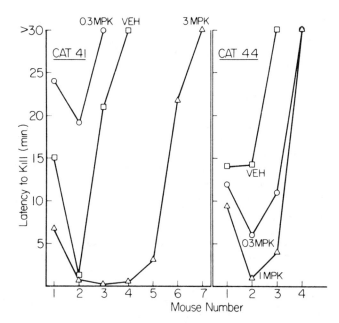

Figure 2. Latencies to kill successive mice during satiation tests under various doses of oxazepam or vehicle. (MPK = mg/kg; VEH = vehicle; >30 = cat did not kill mouse within 30 min.)

(nose contact, mouth contact, bite). As will be shown later, these three forms of contact reveal an increasing strength of contact (lateral striking → vertical striking → orofacial contact). Furthermore, within each category there is an increase in intensity of attack (lateral: touch → swipe; vertical: tap → bat → pin; orofacial: nose contact → mouth contact → bite). However, there are qualitative differences in these changes in intensity. Swipe and bat could result from either increased attack or defense. In contrast, pin and orofacial contact, especially biting, are more pure forms of attack, occurring when defense is reduced.

Lucy represents 1 of the 3 cats that never killed. With vehicle, she did not approach the mice but visually tracked them. If approached by the mouse, she withdrew. In a corner, when she could not withdraw, she tentatively struck the mouse with a forepaw either laterally (touch) or vertically (tap), but if the mouse persisted in approaching, she struck laterally with increased strength (swipe; Figure 3). She did not bat or pin, nor did she approach closely enough with her head to make orofacial contact. In contrast, under diazepam, Lucy followed and approached the mouse and made both forepaw and orofacial contact. However, the gentler, more tentative forms of contact were still predominant (touch, tap, and nose contact; Figure 3). The stronger forms of contact, although present, were not delivered with sufficient strength (e.g., bite) or with sufficient repetitions (e.g., bat) to cause injury to the prey. For example, if she stunned the mouse by batting it on the thorax, she did not follow up with a killing bite to the now stationary mouse. Any movements such as a sudden, deep gasp by the mouse after being batted caused Lucy to withdraw. Therefore, in Lucy the administration of diazepam caused a shift from pure defense, that is from withdrawal and defensive forepaw striking to a blend of attack and defense, in which approach and withdrawal intermingled. Although all forms of contact were present, none were used with sufficient strength to injure the prey. The increased attack with failure to injure or kill mice induced by diazepam resulted in a 95.3% increase in the frequency of contact per mouse (compared with vehicle).

Molly represents 1 of the 3 cats that became killers when given diazepam. When injected with the vehicle, she resembled Lucy under diazepam. She approached the mice and usually initiated gentle contact with them (particularly touch and nose contact; Figure 3). She oscillated between approach and withdrawal, making repeated tentative strikes (touch and swipe) that were insufficient to stun the mice, and withdrew when they moved. Under diazepam, the strength of attack increased, as indicated by the increased relative frequency of

Table 2
Descriptions and Definitions of Behavior Patterns Used by the Cats to Contact the Mouse's Body During Predatory Sequences

---

Paw contact

Swipe: striking the mouse laterally from the side with the palm of the paw perpendicular to the ground. The mouse was moved along the ground by the strike.

Touch: same as the swipe but gentle, so that the mouse was not moved by the strike.

Bat: striking the mouse from above with the palm of the paw parallel to the ground. The mouse was moved or compressed by the strike, but the paw was quickly withdrawn.

Tap: same as the bat but gentle, so that the mouse was not moved or compressed by the strike.

Pin: the initial strike was the same as the bat, but then the paw was held down, pinning the mouse to the ground.

Oral and facial contact

Nose contact: The cat's nose was brought into close proximity or actual contact with the mouse and the nostrils flared. This probably included both tactile and olfactory input from the prey.

Mouthing: The lips of the mouth were brought into contact with the mouse, which could involve gentle touching all along the mouse's body.

Bite: The mouth was closed around the mouse. This could be subdivided into light bites, in which the skin was only grasped and the mouse uninjured, and heavy bites, in which the skin was broken and a firm grip on structures beneath the skin was achieved.

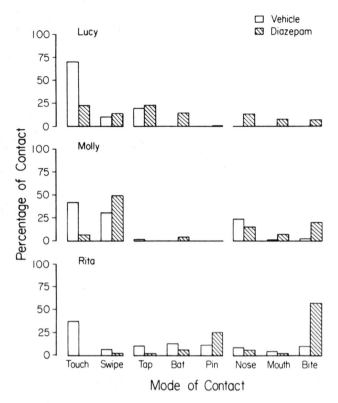

Figure 3. Frequency of occurrence of individual behavior patterns (forms of contact with prey) expressed as a percentage of the total occurrence of all behavior patterns for each cat. (Three cats are represented when given vehicle and diazepam. Lucy never killed mice, Molly only killed mice when given diazepam, and Rita killed mice whether given vehicle or diazepam. All three cats exhibit, at different levels of intensity, increased strength of attack when given diazepam. Data for all vehicle and all diazepam trials are pooled for each cat. [Lucy—10 behavior patterns for 2 vehicle trials and 204 for 2 diazepam trials; Molly—140 for 2 vehicle trials and 152 for 6 diazepam trials; Rita—535 for 6 vehicle trials and 53 for 6 diazepam trials].)

swipe, bat, mouth contact, and bite (Figure 3). Strong strikes with the forepaws were more likely to be repeated, and movement by the mouse when being sniffed or mouthed was more likely to elicit biting instead of withdrawal. After a series of repeated attacks, the mice were killed. In Molly, the increased strength of attack under diazepam was reflected in a 73.5% reduction in the frequency of contact per mouse.

Rita represents 1 of the 4 spontaneous killers. With vehicle, she resembled Molly under diazepam. That is, Rita made a lot of preliminary, tentative contact with mice before killing. Under diazepam, such preliminary contact was greatly reduced, so that she essentially approached, pinned, and bit the mice (Figure 3). Movements by the mice elicited further attack rather than withdrawal, and when withdrawal did occur, it was quickly followed by attack. Thus, under diazepam, Rita exhibited almost pure attack, with little defense, and hence only limited oscillation between approach and withdrawal. The reduced withdrawal led to quicker killing and hence a 91.1% reduction in frequency of contact per mouse. All of

the remaining cats exhibited one of the three patterns of attack described earlier.

Thus the effect of diazepam on any individual cat depended on the baseline pattern of attack. Cats that avoided mice were induced to interact tentatively, cats that interacted tentatively were induced to kill, and cats that killed did so with less preliminary interaction. If play is equated with tentative interaction, in which prey are not killed (e.g., Biben, 1979), then had only nonkillers been studied, one would have concluded that the drug enhanced play. If only killers had been studied, one would have concluded that diazepam decreased play. If, however, play and killing are viewed as being on a continuum of predatory behavior, then each individual cat can be seen to be shifted along the continuum by treatment with diazepam.

The escalation of attack towards killing was further analyzed by scoring the frequency of various behavior patterns during the course of the interactions. From the vehicle controls of the diazepam experiment, 22 sequences leading to killing and 20 sequences not leading to killing were divided into quarters, and the frequency of occurrence of each behavior pattern (listed in Table 2) in each quarter was scored. Division of sequences into quarters was necessary because each sequence contained different frequencies of behavior patterns. A problem existed for those sequences having less than four contacts with the prey, or an odd number. For consistency, we weighted the scoring procedure in favor of the last quarter. This choice was partly arbitrary and partly based on the fact that if a cat contacted the mouse only once, then that represents the strongest final form of contact. Otherwise, a cat biting and killing a mouse with only one bite would have been scored in the first quarter. If only one behavior pattern was scored for any sequence, then it was placed in the last quarter; if two behavior patterns were scored, the last was placed in the final quarter and the second in the third quarter; if five behavior patterns were scored, the last two were placed in the final quarter and one each in the other three quarters, and so on.

At the beginning of interactions leading to killing, the behaviors of touching, tapping, and nose contact occurred significantly more often than other behavior patterns; then they decreased; whereas batting, pinning, and biting increased significantly (Figure 4). Similarly, bites to the neck and base of the skull, compared with bites more posterior on the prey's body, increased significantly towards the end of the sequence, $\chi^2(3, N = 114) = 73.16, p < .001$, for 114 bites. (For sequences not ending in killing, there were no significant changes in the frequency of any of the behavior patterns nor in the placement of the bites during the course of the interaction.) Such escalations during the course of interactions have been reported previously (e.g., Biben, 1979; Leyhausen, 1973) but have been interpreted as signifying a shift from playful to serious predation (Biben, 1979).

Taken together, the results of these experiments revealed two important findings with respect to the conceptualization of predatory play. First, under the influence of anxiolytic drugs, cats that otherwise played with mice showed an escalation of attack towards killing. Second, the escalation of attack was characterized by an increase in the intensity of

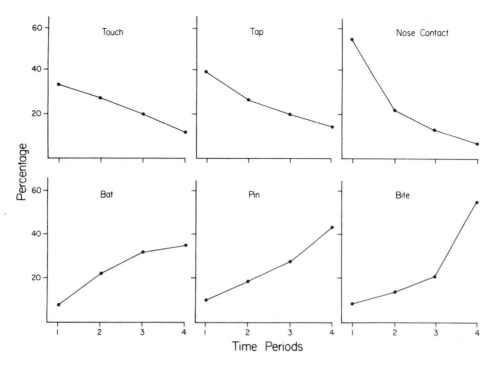

Figure 4. Changes in the frequency of individual behavior patterns during the course of attacks on mice leading to kills. (The periods refer to the first, second, third, and fourth quarters of the sequence [$N = 22$ kills]. Chi-square analysis of raw scores [$p < .01$].)

contact within each of the defined categories of behavior (lateral swipes, vertical strikes, and orofacial contact). In nonkilling cats, because the interaction did not escalate to the point of a killing bite, they appeared to be "locked" into a pattern of repeated contact. The failure to escalate yielded a blend of attack and defense, which made the cat look playful. In light of these findings, we believe that it is more appropriate to view predatory play and attack as points along a continuum rather than as independent categories of behavior.

Discussion

We conclude that the dichotomy between "predatory play" and "predation" is misleading, in that it draws our attention away from the processes that escalate and inhibit predatory attack. The impression of playfulness arises from repeated contact with the mice, such as touching, tapping, and nose contact, as well as some more forceful behavior patterns, such as swiping, batting, and gentle bites, which do not lead to killing. In hesitant killers, such preliminary interaction may escalate toward more intense approach-oriented behavior patterns, such as strongly delivered bats, pins, and killing bites (Figure 4). A failure to escalate keeps nonkilling cats locked into repeated contact because the interaction is not ended by a killing bite. When viewed in isolation, such behavior appears playful. Within the context of a continuous gradient, however, the cat simply fails to escalate sufficiently towards killing. This yields a blend of attack and defense, which makes the cat look playful. As we show in the next section, when the escalation of attack occurs in the absence of defense, the gentle, repeated and noninjurious contact with prey in the early phases of escalation does not look playful at all.

Part 2. Deactivation of Predatory Attack by Lateral Hypothalamic Lesions and its Escalation, in the Absence of Defense, With Recovery

In cats and rats, lateral hypothalamic (LH) lesions have a deactivating effect on the electroencephalogram and on behavior (De Ryck & Teitelbaum, 1978; Shoham & Teitelbaum, 1982; Wolgin & Teitelbaum, 1978; Wright & Craggs, 1979). Such deactivation has been shown to interfere with predatory behavior in the cat (Bandler & Halliday, 1982; Wolgin & Teitelbaum, 1978). Although predatory attack gradually recovers, cats with LH lesions display an enduring deficit in their reaction to threatening stimuli. Such cats are less reactive than normal cats to aggressive swipes by a conspecific (Wolgin & Teitelbaum, 1978). In the following experiment, we show that during recovery, LH cats exhibit a gradual escalation of attack in the absence of defense, and hence, in the absence of play. Administration of diazepam at any point along recovery facilitates this "pure" form of predatory attack. These results reinforce our previous conclusion that the combination of defense with low-intensity attack creates the illusion of playfulness.

Method

Five of the 13 cats at Illinois received large bilateral LH lesions. One male and two females were those previously used in Experiment

2 of Part 1. The spontaneous attack behavior of the 2 other males, which had not previously been tested with diazepam, was filmed and analyzed (see Part 4). Following the lesion, predatory behavior was tested 2-3 times per week beginning the day following the operation, if possible, although spontaneous behavior did not usually begin until the third of fourth day. In addition to spontaneous recovery, 3 of the cats were given diazepam (1.0 or 4.0 mg/kg) at weekly intervals, for up to 6 weeks, and were tested for predation. Testing and recording procedures were the same as those described for Part 1, Experiment 2.

Lateral Hypothalamic Lesions

The cats were anesthesized with sodium thiobarbital, intubated, and then maintained on halothane gaseous anesthesia. Some cats were given acetylpromazine (0.1 mg/kg) as a preanesthestic sedative. They were mounted in a stereotaxic instrument (Kopf Model 1404) and prepared for sterile surgery.

Using a rectal indifferent electrode, LH lesions were made by passing 3.0 mA of anodal current for 90 s through stainless steel electrodes insulated except for 0.5 mm at the conical tip. The coordinates, taken from Snider and Niemer (1961), were 11.0 mmA, 3.0 mmL, and −4.0 mmH. During surgery, cats were given 10 mg/kg of ampicillin and 1 mg/kg dexamethasone. They were allowed to recover in an incubator at 29–30 °C. Some cats were initially unable to maintain their body temperature at room temperature and were kept in the incubator for 1–2 days postoperatively.

Histology

At the completion of the experiment, the cats were injected with an overdose of sodium pentobarbital. They were perfused through the thoracic aorta with isotonic saline followed by 10% formalin. The brains were removed and stored in 10% formalin. Frozen sections 40 μ thick were cut through the extent of the lesion. Every fifth section was mounted on a glass slide and stained with cresyl violet. Selected slides were viewed through a microprojector (Bausch and Lomb), and tracings were made of the brain sections throughout the extent of the lesion. Affected structures were identified with the aid of a stereotaxic atlas (Snider & Niemer, 1961).

In all 5 cats, the lesion was centered at the level of the ventromedial hypothalamus. The lesions were located in the far lateral hypothalamus at the tip of the internal capsule and extended ventrally to the base of the brain (Figure 5).

Results and Discussion

In varying degrees, LH lesions produced the expected deactivating effect on predatory attack in all 5 cats. The male that had avoided mice prior to the lesion completely ignored them postoperatively. The 2 other males that prior to the lesion had exhibited tentative approach (i.e., touching, tapping, and nose contact), also showed no predatory behavior towards the mice after the lesions. One of the females that had occasionally killed mice preoperatively, now made nose contact, and tentatively struck the mice with the forepaws (touching and tapping), but did not pin or bite the prey. The other female that was an efficient killer before the lesion still attacked the mice afterwards, but failed to kill them.

Over the course of the following 4–6 weeks, the cats' predatory behavior recovered to preoperative levels, with the more tentative behaviors reappearing first, followed by more forceful behaviors. Such recovery was often characterized by a series of subtle changes in the cat's response to prey that unfolded over a period of days. These changes are well illustrated by the recovery of the female that was an efficient killer before the lesion. On several occasions in the first 3 days following the lesion, she approached, pinned, and bit the mouse on the thorax. However, instead of shifting the bite to the neck and making a killing bite, she then fell asleep while holding the mouse in her mouth. By the fourth day after the operation, the cat did shift her bite anteriorly, but again failed to deliver the killing bite. By the eighth day, she shifted her bite to the mouse's neck and delivered a killing bite but then fell asleep. At this stage, escalation of attack was contingent on the mouse's movements. Only a moving mouse was attacked, and once the seizing bite was applied, it was shifted to the neck only if the mouse struggled. Similarly, once the bite was shifted to the neck, a killing bite was delivered only when the mouse struggled. It was not until 20 days after the lesion that the whole sequence occurred without movement by the mouse. Therefore, during recovery the activation necessary to shift the behavior closer to killing was transferred from external to endogenous stimuli. A similar transformation occurs during the ontogeny of predation in kittens (Wolgin, 1982).

Throughout recovery, diazepam facilitated attack by inducing a shift along the predatory gradient. For instance, at the stage in which the cat only seized the mouse around the thorax, diazepam induced the cat to readjust its seizing bite to the neck. Similarly, at the stage in which a seizing bite was made to the neck, diazepam induced a killing bite. These diazepam-induced changes did not require movement by the mouse. The other 2 cats given diazepam also exhibited shifts upward along the predatory gradient during various stages of recovery. Thus, at each point in recovery, diazepam increased the intensity of attack, in that the cat progressed closer to killing. The fact that diazepam facilitated attack in cats that did not exhibit defense demonstrates that the drug has activating properties in addition to its well-known anxiolytic effects.[1]

Therefore, the deactivation and reactivation of predation in the LH cat revealed that (a) attack involves an increasing escalation of intensity and (b) attack can occur and grow in strength independently of defense. The failure to withdraw was strikingly apparent in the case of the cat that fell asleep while holding a mouse in its teeth. Although the mouse struggled, its movements failed to elicit release: indeed later in recovery such movements were necessary to elicit more intense attack (i.e., shifting to a killing bite). In contrast, intact

---

[1] It should be noted that in the first 2–3 days after the lesion, diazepam completely deactivated all behavior, including postural support. This deactivating effect of diazepam appears to depend on dopamine depletion (O'Brien et al., 1988; Schallert, Wang, Hsiao, & Whishaw, 1979) and is not seen in diazepam-treated intact cats. Later in recovery, administration of diazepam activated predatory behavior. Thus, even though the cats were ataxic due to the drug (unsteady when moving, frequently falling), they still actively chased the prey.

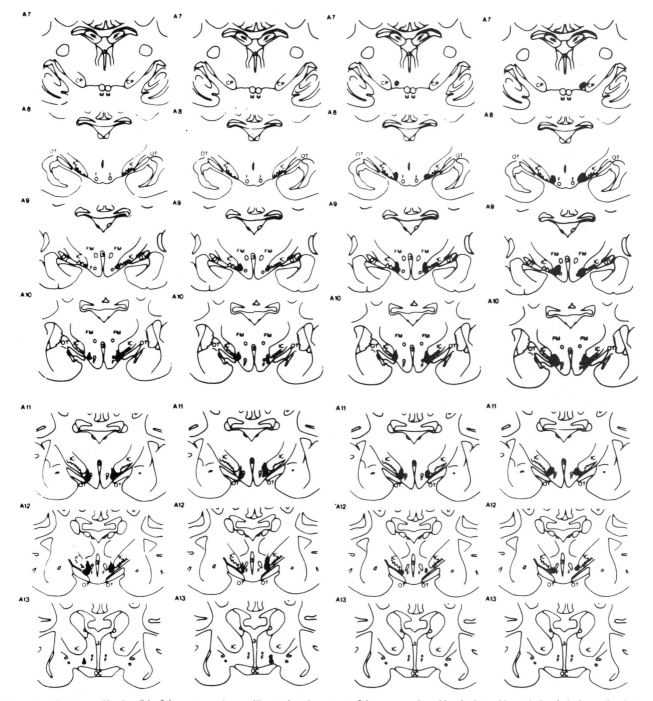

Figure 5. Sections of brain of 4 of the cats are shown, illustrating the extent of damage produced by the lateral hypothalamic lesions. (Sections from lower to upper show the anterior to posterior range of the damage. CP = cerebral peduncles; 3 = ventricle; M = mammillary body; OC = optic chiasm; OT = optic tract; F = fornix; IC = internal capsule; FM = mammillothalamic fasciculus.)

cats showing low-intensity attack react very strongly to such struggles by the prey, by markedly withdrawing the head (see Part 4, and also Wolgin, 1982). Because recovered LH cats failed to show withdrawal, their attack did not appear playful, even though such cats repeatedly attacked mice without causing injury, a cardinal criterion of play for some authors (e.g., Biben, 1979). This further supports our view that play is not merely low-intensity attack but is an artifact of the interaction of attack and defense. As will be shown in the next section, a more extreme case of attack in the absence of defense was shown by both intact and LH cats when predation was induced by pinching the cat's tail.

Part 3. Tail-Pinch-Induced Predation: "Short Circuiting" the Escalation Gradient

Application of mild pressure on the tail has been previously reported to induce predatory attack in otherwise unresponsive cats with damage to the LH (Wolgin & Teitelbaum, 1978). Here we show that tail pinch (TP) can also induce predation in intact cats, and that for both intact and LH-damaged cats, only the terminal components of attack are induced. In this way, pinning (if the prey is evasive) and killing bites occur in the absence of gentler forms of contact.

Method

Animals

Ten of the 17 cats at Illinois were used. All 10 cats were tested for spontaneous predation and again when TP was applied. Five cats then received bilateral lesions of the lateral hypothalamus and were retested for both spontaneous (as described in Part 2 above) and TP-induced predation.

Experimental Procedure

The procedures outlined in the General Methods were used. In this way, each cat was tested for spontaneous attack with 2–6 mice and then for TP-induced attack with another 2–6 mice. To apply tail pressure, the experimenter gripped the tail firmly, about ⅔ from the base, and then squeezed. Although in previous studies (on LH-damaged cats) a clamp was used to apply such pressure (O'Brien, Chesire, & Teitelbaum, 1985; Wolgin & Teitelbaum, 1978), such a device was found to be impractical for intact cats, which oriented and bit at the clamp. Gripping the tail by the hand had the advantage that the experimenter could maneuver away from the cat's attack. After attack on the prey was elicited, no further attacks were directed at the experimenter's hand. Also, a standardized pressure on the tail was a hindrance to the objective of this study. We were not interested in determining how many cats exhibited TP-induced attack for a given tail stimulus, but rather our concern was with the form of the attack. Therefore, whatever the pressure required to stimulate such attack, our analysis focused on the form of the attack itself. As will be shown later, once elicited, TP-induced attack was remarkably similar for all cats, whether brain damaged or intact.

The entire session during which the prey were in the enclosure with the cats was videotaped, and selected sequences were filmed.

Results and Discussion

Four of the 10 intact cats spontaneously killed mice, 3 attacked but did not kill, and 3 neither attacked nor killed. TP induced attack in 7 out of 10 intact cats, and of the 7, 6 killed the mice. After LH lesions, all 5 brain-damaged cats attacked and killed mice when TP was applied, but not spontaneously. This included a male that, prior to the lesion, failed to attack mice even with TP and a female that attacked, but did not kill, when TP was applied.

In Table 3 data for three representative cats before and after the brain lesions are shown. Cats 1 and 2 did not spontaneously kill prey and showed a high frequency of gentle contact (e.g., touch) and little or no strong contact (e.g., bite, pin). Cat 3 used all the behavior patterns, and after escalating its attack, killed all mice presented.

In TP-induced predation, only the behavior patterns directly used for killing were elicited. That is, biting was elicited in all the cats in all trials and pinning was elicited when the prey was behaving evasively (66.7% of 81 examples), χ^2 (1, $N = 81$) = 9.00, $p < .05$, based on raw scores. Of all the attacks scored in all 10 cats, two instances of swiping were observed in only 1 cat, and one instance of batting in 1 other cat. No other behavior patterns occurred when attack was induced by TP (Table 3). Such attack was induced only in the presence of prey. Mouse-sized inanimate objects, a plastic toy, a block of wood, or a rumpled piece of paper were not adequate to elicit attack when TP was applied. Furthermore, during TP-induced attack, the cats did not exhibit defense. Indeed, such cats usually attacked mice that were facing them defensively from a corner and were frequently bitten on the face by the prey.

In sum, TP short-circuits the gradual escalation of attack to killing and directly elicits killing behavior. In TP-induced predation, there is no vacillation between approach and withdrawal; the terminal components of attack were elicited in the absence of defensive behavior by the cat, and thus predatory attack induced by TP did not in the least appear playful. It is interesting that a similar pattern of predatory attack is also elicited by electrical stimulation of the hypothalamus (Berntson, Hughes, & Beattie, 1976).

Part 4. Functional Analysis of Predatory Play

One reason that predatory play and predatory attack are considered motivationally distinct is that playful responses appear to be unrelated in any obvious functional way to the delivery of the killing bite (Biben, 1979). Unlike serious predation, playful attack is characterized by exaggerated and repetitive movements (e.g., repeated swiping, batting, and throwing) even of dead prey. In the following experiment we used the Eshkol-Wachmann Movement Notation (EWMN; Eshkol & Wachmann, 1958) to show that when the movements of the cat are viewed in relation to those of the mouse, such playful behavior patterns are actually quite functional.

Method

All 13 of the cats (9 females and 4 males) at Illinois were used for this portion of the study. Data were obtained from the movies and videotapes taken from the preliminary testing of the cats' responses to mice and also from the filmed sequences from Part 1, Experiment 2.

Thirty sequences of predatory attack, ranging over the whole spectrum of the escalation gradient, were described and analyzed using EWMN. Particular attention was paid to the simultaneous movements of each cat and mouse pair in terms of their orientation and distance from one another (see Moran, Fentress, & Golani, 1981; Pellis, 1982) and to specific movements by the cats in terms of kinematics of the body segments involved (see Pellis, 1981, 1983). Patterns and regularities identified in this way were then checked in other filmed sequences. The method has been described elsewhere (e.g., Golani, 1976; Golani, Wolgin, & Teitelbaum, 1979; Pellis, 1981, 1982), and further results derived from the notation appear elsewhere (Pellis & Officer, 1987).

Table 3
Frequency (%) of Behavior Patterns Used During Predatory Attack, Spontaneous Versus Tail Pinch Induced

Subject	No. of mice attacked	Touch	Tap	Swipe	Bat	Pin	Nose contact	Mouth contact	Bite	Total Obs.
Intact cat/Spontaneous										
1 (f)	2	45.6	0.8	31.5	—	—	20.5	0.8	0.8	127
2 (m)	4	54.1	14.3	18.8	3.0	—	—	—	0.8	133
3 (f)	6	35.0	11.9	6.9	13.9	8.5	9.3	4.1	10.4	555
Intact cat/TP induced										
1	2	—	—	—	—	8.3	—	—	91.7	12
2	12	—	—	—	—	29.0	—	—	71.0	121
3	2	—	—	—	—	33.3	—	—	66.7	15
LH cats*/TP induced										
1	6	—	—	3.4	—	32.8	—	—	63.8	58
2	10	—	—	—	1.0	34.0	—	—	65.0	83
3	6	—	—	—	—	29.7	—	—	70.3	37

Note. TP = tail pinch; LH = lateral hypothalamic; f = female; m = male. * Spontaneous attack by LH cats is described in Part 2 of this article.

In order to evaluate the function(s) of the behavior patterns listed in Table 2, the following variables were quantified: (a) location of contact on the prey's body, (b) horizontal angle between the longitudinal axes of the opponents' bodies at the moment of contact, (c) the relative distance between opponents (i.e., in cat lengths: tip of snout to base of tail), (d) activity of prey (e.g., mobile vs. stationary), and (e) location of prey (e.g., middle of enclosure vs. corner).

Results and Discussion

Inhibition of Attack by Face-to-Face Orientation

No matter how strongly motivated to attack, the cats were always inhibited to some degree by the mouse facing them and actively avoided a face-to-face orientation. For example, in Figure 6, the cat is inhibited from attacking when faced by the mouse (Panel a); in response, the cat moves to the right and prepares to attack (as indicated by the raised forepaw; Panel b); but again is inhibited by the mouse turning its head toward the cat (Panel c); this stalemate persists (Panel d) until the mouse turns further, and runs away from the cat (Panel e), which elicits an attack (Panel f). In this example the mouse was initially wedged against one of the walls of the enclosure, and so the cat could not maneuver around and away from the mouse's face. In other situations in which the cat was free to maneuver around the mouse (i.e., the mouse was in the middle of the enclosure), it did so, thus enabling an attack to proceed from the side or rear.

More timid cats retreated when faced by a mouse; the more timid the cat, the greater the relative distance at which it stopped its approach or withdrew. Very timid cats stopped when they were ½–1 cat length away, whereas bolder cats approached to within ¼ cat length before stopping. Withdrawal was amplified in all cats if the mouse reared into an upright defensive posture when turning to face the cat. The weaker the attack, the more prolonged the stopping or withdrawal. Efficient killers quickly and effectively counteracted the mouse's movement, either by circling to the flank or by delivering a forepaw strike (see later). Consequently, although withdrawal was more obvious in timid cats, it was also present in bolder cats; it was simply harder to notice, because they displayed a more rapid and varied range of responses.

The aversiveness of a face-to-face orientation and the cats' tendency to deliver killing bites from the rear (see later) influenced all aspects of the attack sequences. With increased strength of attack, the cat moved closer to the prey, and as it did so, there was a more pronounced avoidance of the face-to-face orientation. This was reflected in the direction from which the cat contacted the prey: the stronger forms of contact occurred less frequently from the front. From filmed sequences, the angle between the horizontal axes of the bodies of predator and prey was scored in units of 45 degrees at the moment of contact. In all forms of contact, 25–30% occurred when the cat was perpendicular to the mouse. What is re-

Figure 6. The inhibitory effect of the "face-to-face" position is clearly demonstrated in this example. (After the cat began approaching the mouse, the mouse turned to face the cat and the cat stopped, sat down, and looked at the mouse [Panel a]. After a short time, the cat slowly began to attack the mouse [Panel b], but stopped, and leaned to the right as the mouse turned its head to face the cat [Panel c]. After remaining in this face-to-face orientation for several seconds [Panel d], the mouse began to turn away from the cat [Panel e], which was followed immediately by an attack by the cat [Panel f]. Traced from 16-mm film taken at 24 fps.)

vealing is the likelihood of attack from the frontal versus the rear angles (i.e., angles anterior to perpendicular and angles posterior to perpendicular). Touch, tap, swipe, nose contact, and mouth contact occurred as frequently from the front as from the rear (Figure 7). Bat, pin, light bite, and heavy bite, on the other hand, occurred significantly less often from the front (Figure 7). Therefore, cats avoided strong contact from the front but were less discriminating with gentle contact. When contacting the prey gently, either from the front, side, or rear, the cats would stop about ¼–½ cat length away and then stretch either the head or the paw toward the mouse. However, even from the front, some forms of contact had important functions (e.g., swipe and light bite) as shown below.

Functional Context of Behavior Patterns

The nine different modes of making contact with mice (Table 2) used in the previous sections are here further analyzed.

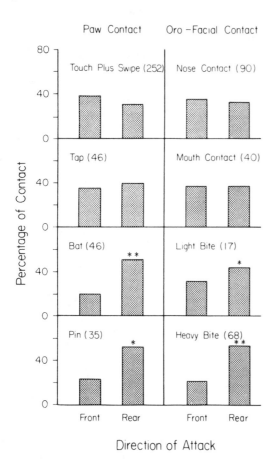

Figure 7. Comparison of contact from frontal versus rear horizontal angles between the longitudinal axes of cat and mouse for each behavior pattern. (The remaining percentage of contact is accounted for by contact made when perpendicular to the prey, which is not shown in this figure. Shown in the parentheses are the number of times each behavior was scored from the front or rear. Chi-square analyses of the raw scores were carried out for each behavior pattern [$*p < .05$; $**p < .01$].)

Table 4
Behavior of the Mouse Compared With Type of Paw Use by the Cat to Make Contact With the Mouse

Behavior pattern	N	Mouse Moving (in %)	Mouse Stationary (in %)
Swipe	324	28.7	71.3*
Bat	109	53.2	46.8
Pin	61	75.8*	24.2

* Chi-square analysis of the raw scores showed that swiping occurred more frequently when the mouse was stationary and that pinning occurred more frequently when the mouse was moving ($p < .01$). Bats occurred at a similar frequency in both cases ($p > .05$).

Forepaw contact. Three major types of paw contact occurred during the interactions: the swipe, the bat, and the pin. The swipe and the bat also occurred in inhibited forms, which we designated as touch and tap, respectively (Table 2). Other forms of paw contact, such as the two-paw grasp and kicking with the hindpaws, occurred only rarely, because they are more commonly used when attacking prey larger than mice (Leyhausen, 1979).

The swipe, which displaced the prey horizontally, occurred most often with stationary mice (Table 4) at maximal distance (½ cat length or more; Figure 8). Even efficient killers initiated contact with the prey with a swipe. Swiping appeared to probe the mouse for its response to being attacked: after a sequence of swipes, a mouse that remained stationary was bitten, whereas a mouse that ran or behaved defensively was first batted and pinned and then bitten. The swipe was also used, however, to position the mouse for an attack. For example, if a mouse stood in a corner, facing the approaching cat, the cat used a swipe to strike the mouse across the side of the face, knocking it to one side, and allowing a biting attack to be directed from an oblique frontal angle (for an example, see Figure 1C in Pellis & Officer, 1987). Thus, unlike bats or pins, swipes occurred as frequently from the front as from the rear (Figure 7) and were often directed to the head (Figure 9).

The role of batting seemed to be that of stunning and disorienting the mouse, because powerful downward strikes on the chest at times could leave a mouse gasping. This function is reflected in the high frequency of batting the mouse's thorax (Figure 9). Batting seemed to occur more frequently after the prey began to respond to contact by the cat. Such "softening" up by batting was particularly evident when attacking an evasive mouse. More hesitant bats (i.e., taps) were delivered when the mouse did not respond to the cat's probing contact. That taps were more exploratory is indicated by their greater frequency of delivery from the front compared with bats (Figure 7). The pin occurred in the most specific functional context, that of holding the prey still while a bite was delivered. In this way, mice, especially when running away (Table 4), were held down by the pelvis (Figure 9) and then bitten (see below).

Facial and oral contact. Facio-oral contact involved nose contact, mouthing, and biting. These rarely occurred when the mouse was moving or standing defensively. However, if the mouse escaped while being bitten, the cat sometimes attempted to bite the moving prey. More typical, the moving mouse was first pinned and then bitten.

Two forms of bite, light and heavy, were categorized (see Table 2). Heavy bites were of two kinds: seizing bites, used to hold and carry the prey, and killing bites, which caused tissue damage. Heavy bites were most frequently directed to the neck or chest (Figure 9). The initial heavy bite was usually a seizing bite to the thorax. Once the prey was subdued, the bite was shifted anteriorly to the neck or base of the skull where a killing bite was delivered. Light bites were usually delivered to the thorax (Figure 9). Although light bites sometimes appeared to be inhibited heavy bites (e.g., during aborted frontal attacks), some instances of their use suggested that they served a specific functional purpose: light bites were often employed when a mouse was wedged facing into a corner. For a firm bite to be delivered, the cat's head had to be at right angles to the mouse's body, so in this case the mouse's position made it difficult to orient a bite. At such times, the cat bit the mouse lightly on the pelvis or chest, lifted it out of the corner, and then carried it to a more appropriate place before renewing the attack. While carrying the mouse, the cat sometimes used a head shake to fling it away (see Leyhausen, 1979) and then delivered a heavy bite to the neck or head while the mouse was still disoriented after landing.

Nose contact and mouthing involved more generalized contact with the mouse's body (Figure 9). The functional roles of both are unclear. However, some aspects of their occurrence suggest that they are exploratory, testing the prey before attacking: nose contact occurred early in the predatory attack

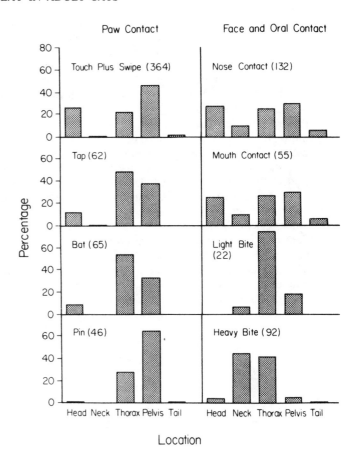

Figure 9. Location of contact on the mouse's body by different behavior patterns. (The numbers in parentheses are the number of times each behavior was scored. As in Figure 7, swipe and touch were lumped together.)

sequence when the mouse was typically stationary. The cat approached, stretched out its neck, and made nose contact, often along the whole length of the prey's body. After such contact the mouse was swiped once or twice and, if it remained motionless, was again touched with the nose. Similarly, mouthing was directed to stationary mice. If the mouse moved while being mouthed, it was immediately bitten.

The preceding analysis provides new insights into several features of playful attack. Consider first the feature of repeated contact without such contact escalating to killing. If a cat is motivated to attack a mouse, but is so timid that it does not approach any closer than ½ cat length away, it will swipe repeatedly at the mouse. Biben (1979) interpreted such repeated, "irrelevant" attacks as being play rather than serious aggression. In our schema, the cat is locked into a level of attack from which it cannot escalate further; it thus uses those behavior patterns available to it at that distance. However, as we showed earlier, if such a cat is given an injection of diazepam, it will approach to within ¼ cat length and attack more forcefully.

An aspect of playful attack that is difficult to explain from a functional perspective is play with dead prey (e.g., Biben, 1979; Leyhausen, 1973). Although we observed only six in-

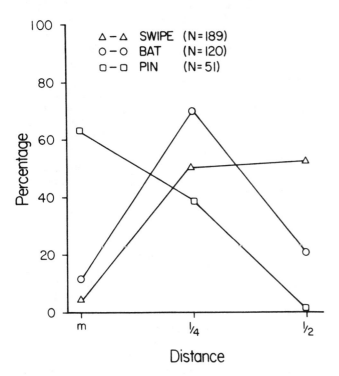

Figure 8. Type of forepaw contact used by the cats in relation to distance from the mouse. (The scale of distance is in terms of fractions of the length of the cat [snout to base of tail]. M = minimal distance.)

stances of such postkill play, the behavior of the cats with dead mice differed in a number of ways from their supposedly playful behavior with live mice. Compared to sequences when cats did not kill the mouse, after the kill there was an increase in swiping and mouthing (Figure 10) and also in gently biting and then throwing the mice. The way these behavior patterns were used suggests that the cats may have failed to recognize that the mice were dead. As noted previously, swiping is used more often with stationary mice (Table 4) and seems to be used to test the prey for a response. Similarly, when attacking live, stationary mice, both mouthing and throwing quickly led to biting if the prey moved. Thus these behaviors appear to be prevalent in cases in which the cat is testing the prey's reactions.

Two examples illustrate further that such postkill interactions may result from a cat's uncertainty as to whether the mouse is dead. In the first example, a cat repeatedly swiped, mouthed, and bit a dead mouse. After several minutes, the mouse was removed and a small incision was made on the dorsal surface of the neck. Immediately after the mouse was returned to the enclosure, the cat approached, sniffed it, licked the blood from the incision, and began eating the mouse without any further playful interaction. In a second example, an efficient killer approached, pinned, and delivered a killing bite. The cat dropped the mouse, walked away, then approached, sniffed it, and began to swipe at it. The mouse was repeatedly swiped, mouthed, bitten, thrown, and carried. This went on for over 10 min. Then, all of a sudden, the supposedly dead mouse began running. The cat immediately ran up to it, pinned it on the thorax, and delivered a killing bite to the head. On releasing the mouse, the cat looked at it, swiped it a couple of times, and walked off.

These observations support the view that prolonged interactions with dead prey reflect the cats' uncertainty as to whether the prey is dead. A similar explanation may account for play with inanimate objects. We have observed many of the behavioral components of attack and defense in the cats' interactions with a string or a ball. Consequently, these responses appear to be elicited by specific features (i.e., "sign stimuli") shared by these stimulus objects. As in the case of dead prey described earlier, sustained play with such stimuli may result from the absence of a suitable "terminating stimulus" for a consummatory response. This may also be true for kittens that also direct predatorylike behavior to inanimate objects (West, 1977). Kittens most frequently play with inanimate objects from 8–12 weeks of age (Barrett & Bateson, 1978) after they are fully capable of killing mice (Baerends-van Roon & Baerends, 1979; Leyhausen, 1979). Furthermore, "object play" in older kittens is more readily elicited by objects that provide stimuli resembling prey (Egan, 1976).

Another characteristic of playful attack is the presence of exaggerated, "out-of-context" movements (Loizos, 1966; Marler & Hamilton, 1966). Consider the situation in Figure 11A. In Figure 11 (Section a), the cat is looking away from the mouse. As she turns to face it, the mouse simultaneously turns toward the cat, creating a face-to-face orientation at close range. In response, the cat withdraws her head as she turns, with the result that her head is cocked obliquely downward (Section b). This "bizarre," playful head movement, then, is simply a reaction to the proximity of the mouse and serves to maintain a less threatening distance from the mouse's face. This is clearly shown by the cat's subsequent behavior. The cat shifts her forequarters slightly to one side of the mouse, but still holds her head withdrawn (Section c), and once facing the side of the mouse's head, she prepares to attack (Section d). The mouse in this example was in a corner and so the cat could not maneuver to the mouse's rear. The cat's head movement in this instance may be compared to one (Figure 1A in Pellis & Officer, 1987) in which another cat cautiously approached a mouse from a frontal angle in the middle of the enclosure. About ¼ cat length away, the cat stopped and began to extend her head forward toward the mouse. As she did so, however, the mouse turned its face toward her, and the cat withdrew her head, by lifting the neck and shifting her body weight backwards, until her head was about ½ length away. The difference between these two examples is that in the latter, the cat approached with the head, neck, and body all in one line, so that the withdrawal of the head was achieved by moving in the same vertical plane. In the former, the cat had to rotate the head toward the mouse

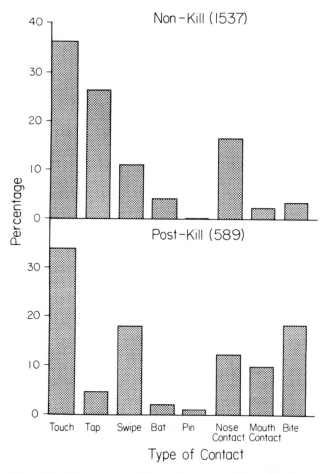

Figure 10. The frequency of behavior patterns during postkill interactions and during interactions not leading to killing. (Mice were the prey in all cases. Data for postkill were derived from 7 interactions; data for nonkill were derived from 20 interactions. The *N* signifies the total number of behavior patterns scored.)

as well as to move the neck laterally, thus placing the head and neck out of alignment with the body. To withdraw the head further from the mouse, rotatory and lateral head and neck movements were made. When the sequence of behavior is viewed quickly, without simultaneously scoring the movements of the mouse, the withdrawal of the cat's head in the first example (with rotation) appears bizarre and playful, whereas that in the second example does not.

Of all the cats tested, including the efficient killers, only one cat, a male, attacked adult rats, though he never killed them. Most laboratory cats do not attack rats spontaneously (Flynn, 1967). When they do, playful behavior is more likely to occur than when attacking mice (Biben, 1979). In this cat's interactions with rats, exaggerated playful movements were especially frequent. However, as in the example described earlier, many of these exaggerated movements had functional significance. In Figure 11B, while interacting with a rat, the cat performed two seemingly bizarre head movements. In the first tracing (Section a), he bit the rat gently on the back, then released the bite, and began to turn his head to the right (Section b). Simultaneously, the rat began to walk forward. As the cat's head continued to turn away, his forepaws grasped the rat, and pulled it back and to the right (Section c). Once the rat was repositioned with the left forepaw (Section c), the cat turned his head back toward the rat and, as indicated by the raised right forepaw, prepared to attack (Section d). As he began to lunge, he turned his head horizontally to the left to allow a sideways bite. As he did so, the rat turned its head toward the cat, and so coupled with the horizontal head movement, the cat also raised his neck and withdrew his head from the rat's head (Section e). The rat then turned its head to the right, and the cat lunged and delivered a bite at the junction of the neck and thorax (Section f). When viewed at normal speed and without simultaneously recording and correlating with the movements of the rat, these head movements by the cat seem to be exaggerated.

The second head movement by the cat in Figure 11B (Sections d, e, and f), like that shown in Figure 11A, may be understood as the coupling of a maneuver to attack with withdrawal from the prey's face. However, the cat's first head movement (Figure 11B, Sections a, b, and c) occurs independently of movement by the rat's head. In cats attacking tentatively, any movement by the prey can elicit withdrawal, and in this case, the rat began to move forward (see also Figure 8.9 in Wolgin, 1982, p. 270). During very tentative attacks, even making nose contact with a stationary prey can elicit an initial withdrawal before continuing the attack. In Figure 11C, for instance, the male cat's body is facing away from the observer, and the rat is being held between the forepaws, facing away from the cat. The cat had just ceased mouthing the rat and is facing to the right (Section a). The cat then turns to face the left, but does so by first raising the head (Section b), then rotating the neck clockwise (Section c), and then lowering the head to the left (Section d). This complicated turn thus avoids visual, and perhaps also vibrissae, contact with the rat. However, there were cases in which the withdrawal from the prey was independent of the prey's behavior and so extreme that such an explanation seems inadequate. For example, in Figure 12, three instances of exaggerated withdrawal are illustrated for the male during an interaction with a juvenile rat (150 g). In all three cases, the cat reared upward as he withdrew from the rat but did so with increasing strength. In Figure 12, Panel A, when he rears, the right forepaw barely leaves the ground; in Panel B, both forepaws raise high off the ground; and finally, in Panel C, he not only rears high with the forequarters, but he also releases his hindpaws from contact with the ground, jumping into the air. The rat, in each case, remains stationary throughout the course of the cat's movements. Even more perplexing from the point of view of ascribing these behaviors to a simple withdrawal is that the most exaggerated withdrawal occurs when the head is farthest from the prey. It must be noted, however, that these highly exaggerated withdrawals only occurred when the cat attacked rats (i.e., large prey), and then only toward the end of a protracted interaction with the prey. Therefore, it may be possible to explain these movements as resulting from an escalation of defense in the absence of any further escalation of attack. Alternatively, such movements may represent a residual category of behaviors, which do not reflect a conflict between attack and defense, but rather, represent truly playfully motivated movements. Finally, such leaps over and away from the prey may result from hyperarousal, with the behavior thus having little functional relation to the prey (see Fentress, 1976, for discussion of such phenomena in different types of behavior). This last possibility is supported by the fact that these leaps only occurred toward the end of a long interaction with large prey. Further analysis is necessary to distinguish between these possibilities. However, it is clear from our analysis that many, perhaps most,

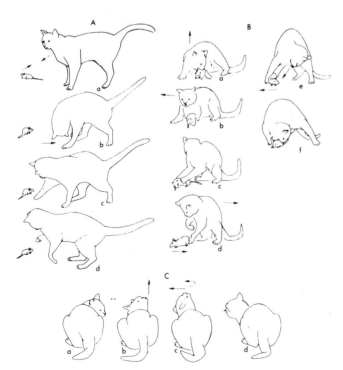

Figure 11. Three examples of seemingly bizarre movements by cats during interactions with prey; in Panel A with a mouse, and in Panels B and C with a rat. (Drawn from 16-mm movie film taken at 24 fps.)

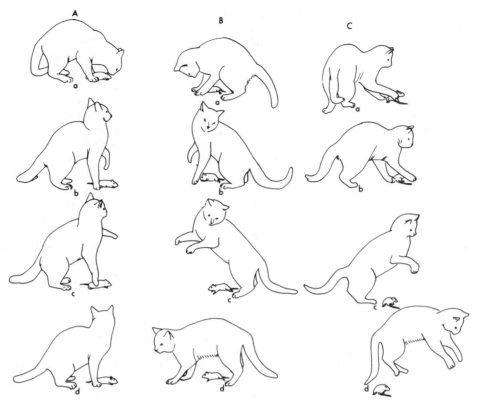

Figure 12. Three examples of exaggerated withdrawal by one of the male cats when interacting with a juvenile rat. (Drawn from 16-mm movie film taken at 24 fps.)

movements that appear to be exaggerated are in reality highly relevant to, and functional in, the ongoing sequence of interaction with the prey.

General Discussion

Our analysis of predatory play in the adult cat suggests that it is not fundamentally different from serious predation. Rather, the data indicate that predation involves a gradient ranging from pure defense at one extreme to pure attack at the other. This gradient is a by-product of two separate gradients, one of defense and one of attack, the interaction of which determine the behavior exhibited by any individual cat. This interaction is illustrated graphically in Figure 13, where, from left to right, defense decreases in strength, and attack increases. The predatory defense-attack gradient fits cats that avoid prey, play with prey, or kill prey (Figure 13).

A completely defensive cat is generally not considered playful (far left of Figure 13), nor is one that attacks forcefully and immediately delivers a killing bite (far right of Figure 13). Only those cats that vacillate between defense and attack are typically viewed as being playful (middle area of Figure 13). As Wolgin (1982, p. 269) noted, it is the superimposition of approach and withdrawal that creates the illusion of playfulness, that is, exaggerated and seemingly out-of-context movements, repeated performance of behavior patterns lacking apparent goal direction, and disordered sequencing of motor patterns (Biben, 1979; Loizos, 1966; Marler & Hamilton, 1966; Poole & Fish, 1975). For example, face-to-face orientation at close quarters (<¼ cat length) inhibits attack at any point along the gradient and evokes exaggerated and distorted playful movements, especially of the head, which seem to have a defensive function. Both efficient killers and playful cats perform such exaggerated movements. In efficient killers, they are usually followed quickly by an attack that results in killing. In playful cats, however, such movements clearly interrupt the cat's attack, conveying the impression that the cat is playing with the prey and not seriously attempting to kill it (see Biben, 1979; Leyhausen, 1973). The necessity of the superimposition of approach and withdrawal in creating the impression of playfulness is well illustrated in recovering LH cats. These animals repeatedly approached, struck with the forepaws, and touched the mice with their noses, lips, and mouths, but did not kill the prey. However, in the absence of defense, these repeated contacts did not appear at all playful.

It must be pointed out that the gradients presented in Figure 13 may not, in reality, be monotonic nor reciprocal. For example, it is possible that play may occur when both attack and defense are high, not just when both are intermediate; and it is also possible that avoidance may occur when both attack and defense are low, not just when defense is high and attack is low. Nevertheless, the proposed model (Figure 13) illustrates how playful appearing behavior may arise without the need for a separate motivational category. Furthermore, the behavioral data of Adamec and his colleagues (Adamec, Stark-Adamec, & Livingston, 1981) "... indicate an inverse

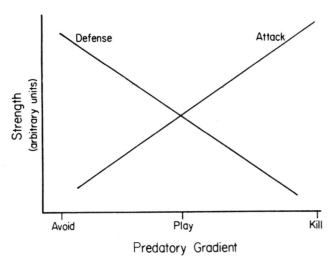

Figure 13. Hypothetical model of the interaction between gradients of defense and attack, which produce the predatory gradient analyzed in this article. (The relative strength of attack and defense is represented on the ordinate, whereas predatory behavior is represented on the abscissa. At the left, the cats avoid prey; at the right, they kill without hesitation. In the midrange, the cats vacillate between attack and defense, yielding seemingly "playful" behavior. As cats escalate toward increased attack or increased defense they appear less and less playful.)

relationship between attack and defense, with the least aggressive cats being the most defensive" (p. 398).

The currently held view is that play has unique behavioral characteristics and that it involves a motivation distinct from other forms of behavior (e.g., predation, intraspecific aggression, sexual behavior, etc.; Bekoff, 1974; Bekoff & Byers, 1981; Einon, 1983; Fagen, 1981; Pellis, 1981, 1983). Our findings show that such playful characteristics can arise without having to invoke separate motivational states for predatory play and for serious predation (see also Leyhausen, 1979). The interaction of two behavioral systems, attack and defense, can account for a wide range of responses toward prey, including avoidance, play, and killing (Figure 13).

The concept of a predatory gradient provides an important framework for understanding the effects of benzodiazepines on aggression. Benzodiazepines are well known clinically for their anxiolytic effects (Rosenbaum, 1982). In the laboratory, these drugs are particularly effective in disinhibiting behavior that is both reinforced and punished (the "conflict" paradigm; cf. Sepinwall & Cook, 1978). Because cats given oxazepam and diazepam showed less defensive behavior toward prey, the escalation of attack induced by these drugs may be due, at least in part, to their anxiolytic properties (see also Apfelbach, 1978; Leaf et al. 1975). However, diazepam also facilitated the attack of cats with lateral hypothalamic lesions, which show little, if any, defensive behavior. In such cats, diazepam induced a more intense form of attack, similar to that evoked by tail pinch. These results suggest that benzodiazepines also directly facilitate attack through their activational properties (Winters & Kott, 1979).[2]

The important role of activation in the escalation of attack is also supported by the results of developmental studies. Leyhausen (1973) noted that "for the first elicitation of the killing bite . . . the exciting, self-intensifying performance of the associated instinctive movements is a necessary condition" (p. 212). Even in more experienced kittens, killing often occurs only after prolonged, energetic play (Wolgin, 1982). These observations suggest that the escalation of play to killing results from the activation engendered by repeated interaction with the prey.

Finally, it is clear that a meaningful analysis of predatory attack requires a simultaneous frame-by-frame correlation of the behavior of both the predator and the prey. Such analyses are equally important for understanding other forms of aggression. For example, play fighting can at times escalate to serious fighting (e.g., Bekoff, 1981), especially in males at puberty (e.g., Meaney & Stewart, 1981; Owens, 1975). At present, however, there are no adequate criteria by which play can be differentiated from its serious counterparts (except perhaps for some reports of supposed play-signals; Bekoff, 1972; Rasa, 1984; but such behavior is yet to be adequately analyzed; see Symons, 1978). Recently, some species have been shown to attack and defend different targets of the body during play-fighting compared with serious fighting (Pellis & Pellis, 1987, 1988). However, such differentiation of play-fighting from serious fighting based on targets of attack and defense is not universal; many species attack the same targets during both forms of fighting (e.g., Fox, 1969; Henry & Herrero, 1974; Pellis, 1981, 1983). Therefore, in many cases the comparison of playful and serious behaviors requires an analysis of the movements performed by each for an adequate differentiation. Based on our analysis of predatory play, we believe that more refined analyses of behavior, in which the simultaneous movements of both animals are recorded and notated will provide a more meaningful baseline for the analysis of drug effects (e.g., Miczek & Krsiak, 1981; Rodgers & Waters, 1985) and help to identify the functional relevance of playful responses to the ongoing sequence of interaction.

---

[2] It is generally acknowledged that there is an optimal level of activation for the efficient performance of any behavior. The relation between activation and performance is often represented as an inverted U-shaped function (e.g., Hebb, 1955). Consequently, under conditions of excessive activation, predatory attack might be expected to become less efficient. This point may rationalize the otherwise paradoxical finding that in efficient killers, diazepam induces a playful pattern of attack (Langfeldt, 1974).

References

Adamec, R. (1975). The behavioral basis of prolonged suppression of predatory attack in cats. *Aggressive Behavior, 1,* 297–314.

Adamec, R. (1978). Normal and abnormal limbic system mechanisms of emotive biasing. In K. E. Livingston & O. Horynkiewicz (Eds.), *Limbic mechanisms* (pp. 405–455). New York: Plenum Press.

Adamec, R., & Stark-Adamec, C. (1983). Limbic control of aggression in the cat. *Progress in Neuro-Psychopharmacology and Biological Psychiatry, 7,* 505–512.

Adamec, R. E., Stark-Adamec, C., & Livingston, K. E. (1980a). The development of predatory aggression and defense in the domestic cat (*Felis catus*). 1. Effects of early experience on adult patterns of aggression and defense. *Behavioral and Neural Biology, 30,* 389–409.

Adamec, R. E., Stark-Adamec, C., & Livingston, K. E. (1980b). The development of predatory aggression and defense in the domestic cat (*Felis catus*). 2. Development of aggression and defense in the first 164 days of life. *Behavioral and Neural Biology, 30,* 410–434.

Adamec, R. E., Stark-Adamec, C., & Livingston, K. E. (1980c). The development of predatory aggression and defense in the domestic cat (*Felis catus*). 3. Effects on development of hunger between 180 and 365 days of age. *Behavioral and Neural Biology, 30,* 435–447.

Adamec, R. E., Stark-Adamec, C., & Livingston, K. E. (1981). Neural and developmental determinants of the balance of attack and defense in the cat. In P. F. Brain & D. Benton (Eds.), *The biology of aggression* (pp. 397–404). Alphen aan den Rijn, The Netherlands: Sijthoff & Noordhoff.

Apfelbach, R. (1978). Instinctive predatory behavior of the ferret (*Putorius putorius furo* L.) modified by chlordiazepoxide hydrochloride (Librium). *Psychopharmacology, 59,* 179–182.

Baerends-van Roon, J. M., & Baerends, G. P. (1979). *The morphogenesis of the behaviour of the domestic cat, with a special emphasis on the development of prey catching.* Amsterdam: North-Holland Publishing Co.

Bandler, R., & Halliday, R. (1982). Lateralized loss of biting attack-patterned reflexes following induction of contralateral sensory neglect in the cat: A possible role for the striatum in centrally elicited aggressive behaviour. *Brain Research, 242,* 165–177.

Barrett, P., & Bateson, P. (1978). The development of play in cats. *Behaviour, 66,* 106–120.

Bekoff, M. (1972). The development of social interaction, play, and metacommunication in mammals: An ethological perspective. *Quarterly Review of Biology, 47,* 412–434.

Bekoff, M. (1974). Social play and play-soliciting by infant canids. *American Zoologist, 14,* 323–340.

Bekoff, M. (1981). Development of agonistic behavior: Ethological and ecological aspects. In P. F. Brain & D. Benton (Eds.), *Multidisciplinary approaches to aggression research* (pp. 161–178). Amsterdam: Elsevier/North-Holland Biomedical Press.

Bekoff, M., & Byers, J. A. (1981). A critical reanalysis of the ontogeny and phylogeny of mammalian social and locomotor play: An ethological hornet's nest. In K. Immelmann, G. W. Barlow, L. K. Petrinovich, & M. Main (Eds.), *Behavioral development: An interdisciplinary approach* (pp. 196–337). Cambridge: Cambridge University Press.

Berntson, G. G., Hughes, H. C., & Beattie, M. S. (1976). A comparison of hypothalamically induced biting attack with natural predatory behavior in the cat. *Journal of Comparative and Physiological Psychology, 90,* 167–178.

Biben, M. (1979). Predation and predatory play behaviour of domestic cats. *Animal Behaviour, 27,* 81–94.

Della-Fera, M. A., Baile, C. A., & McLaughlin, C. L. (1979). Feeding elicited by benzodiazepine-like chemicals in puppies and cats: Structure activity relationships. *Pharmacology Biochemistry and Behavior, 12,* 195–200.

De Ryck, M., & Teitelbaum, P. (1978). Neocortical and hippocampal EEG in normal and lateral hypothalamic-damaged rats. *Physiology and Behavior, 20,* 403–409.

Egan, J. (1976). Object-play in cats. In J. S. Bruner, A. Jolly, & K. Sylva (Eds.), *Play: Its role in development and evolution* (pp. 161–165). Harmondsworth, Middlesex, England: Penguin.

Einon, D. F. (1983). Play and exploration. In J. Arther & L. I. A. Birke (Eds.), *Exploration in animals and humans* (pp. 210–229). Wokingham, Berkshire, England: Van Nostrand Reinhold (UK).

Eisenberg, J. F., & Leyhausen, P. (1972). The phylogenesis of predatory behaviour in mammals. *Zeitschrift fur Tierpsychologie, 30,* 59–93.

Eshkol, N., & Wachmann, A. (1958). *Movement notation.* London: Weidenfeld and Nicholson.

Fagen, R. (1981). *Animal play behavior.* Oxford: Oxford University Press.

Fentress, J. C. (1976). Dynamic boundaries of patterned behavior: Interaction and self-organization. In P. P. G. Bateson & R. A. Hinde (Eds.), *Growing points in ethology* (pp. 135–169). Cambridge: University of Cambridge Press.

Flynn, J. P. (1967). The neural basis of aggression in cats. In D. C. Glass (Ed.), *Neurophysiology and emotion* (pp. 40–60). New York: Rockefeller University Press.

Fox, M. W. (1969). The anatomy of aggression and its ritualization in Canidae: A developmental and comparative study. *Behaviour, 35,* 242–258.

Golani, I. (1976). Homeostatic motor processes in mammalian interactions: A choreography of display. In P. P. G. Bateson & P. H. Klopfer (Eds.), *Perspectives in ethology* (Vol. 2, pp. 69–134). New York: Plenum Press.

Golani, I., Wolgin, D., & Teitelbaum, P. (1979). A proposed natural geometry for recovery from akinesia in the lateral hypothalamic rat. *Brain Research, 164,* 237–267.

Hebb, D. O. (1955). Drives and the CNS (conceptual nervous system). *Psychological Review, 62,* 243–254.

Henry, J. D., & Herrero, S. M. (1974). Social play in the American black bear, its similarity to canid social play and an examination of its identifying characteristics. *American Zoologist, 14,* 371–389.

Langfeldt, T. (1974). Diazepam-induced play behavior in cats during prey killing. *Psychopharmacology, 36,* 181–184.

Leaf, R. C., Wnek, D. J., Gay, P. E., Corcia, R. M., & Lamon S. (1975). Chlordiazepoxide and diazepam induced mouse killing by rats. *Psychopharmacology, 44,* 23–28.

Leaf, R. C., Wnek, D. J., & Lamon, S. (1984). Oxazepam induced mouse killing in rats. *Pharmacology Biochemistry and Behavior, 20,* 311–313.

Leyhausen, P. (1973). On the function of the relative hierarchy of moods (As exemplified by the phylogenetic and ontogenetic development of prey-catching in carnivores). In K. Lorenz & P. Leyhausen (Eds.), *Motivation of human and animal behavior: An ethological view* (pp. 144–247). New York: Van Nostrand Reinhold.

Leyhausen, P. (1979). *Cat behavior: The predatory and social behavior of domestic and wild cats.* New York: Garland STPM Press.

Loizos, C. (1966). Play in mammals. *Symposia of the Zoological Society of London, 18,* 1–9.

MacDonnell, M. F., & Flynn, J. P. (1966a). Control of sensory fields by stimulation of hypothalamus. *Science, 152,* 1406–1408.

MacDonnell, M. F., & Flynn, J. P. (1966b). Sensory control of hypothalamic attack. *Animal Behavior, 14,* 339–405.

Marler, P., & Hamilton, W. J. III. (1966). *Mechanisms of animal behavior.* New York: Wiley.

Meaney, M. J., & Stewart, J. A. (1981). A descriptive study of social development in the rat (*Rattus norvegicus*). *Animal Behaviour, 29,* 34–45.

Mereu, G. P., Fratta, W., Chessa, P., & Gessa, G. L. (1976). Voraciousness induced in cats by benzodiazepines. *Psychopharmacology, 47,* 101–103.

Miczek, K. A., & Krsiak, M. (1981). Pharmacological analysis of attack and flight. In P. F. Brain & D. Benton (Eds.), *Multidisciplinary approaches to aggression research* (pp. 341–354). Amsterdam: Elsevier/North-Holland Biomedical Press.

Moran, G., Fentress, J. C., & Golani, I. (1981). A description of relational patterns of movement during "ritualized fighting" in wolves. *Animal Behaviour, 29,* 1146–1165.

O'Brien, D. P., Chesire, R., & Teitelbaum, P. (1985). Tail-pinch vs. vestibular activation in lateral hypothalamic akinesia. *Physiology and Behavior, 34,* 811–814.

O'Brien, D. P., Pellis, S. M., Pellis, V. C., Schallert, T., Whishaw, I. Q., & Teitelbaum, P. (1988). *Diazepam deactivation of haloperidol-induced catalepsy in the rat: A behavioral and electroencephalographic study.* Manuscript in preparation.

Owens, N. W. (1975). Social play in free living baboons *Papio anubis. Animal Behaviour, 23,* 387–408.

Pellis, S. M. (1981). A description of social play by the Australian magpie *Gymnorhina tibicen,* based on the Eshkol-Wachmann notation. *Bird Behaviour, 3,* 61–79.

Pellis, S. M. (1982). An analysis of courtship and mating in the Cape Barren goose *Cereopsis novaehollandiae* Latham based on the Eshkol-Wachmann Movement Notation. *Bird Behaviour, 4,* 30–31.

Pellis, S. M. (1983). Development of head and foot coordination in the Australian Magpie *Gymnorhina tibicen,* and the function of play. *Bird Behaviour, 4,* 57–62.

Pellis, S. M., & Officer, R. C. E. (1987). An analysis of some predatory behaviour patterns in four species of carnivorous marsupials (Dasyuridae), with comparative notes on the eutherian cat *Felis catus. Ethology, 75,* 177–196.

Pellis, S. M., & Pellis, V. C. (1987). Play-fighting differs from serious fighting in both target of attack and tactics of fighting in the laboratory rat *Rattus norvegicus. Aggressive Behavior, 13,* 227–242.

Pellis, S. M., & Pellis, V. C. (1988). Play-fighting in the Syrian golden hamster *Mesocricetus auratus* Waterhouse, and its relationship to serious fighting during post-weaning development. *Developmental Psychobiology, 21,* 323–337.

Poole, T. B., & Fish, J. (1975). An investigation of playful behavior in *Rattus norvegicus* and *Mus musculus* (mammalia). *Journal of Zoology, 175,* 61–71.

Rasa, O. A. E. (1984). A motivational analysis of object play in juvenile dwarf mongooses (*Helogale undulata rufula*). *Animal Behaviour, 32,* 579–589.

Rodgers, R. J., & Waters, A. J. (1985). Benzodiazepines and their antagonists: A pharmacological analysis with particular reference to effects on "aggression." *Neuroscience Biobehavioral Reviews, 9,* 21–35.

Rosenbaum, J. F. (1982). The drug treatment of anxiety. *New England Journal of Medicine, 306,* 491–504.

Schallert, T., Wang, C. H., Hsiao, S., & Whishaw, I. Q. (1979). Differential effects of antianxiety and antipsychotic drugs on two distinct systems of arousal. *Society for Neuroscience Abstract, 5,* 2257.

Sepinwall, J., & Cook, L. (1978). Behavioral pharmacology of antianxiety drugs. In L. L. Iverson, S. D. Iversen, & S. H. Snyder (Eds.), *Handbook of psychopharmacology* (Vol. 13, pp. 345–392). New York: Plenum Press.

Shoham, S., & Teitelbaum, P. (1982). Subcortical waking and sleep during lateral hypothalamic "somnolence" in rats. *Physiology and Behavior, 28,* 335–347.

Snider, R. S., & Niemer, W. T. (1961). *A stereotaxic atlas of the cat brain.* Chicago: University of Chicago Press.

Symons, D. (1978). *Play and aggression: A study of rhesus monkeys.* New York: Columbia University Press.

West, M. J. (1977). Exploration and play with objects in domestic kittens. *Developmental Psychobiology, 10,* 53–57.

Winters, W. D., & Kott, K. S. (1979). Continuum of sedation, activation, and hypnosis or hallucinosis: A comparison of low dose effects of pentobarbital, diazepam and gamma-hydroxybutyrate in the cat. *Neuropharmacology, 18,* 887–894.

Wolgin, D. L. (1982). Motivation, activation, and behavioral integration. In R. L. Isaacson & N. E. Spear (Eds.), *The expression of knowledge* (pp. 243–290). New York: Plenum Press.

Wolgin, D. L., & Servidio, S. (1979). Disinhibition of predatory attack in kittens by oxazepam. *Society for Neuroscience Abstracts, 5,* 667.

Wolgin, D. L., & Teitelbaum, P. (1978). Role of activation and sensory stimuli in the recovery from lateral hypothalamic damage in the cat. *Journal of Comparative and Physiological Psychology, 92,* 474–500.

Wright, J. J., & Craggs, M. D. (1979). Intracranial self-stimulation, cortical arousal and the sensorimotor neglect syndrome. *Experimental Neurology, 65,* 42–52.

Received December 4, 1986
Revision received July 28, 1987
Accepted August 5, 1987 ∎

Glossary

anxiolytic - antianxiety

benzodiazepines - a class of drugs, such as oxazepam and diazepam, that are commonly prescribed for their antianxiety effects

consummatory - any behavior that consummates or ends an adaptive behavioral sequence

cresyl violet - a dye that preferentially stains the cell bodies of the neurons in a slice of neural tissue

diazepam - a benzodiazepine drug; one of the most commonly prescribed antianxiety drugs

Eshkol-Wachman Movement Notation - a system for recording behavioral sequences in terms of a set of written symbols that represent the movements of individual body segments

orofacial - pertaining to the mouth and face

oxazepam - a benzodiazepine drug

stereotaxic atlas - an atlas of the three dimensional structure of the brain; it tells the surgeon how far to move the electrode tip from a prescribed reference point (along three axes: front-back, up-down, medial-lateral) so that the electrode tip ends up in the intended structure

stereotaxic instrument - a device used for certain kinds of brain surgery; it has two parts: a headholder for holding the subject's head perfectly still in the correct orientation and an electrode manipulator for precisely positioning the tip of an electrode at a prescribed location in the brain

Tween 80 - when a drug does not readily dissolve, it is crushed into small particles, mixed with saline, and injected as a suspension; a small amount of Tween 80 added to the mixture helps keep the particles from settling out of the suspension

Essay Study Questions

1. What is the usual definition of play behavior?

2. Describe the observations of Adamec and his colleagues that provided the stimulus for the research of Pellis et al.

3. Compare the effects of diazepam on the behavior of Lucy, Molly, and Rita. What does this comparison suggest about the predatory behavior of cats?

4. Avoidance, play, and attack are best viewed as points along a continuum. Explain, and support with experimental results.

5. What is the effect of bilateral lateral hypothalamic lesions on predatory behavior in the cat?

6. What is the effect of a tail pinch on feline predatory behavior?

7. What is Eshkol-Wachmann Movement Notation (EWMN), and how did Pellis and his collaborators use it to increase their understanding of cat predatory "play" behavior? What did they conclude?

Multiple-Choice Study Questions

1. Activities that are commonly classed as play behavior
    a. appear to have no functional significance.
    b. seem exaggerated.
    c. seem to be organism-dominated.
    d. are flexible.
    e. all of the above

2. Which of the following is a benzodiazepine?
    a. Tween 80
    b. propylene glycol
    c. sodium pentobarbital
    d. cresyl violet
    e. none of the above

3. Which of the following drugs is an anxiolytic drug?
    a. oxazepam
    b. dexamethazone
    c. diazepam
    d. both a and c
    e. none of the above

4. Molly did not spontaneously kill mice, but she "played" with them. After an injection of diazepam, she
    a. fled from them.
    b. killed them.
    c. did not approach them but was very relaxed in their presence.
    d. played with them less.
    e. played with them slightly more, but she never killed them.

5. In one experiment, the effect of diazepam on the recovery from bilateral lateral hypothalamic lesions was assessed. The results of this experiment suggested that
    a. diazepam is an anxiolytic.
    b. diazepam's ability to suppress defensive behavior cannot totally account for its ability to facilitate killing.
    c. diazepam facilitates regeneration.
    d. diazepam's anxiolytic effect increases killing.
    e. diazepam's anxiolytic effect decreases killing.

6. The main conclusion of the study that employed Eshkol-Wachmann Movement Notation was that
    a. tail pinch increases mouse killing.
    b. cat predatory behavior is similar in many respects to classical dance.
    c. the seemingly exaggerated and inappropriate movements of the cat during predatory "play" are actually adaptive.
    d. the movements that compose cat predatory "play" are not adaptive.
    e. the movements that compose cat predatory "play" are, in fact, maladaptive.

The answers to the preceding questions are on page 290.

Food-For-Thought Questions

1. The main advantage of the Eshkol-Wachmann Movement Notation System is that in using it the experimenter is forced to look carefully and objectively at the behavior that she or he is studying. As in the present case, this often leads to important insights. Interestingly, once these insights have occurred, the fundamental behavioral principles that were initially hidden in the intricacy of the behavioral sequences become obvious, even to casual inspection. Discuss.

2. If you have ever played with a cat, you will recognize that many of the behaviors observed by Pellis et al. can be elicited in cats by a human hand or a moving inanimate object (such as a piece of string). On the basis of Pellis' research, devise a set of simple objective tests that might be used to determine where your cat, or the cat of a friend, would lie on Pellis' avoidance-play-kill continuum. If you have access to several cats, you might wish to see how well your test battery distinguishes between them.

ARTICLE 9

Gastric Emptying Changes Are Neither Necessary Nor Sufficient for CCK-Induced Satiety

K.L. Conover, S.M. Collins, and H.P. Weingarten

Reprinted from American Journal of Physiology, 1989, volume 256, R56-R62

Peptides are short chains of amino acids. When peptides were initially discovered in the nervous system, they were assumed to be inactive, intermediate steps in the manufacture of proteins, which are long chains of amino acids. However, in the last two decades peptides have been shown to have a variety of important direct effects on neural activity and behavior.

There has been considerable interest in the peptide CCK (cholecystokinin) because of the ability of CCK injections to reduce the size of subsequent meals. Food in the upper intestine triggers the release of CCK from the intestine wall, and the CCK causes the pyloric sphincter to contract thus retarding the movement of food from the stomach until the nutrients are absorbed from the food that is already in the intestine. Accordingly, injections of CCK prior to a meal have been assumed to reduce meal size by increasing stomach distension. In opposition to this widely accepted theory, the following two experiments by Conover, Collins, and Weingarten suggest that the satiety-accelerating effect of CCK is not mediated by its inhibitory effect on stomach emptying.

In Experiment 1, an injection of CCK-8 (an easy-to-synthesize form of the CCK molecule) immediately before a meal reduced both meal size and gastric emptying, however an injection 15 minutes before a meal reduced meal size without reducing gastric emptying. In Experiment 2, the effects of three peptides were compared: pentagastrin influenced neither food intake nor gastric emptying; bombesin caused a large sustained reduction in eating but only a transient delay in gastric emptying; and secretin did not reduce eating despite the fact that it inhibited stomach emptying to the same degree as did CCK-8.

On the basis of their experiments, Conover, Collins, and Weingarten suggested that the satiety-inducing effects of CCK may be mediated by its direct effects on the tone of the stomach wall or on the vagus nerve, which carries signals from the stomach to the brain. They favored these interpretations because at the time that their manuscript was being written, the evidence seemed to indicate that the ability of CCK to reduce meal size was mediated by its peripheral effects. However, it has recently become clear that CCK can enter the brain and influence behavior by binding to specialized receptor molecules in neural membranes.

Gastric emptying changes are neither necessary nor sufficient for CCK-induced satiety

KENT L. CONOVER, STEPHEN M. COLLINS, AND HARVEY P. WEINGARTEN
Departments of Psychology and Medicine, McMaster University, Hamilton, Ontario L8S 4K1, Canada

CONOVER, KENT L., STEPHEN M. COLLINS, AND HARVEY P. WEINGARTEN. *Gastric emptying changes are neither necessary nor sufficient for CCK-induced satiety.* Am. J. Physiol. 256 (Regulatory Integrative Comp. Physiol. 25): R56–R62, 1989.—If gastric emptying plays a significant role in the satiety produced by exogenous cholecystokinin octapeptide (CCK-8) then *1*) the effects on emptying and feeding should share similar kinetics and *2*) peptides that inhibit emptying should also inhibit feeding. In the first experiment, CCK-8 (5.6 µg/kg) injected immediately before the introduction of an intragastric load (10 ml saline) or presentation of a test meal (15% sucrose) produced a rapid inhibition of both emptying and feeding. In contrast, the same dose administered 15 min before testing caused no inhibition of emptying, even though it retained the ability to reduce meal size. In *experiment 2*, the abilities of the peptides pentagastrin (100 µg/kg), bombesin (8 and 16 µg/kg), and secretin (2.86, 14.3, and 28.6 µg/kg) to reduce food intake and inhibit emptying were tested. Pentagastrin influenced neither food intake nor gastric emptying. Bombesin caused a small transient delay in emptying but a large and sustained eating suppression. However, a high dose of secretin caused no significant reduction of food intake, in spite of the fact that it inhibited emptying to the same degree as 1.4 µg/kg CCK-8, which does reduce intake. These results suggest that the inhibition of emptying by CCK is not sufficient to explain the satiety effect of CCK-8.

feeding; behavior; kinetics; peptides; cholecystokinin; gastrin; bombesin; secretin

---

ALTHOUGH THE COOH-TERMINAL octapeptide of cholecystokinin (CCK-8) suppresses meal size (15, 16, 33, 34), the mechanism mediating CCK-induced satiety is not understood. Moran and McHugh (21, 23, 28) have proposed that the satiety produced by CCK-8 depends on the peptide's ability to inhibit gastric emptying. This hypothesis is supported by demonstrations that doses of CCK-8 that suppress meal size also slow gastric emptying in monkeys (23) and rats (1, 2, 7, 11). Although the gastric emptying hypothesis requires these correlational data, findings of this nature do not demonstrate conclusively that the satiety effect of the peptide is mediated exclusively by the inhibition of gastric emptying.

This study uses two approaches to determine the degree to which CCK-induced changes in gastric emptying mediate its satiety action. In *experiment 1*, the kinetics of CCK-8's suppression of emptying and feeding are compared. In spite of the short half-life CCK in plasma [17 min in rats (18)] and in vivo [2–3 min in humans and dogs (38)], exogenous CCK-8 reduces feeding even when administered 15 min before initiation of a test meal (6, 13). If suppression of emptying is critical to the satiety effect of CCK, then inhibition of emptying must also occur 15 min after the administration of CCK.

In *experiment 2* we examine the ability of other gut peptides to affect emptying and satiety. If CCK produces satiety because it slows emptying, then any peptide that inhibits emptying to a degree similar to CCK should produce reduction of food intake of similar magnitude.

METHODS

Subjects

Male Long-Evans hooded rats were bred in the McMaster Psychology Department colony from stock originating from Blue Spruce Farms (Altmont, NY). They weighed 375–500 g at the time of testing. They were housed in individual hanging cages in a colony room maintained at 26°C with a 14:10 h light-dark cycle. Water was available ad libitum and food was provided according to the experimental protocol.

Gastric Cannula Surgery

For both emptying and feeding studies, access to the stomach was provided by a stainless steel gastric cannula implanted surgically (42), under pentobarbital sodium (45 mg/kg) anesthesia. Rats recovered from surgery for a minimum of 14 days.

Injections

Cholecystokinin octapeptide was a kind gift of S. J. Luciana, Squibb Institute for Medical Research, Princeton, NJ (SQ 19 844, no. NN025NC). Pentagastrin (Peptavlon) was purchased from Ayerst Laboratories, Montreal, Canada. Cimetidine (Tagamet) was purchased from Smith Kline & French Laboratories, Canada. Bombesin (B 5508) and secretin (S 5014) were purchased from Sigma (St. Louis, MO). All injections were made up to a volume of 1 ml with 0.15 M saline and were injected intraperitoneally. Control injections consisted of 1 ml 0.15 M saline.

Gastric-Emptying Procedure

Subjects were placed in individual cages (10 × 15 × 25 mm), with a slot in the floor through which the gastric catheter could hang. Rats were 6-h food deprived before each session, and all tests took place between 1500 and 1700 h.

Before testing, the stomach was cleaned by saline lavage, the gastric catheter was installed, and fluid remaining in the stomach was aspirated. The test load of 10 ml 0.15 M saline was infused directly intragastrically through the gastric cannula at the rate of 10 ml/min. The introduction of the test load into the stomach was designated as *time 0*; the volume of test load remaining in the stomach was measured at 2, 5, 10, 15, 20, 25, 30, 35, and 40 min.

Gastric emptying was monitored using a double sampling method (8). With this procedure each emptying test yielded a complete descripton of the time course of emptying from which half-emptying times could be derived. There were six subjects in each experiment with each subject serving as its own control. Treatment conditions were administered in a random order.

Satiety Procedure

Experimental conditions for satiety tests paralleled those used for emptying tests. Thus animals were 6-h food deprived before each trial with tests performed between 1500 and 1700 h. The gastric cannula was opened and the stomach was flushed with saline lavage. The cannula was reclosed and the animals were returned to their home cages for feeding tests. Animals were fed 15% (wt/vol) sucrose solutions. Experiments were not initiated until the subjects' intakes following control intraperitoneal administration of 1 ml of 0.15 M saline were stable. Sucrose consumption was measured at 5, 10, 15, 20, 25, and 30 min.

Data Analysis

Half-emptying times ($t_{1/2}$) were defined as the time needed for half of the test load to leave the stomach and were derived by fitting the power exponential function to the individual test load time-course data (10). In addition to $t_{1/2}$, the power exponential has the parameter, β, which describes the shape of the emptying curve relative to a simple exponential with the same $t_{1/2}$. When $\beta = 1$, the emptying curve is perfectly exponential. A $\beta < 1$ indicates a faster than exponential initial phase of emptying (i.e., approximately the first 10 min) followed by a slower than exponential phase of late emptying. This type of emptying is characteristic of subjects with impaired pyloric function, as in the case of vagotomy and pyloroplasty (10). A $\beta > 1$ indicates initial emptying slower than exponential followed by a later phase of emptying more rapid than exponential. In this study a $\beta > 1$ is best interpreted as a lag or suppression of emptying in the early phase of gastric emptying. The goodness of fit of the power exponential function to the time courses was measured by the coefficient of determination (R^2); the median R^2 is reported for each treatment. As R^2 approaches 1, the fit of the curve to the data becomes perfect.

To test the effects of dose CCK-8 and injection time on emptying inhibition (*experiment 1*), two-way repeated measures analysis of variance (ANOVAs) (dose CCK-8 by injection time) were performed on the $t_{1/2}$ and β estimates and upon the test load remaining values for 2, 5, and 10 min. Multiple comparisons of the $t_{1/2}$ and β estimates were performed using the Newman-Keuls procedure. To maximize statistical sensitivity to emptying inhibition, the mean remaining test volumes were compared using Dunnett's test (one-tailed) for experimental treatment vs. the control of intraperitoneal saline.

The effects of pentagastrin and cimetidine on $t_{1/2}$, β, and gastric secretion (*experiment 2*) were assessed using two-way repeated measure analysis of variances (ANOVAs). Dunnett's two-tailed test was used for post hoc comparisons of the treatment means. One-way repeated measures ANOVAs were used to analyze the effect of bombesin and secretin on $t_{1/2}$ and β, and multiple comparisons were performed by the Newman-Keuls test.

Feeding data were analyzed with one-way repeated measure ANOVAs on the cumulative sucrose consumptions at 5, 15, and 30 min. Multiple comparisons were analyzed by the Newman-Keuls method.

Peptide Administration Protocols

Experiment 1: kinetics of emptying and feeding inhibition. Cholecystokinin inhibits feeding even when injected 15 min before the initiation of eating (6, 13). We tested whether the effects of CCK-8 on emptying possessed similar kinetics by injecting rats with CCK-8 (5.6 or 11.2 µg/kg) either immediately or 15 min before an intragastric test load of 0.15 M saline.

The satiety effect of CCK-8 under these experimental conditions was assessed by comparing the intakes of 15% sucrose under saline control trials with intakes after injections of 5.6 µg/kg CCK-8 either immediately or 15 min before meal presentation.

Experiment 2: effect of other gut peptides on emptying and feeding. For emptying and feeding studies, all injections were administered immediately before the intragastric load or meal, respectively. To examine the effect of pentagastrin on emptying, subjects received two injections at *time 0* consisting of *1*) saline and saline; *2*) saline and pentagastrin; *3*) cimetidine and saline; or *4*) cimetidine and pentagastrin. Doses used were 100 µg/kg pentagastrin and 5 mg/kg cimetidine. Feeding studies used the identical dose of pentagastrin. To examine the ability of bombesin to inhibit gastric emptying, rats were injected with either saline or that peptide at doses of 8 and 16 µg/kg. Feeding studies with bombesin used the dose of 8 µg/kg only. Studies on the effect of secretin on emptying used 2.86, 14.3, and 28.6 µg/kg, corresponding to doses of 10, 50, and 100 clinical units/kg. Feeding studies only used secretin doses of 14.3 and 28.6 µg/kg.

As indicated, the satiety effect of these peptides was studied under conditions similar to those of the gastric emptying procedure. Each day of peptide administration was preceded and followed by a control saline trial. The satiety effect of the peptides was measured by comparing sucrose consumption on peptide days with the average on saline days.

RESULTS

Experiment 1: Kinetics of Emptying and Feeding Inhibition

Examination of Fig. 1 and statistical analysis of corresponding $t_{1/2}$s (Table 1) revealed that CCK-8 significantly inhibited emptying [$F(2,10) = 42.66, P < 0.001$] when injected at *time 0*. Multiple comparisons determined that when injected at *time 0*, both 5.6 μg/kg ($q_3 = 5.81, P < 0.01$) and 11.2 μg/kg ($q_5 = 6.18, P < 0.01$) CCK-8 significantly raised mean $t_{1/2}$ over that for saline. Emptying inhibition in the early phases of emptying was also indicated by significantly increasing b values with dose [$F(2,10) = 14.55, P < 0.001$; see Table 1]. Post hoc comparisons revealed that βs following both 5.6 ($q_4 = 5.81, P < 0.01$) and 11.2 μg/kg were significantly higher than control values (respectively: $q_4 = 5.81$, $q_5 = 6.81$, $Ps < 0.01$).

Injecting the peptide 15 min before the test load significantly reduced the ability of the peptide to inhibit gastric emptying. The mean $t_{1/2}$ after 5.6 μg/kg CCK-8 administered at −15 min was not different from the saline mean at −15 min ($q_3 = 3.90$, NS). Although 11.2 μg/kg CCK-8 injected at −15 min significantly increased half-emptying time compared with saline ($q_3 = 5.30, P < 0.01$), the degree of inhibition was significantly less than when the identical dose was injected at 0 min ($q_3 = 5.30, P < 0.01$). The reduced ability of CCK-8 to retard emptying when injected 15 min before the intragastric load was also evident in the analysis of β, which revealed that the βs for CCK-8 at both 5.6 and 11.2 μg/kg were not significantly greater than that for saline.

To be confident that 5.6 μg/kg CCK-8 injected at −15 min did not inhibit emptying, the volumes of test load remaining at times 2, 5, and 10 min after peptide administration were compared with those following saline, using Dunnett's one-tailed test (see Table 2). Stomach values after CCK-8 5.6 μg/kg at −15 min were similar to control values at 2, 5, and 10 min, reinforcing the conclusion that 5.6 μg/kg CCK-8 does not inhibit emptying when administered 15 min before the IG load.

Analysis of sucrose consumption at 5 min (see Fig. 2 and Table 3) revealed that CCK-8 significantly reduced meal size when injected at 0 min ($q_4 = 1.20, P < 0.01$), indicating that the peptide acts rapidly. CCK-8 injected coincident with meal initiation also reduced intake at 15 ($q_4 = 2.54, P < 0.01$) and 30 ($q_4 = 2.11, P < 0.01$) min.

Although the suppression of feeding was significantly reduced following CCK-8 at −15 min compared with 0 min (q_2s = 0.95 and 2.01 at 5 and 15 min, respectively, $Ps < 0.01$), CCK-8 administered at −15 min (Table 3) did result in significant reductions of intake at 5, 15, and 30 min (q_3s = 1.10, 2.34, and 1.84, respectively, $Ps < 0.01$), indicating that the satiety effect can persist over a 15-min interval.

The results of *experiment 1* indicate different kinetics of the CCK-induced emptying and feeding effects. The time frame over which CCK influences emptying is brief and consistent with the demonstrated short half-life of the peptide (18, 38). In comparison, the ability of CCK-8 to suppress feeding survives longer intervals.

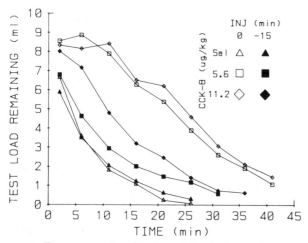

FIG. 1. Time courses for emptying of 10-ml saline (sal) test loads after cholecystokinin octapeptide (CCK-8, 5.6 or 11.2 μg/kg) administered coincident with (0 min) or 15 min before (−15 min) infusion of the test load. Test load remaining is volume remaining in stomach corrected for gastric secretion.

TABLE 1. *Effect of injection time on inhibition of gastric emptying by CCK-8*

	Saline	CCK-8, μg/kg	
		5.6	11.2
Injection time, 0 min			
$t_{1/2}$	3.9±0.4	20.8±2.3*†	23.5±2.4*†
$β^3$	0.9±0.06	1.7±0.22*†	1.8±0.22*†
Median R^2	0.99	0.92	0.88
Injection time, −15 min			
$t_{1/2}$	3.1±0.4	5.3±1.2	10.4±1.4*
β	0.7±0.06	0.7±0.08	1.1±0.06
Median R^2	0.98	0.99	0.98

Values are means ± SE. Mean $t_{1/2}$ is the time at which half the test load had emptied. Injection time represents time at which cholecystokinin octapeptide (CCK-8) was injected relative to infusion of gastric test load. See text for interpretation of β values. * $P < 0.01$ when compared with saline; † $P < 0.01$ when compared with corresponding dose of CCK-8 administered at −15 min.

Experiment 2: Effect of Other Gut Peptides on Emptying and Feeding

Figure 3 and statistical analysis indicated that 100 μg/kg pentagastrin significantly increased gastric secretion during the first 20 min [$F(1,5) = 9.67, P < 0.03$], an effect that was blocked by cimetidine [$F(1,5) = 27.98, P < 0.005$]. Nevertheless, 100 μg/kg pentagastrin had no effect on the time course of gastric emptying, verified by the absence of significant changes in $t_{1/2}$ [$F(1,5) = 0.02$, NS, or β, $F(1,5) = 0.27$, NS compared with saline trials]. The amount eaten after 100 μg/kg pentagastrin was also not significantly different from control (Table 4). Bombesin (8 and 16 μg/kg) had only a minor impact on the time course of gastric emptying (Fig. 4). Although $t_{1/2}$ values after administration of bombesin were not significantly different from those after saline [$F(2,10) = 1.19, P > 0.05$], there was a small, but significant, increase in β at both the 8 μg/kg ($q_2 = 0.17, P < 0.05$) and 16 μg/kg ($q_3 = 0.21, P < 0.05$), doses. This increase in β, ~20%, is

TABLE 2. *Effect of CCK-8 on test load remaining in stomach at times indicated after intragastric infusion*

Time, min	Saline	CCK-8, µg/kg		Saline	CCK-8, µg/kg		$F(2,10)$	P
		5.6	11.2		5.6	11.2		
	Injection time, 0 min			*Injection time, −15 min*				
2	6.7±0.3	8.6±0.3†	8.3±0.2†	5.9±0.5	6.8±0.4	8.0±0.3*	3.22	0.082
5	3.6±0.2	8.9±0.5†	8.2±0.2†	3.5±0.2	4.6±0.7	7.2±0.6†	24.62	0.001
10	1.8±0.3	7.9±0.8†	8.4±0.5†	2.0±0.2	2.9±0.5	4.8±0.5†	25.80	0.001

Values are means ± SE of test loads remaining. The time at which the test volumes remaining were measured relative to the onset of infusion. F ratios are for injection by time of dose interaction for mean test volumes remaining at 2, 5, and 10 min. Time represents the time at which cholecystokinin (CCK-8) was injected relative to infusion of the test load. Significantly greater than saline at 0 min: * $P < 0.05$; † $P < 0.01$.

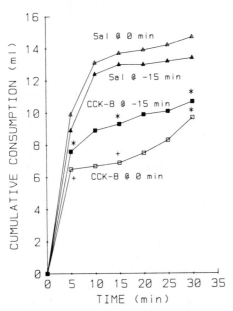

FIG. 2. Cumulative consumption of 15% sucrose after 5.6 µg/kg cholecystokinin octapeptide (CCK-8), administered coincident with (0 min) or 15 min before (−15 min) meal presentation.

TABLE 3. *Effect of injection time on CCK-8-induced satiety*

	Consumption, ml		
	5 min	15 min	30 min
	Injection time, 0 min		
Saline	10.1±0.5	13.7±1.0	14.7±0.8
CCK (5.6 µg/kg)	6.5±0.5*†	6.9±0.6*†	9.7±0.8*
	Injection time, −15 min		
Saline	9.2±0.5	13.2±0.8	13.6±0.9
CCK (5.6 µg/kg)	7.6±0.4*	9.3±0.7*	10.6±0.8*

Values for consumption of 15% sucrose are means ± SE. Injection is time at which cholecystokinin (CCK-8) was injected relative to presentation of the test meal. * $P < 0.01$ when compared with corresponding saline condition; † $P < 0.01$ when compared with corresponding dose of CCK-8 when administered at −15 min.

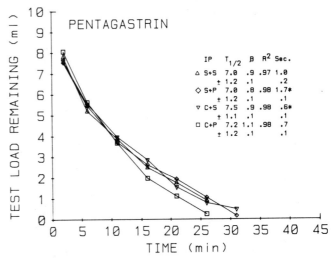

FIG. 3. Time courses for emptying of a 10-ml saline intragastric test load in the following test conditions: S+S, saline + saline; S+P, saline + pentagastrin; C+S, cimetidine + saline; or C+P, cimetidine + pentagastrin. See text for doses and explanation of $t_{1/2}$, β, and R^2 measures. Gastric secretion was measured in milliliters over the 1st 20 min. * $P < 0.05$ Dunnett's two-tailed test.

TABLE 4. *Effect of pentagastrin, bombesin, and secretin on consumption of 15% sucrose*

	Consumption, ml		
	5 min	15 min	30 min
Saline	8.0±0.7	10.3±0.7	10.6±0.7
Pentagastrin (100 µg/kg)	7.3±0.7	10.1±0.7	10.7±0.7
Saline	7.7±0.6	10.6±0.8	10.9±0.8
Bombesin (8 µg/kg)	5.1±0.6*	5.4±0.7*	5.7±0.7*
Saline	8.1±0.6	11.0±0.7	11.4±0.6
Secretin (14.3 µg/kg)	7.1±0.6	10.5±0.8	11.7±0.5
Secretin (28.6 µg/kg)	5.3±0.7†	8.1±0.6†	9.3±0.6†

Values for consumption of 15% sucrose are means ± SE. Peptides were injected immediately before presentation of the test meal. * $P < 0.01$ compared with corresponding mean for saline. † $P < 0.01$ compared with corresponding means for saline and smaller dose of secretin.

small compared with the 100% increase produced by CCK-8 in *experiment 1*, suggesting that bombesin produces only a small and transient retardation of emptying. In spite of the small emptying effects, even the lower dose of bombesin, 8 µg/kg, profoundly suppressed eating (Table 4) measured at 5, 15, and 30 min after meal initiation [$F(1,14) = 19.10$, 48.58, and 46.79, respectively, all $P < 0.001$]. The percent suppression produced by bombesin, 49% at 15 min, is similar to the level of inhibition produced by CCK-8.

Secretin significantly inhibited gastric emptying in a dose-related manner, indicated by an increase in $t_{1/2}$ with dose, $F(3,15) = 31.94$, $P < 0.001$ (Fig. 5). Secretin, 14.3 and 28.6 µg/kg, produced elevated but equal increases in $t_{1/2}$ over saline ($q_3 = 4.84$ and $q_4 = 5.25$, respectively, $P < 0.01$). β Values were unaffected by secretin [$F(3,15) = 2.88$, $P > 0.05$)]. The lowest dose of secretin used, 2.86 µg/kg, did not alter the time course of emptying.

Multiple comparisons revealed that only secretin at 28.6 µg/kg significantly reduced sucrose consumption

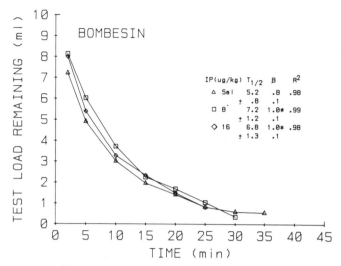

FIG. 4. Time courses for emptying of a 10-ml saline (sal) test load after intraperitoneal (IP) injection of saline or bombesin (8 and 16 µg/kg).

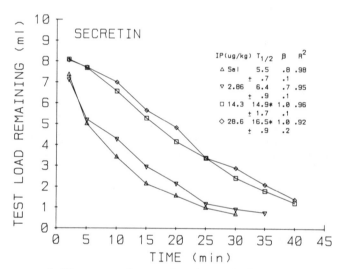

FIG. 5. Time courses for emptying of a 10-ml saline (sal) test load after IP injection of saline or secretin (2.86, 14.3, and 28.6 µg/kg). See text for doses and explanation of $t_{1/2}$, β, and R^2 measures.

(Table 4). Secretin, 14.3 µg/kg, had no significant effect on sucrose intake, even though this dose retards emptying to the same degree as the highest dose of secretin and 1.4 µg/kg CCK-8 (7).

DISCUSSION

The purpose of this study was to examine the importance of gastric emptying in the mediation of peptide-induced satiety. The motivation for these examinations is the necessity of evaluating the most viable hypothesis regarding CCK's mechanism of behavioral action, i.e., that its satiety is mediated by its ability to retard the rate of gastric emptying (21, 23, 28). The results of the current and previous studies (2, 7, 11) demonstrating that CCK-8 suppresses gastric emptying in a dose-dependent manner are consistent with this hypothesis. However, the manipulations employed in this study, which investigate CCKs response kinetics and the ability of other peptides to modulate emptying and eating, demonstrate several important dissociations between the emptying response and satiety. These results are summarized in Table 5 and are discussed in more detail below.

CCK-8 injected coincident with meal initiation produces a dose-dependent inhibition of gastric emptying and meal size. These data are consistent with previous reports of an inhibition of gastric emptying with this peptide (1, 2, 7, 11, 23). This correlation between CCK's effects on emptying and satiety are congenial with the gastric emptying hypothesis. Similarly, the correlation reported in this study between the inability of the structurally related peptide, that pentagastrin to suppress both emptying and eating, is consistent with the predictions of the gastric emptying hypothesis.

The kinetics of the CCK-induced emptying response, however, do not coincide with the kinetics of CCK-induced satiety. Specifically, CCK-8 injected 15 min before the emptying test does not affect the rate of gastric emptying. This observation is consistent with the demonstrated short half-life of the peptide (18, 38). In contrast, administering CCK-8 15 min before a meal preserved a significant and large suppression of food intake. The suppression of feeding behavior when there is no corresponding change in the rate of gastric emptying suggests that changes in gastric emptying are not necessary for CCK-induced satiety. One possible interpretation of these data is that there exists a subpopulation of peripheral CCK receptors responsible for satiety, which are independent of a receptor population mediating CCK-induced emptying changes. Alternatively, the same population of receptors may mediate both responses. However, the time course of the induced emptying changes are comparatively brief, whereas the effect of that peptide on food intake is considerably longer.

The data comparing the effects of bombesin on emptying and satiety further reveal the dissociation between changes in gastric emptying and food intake produced by peptides. Bombesin in a dose of 8 µg/kg produced a profound and reliable suppression of feeding. This finding is consistent with numerous similar demonstrations in the field (6, 19, 37, 41). In contrast, this dose of bombesin produced only a trivial effect on the rate of gastric emptying. The effects of bombesin on gastric emptying reported in the literature have been equivocal. Some (11, 29, 30) have found a decrease in gastric emptying with intraperitoneal bombesin; others (26) have failed to find any effect. It is not clear why different investigators have found different results, although considerations such as dose, species, time of peptide injection

TABLE 5. *Comparison of the effects of gastrointestinal peptides on satiety and gastric emptying*

Treatment	Induces Satiety	Inhibits Emptying
CCK-8 (5.6 µg/kg at 0 min)	Yes	Yes
Pentagastrin (100 µg/kg)	No	No
CCK-8 (5.6 µg/kg at −15 min)	Yes	No
Bombesin (8 µg/kg)	Yes	No
Secretin (14.3 µg/kg)	No	Yes

CCK-8, cholecystokinin octapeptide.

relative to test load, and sensitivity of the preparation used to measure gastric emptying, are all candidates. The current study was designed deliberately so that the protocols used to assess the effects of the peptide on emptying and satiety would be identical. Thus the currently demonstrated dissociation between the effects of bombesin on emptying and satiety demonstrate that changes in stomach emptying are not necessary for peptide-induced satiety. This conclusion with respect to bombesin is not novel, as investigators have already speculated that the mechanism underlying bombesin-induced satiety differs from that of CCK-induced satiety (15).

The secretin data demonstrate that changes in gastric emptying are not sufficient for the induction of satiety. With the use of our preparation, secretin inhibited gastric emptying in a dose-dependent manner, a finding that is consistent with several other reports in the literature relating secretin to gastric emptying (4, 11, 40). In our hands, a dose of 14.3 µg/kg secretin produced an inhibition of gastric emptying identical to that produced by a dose of 1.4 µg/kg CCK-8 (7). However, the critical observation is that the same dose of CCK-8 resulted in a significant inhibition of food intake, whereas 14.3 µg/kg secretin, a dose that retarded gastric emptying to a rate equivalent to that of 1.4 µg/kg CCK-8 (7), produced no satiety whatsoever. These results are consistent with previous reports that secretin does not affect meal size (11, 31).

In summary, these data demonstrate that changes in gastric emptying are neither necessary nor sufficient for peptide-induced satiety. The degree to which changes in gastric emptying contribute to the satiety action of CCK is still somewhat open. The observation that gastric emptying is neigher necessary nor sufficient for peptide-induced satiety does not suggest that in the case of CCK-8 there is no contribution of inhibition of gastric emptying to the satiety effect of the peptide. The current results suggest, however, that the changes in gastric emptying cannot represent the sole mechanism whereby CCK induces satiety in the rat and support other recent reports (7, 22).

In addition to independently verifying previous results, the present study provides new data that address the issue of how CCK-8 induces satiety. The finding that doses of secretin and CCK-8 matched for the inhibition of emptying have differing effects on feeding (*experiment 2*) suggests that comparison of the gastrointestinal responses to these peptides may provide a clue to the basis of CCK's satiety effect. Thus the proposal that contraction of the pyloric sphincter is critical (28) seems unlikely because both peptides induce constriction of the pylorus (12, 25). Moreover, the fact that these hormones produce similar decreases in intragastric pressure (9, 39) indicates that relaxation of the proximal stomach may not account for the behavioral difference. The hormones do vary in their affect on motility of isolated antral muscle, in that CCK-8 causes contraction (3, 14, 24), whereas secretin does not (3, 5, 32). However, antral contractions are not sufficient for the satiety effect of CCK, because gastrin also increases antral motility (3, 14, 32) but has no effect on feeding [*experiment 2*, (20)]. Similarly, the fact that

CCK and secretin have opposite effects on intestinal intraluminal pressure (17, 27) cannot account for the CCK-induced satiety, because gastrin produces a pressure increase comparable to CCK (35) without affecting behavior. While tentative, this analysis does not imply that emptying changes are central to feeding control by CCK. Possible mechanisms of CCK-induced satiety may involve a primary effect of the peptide increasing tone in the gastric wall, thereby stimulating vagal afferent fibers or, alternatively, a direct interaction between CCK and its receptors on afferent vagal nerve fibers (43).

We thank Drs. M. Ondetti and S. Luciana, Squibb Institute of Medical Research, for providing the CCK-8.

This research was supported by the Medical Research Council of Canada and the McMaster Science and Engineering Research Board. These studies represent part of a doctoral dissertation by K. L. Conover submitted to the Department of Psychology, McMaster University in partial fulfillment of requirements for the PhD degree.

Address for reprint requests: H. P. Weingarten, Dept. of Psychology, McMaster University, Hamilton, Ontario L8S 4K1, Canada.

Received 26 January 1988; accepted in final form 11 August 1988.

REFERENCES

1. ANIKA, M. S. Effects of cholecystokinin and caerulein on gastric emtpying. *Eur. J. Pharmacol.* 85: 195–199, 1982.
2. AVILON, A. A., J. D. FALASCO, G. P. SMITH, AND J. GIBBS. Proglumide antagonized the effect of exogenous CCK on gastric emptying, but does not antagonize the effects of L-phenylalanine or corn oil (Abstract). *Proc. Annu. Meet. East. Psychol. Assoc.* 57: 72, 1986.
3. CAMERON, A. J., M. B. SIDNEY, F. PHILLIPS, H. J. WILLIAM, AND D. M. SUMMERSKILL. Comparison of effects of gastrin, cholecystokinin-pancreozymin, secretin, and glucagon on human stomach muscle in vitro. *Gastroenterology* 59: 539–545, 1970.
4. CHEY, W. Y., S. HITANANT, J. HENDRICKS, AND S. H. LORBER. Effect of secretin and cholecystokinin on gastric emptying and gastric secretion in man. *Gastroenterology* 58: 820–827, 1969.
5. CHEY, W. Y., S. KOSAY, J. HENDRICKS, S. BRAVERMAN, AND S. H. LORBER. Effect of secretin on motor activity of stomach and Heidenhain pouch in dogs. *Am. J. Physiol.* 217: 848–852, 1969.
6. COLLINS, S., D. WALKER, P. FORSYTH, AND L. BELBECK. The effects of proglumide on cholecystokinin-, bombesin-, and glucagon-induced satiety in the rat. *Life Sci.* 32: 2223–2229, 1983.
7. CONOVER, K. L., S. M. COLLINS, AND H. P. WEINGARTEN. A comparison of cholecystokinin-induced changes in gastric emptying and feeding in the rat. *Am. J. Physiol.* 255 (*Regulatory Integrative Comp. Physiol.* 24): R21–R26, 1988.
8. CONOVER, K. L., H. P. WEINGARTEN, AND S. M. COLLINS. A procedure for within-trial repeated measurement of gastric emptying in the rat. *Physiol. Behav.* 39: 303–308, 1987.
9. DINOSO, V., W. Y. CHEY, J. HENDRICKS, AND S. H. LORBER. Intestinal mucosal hormones and motor function of the stomach in man. *J. Appl. Physiol.* 26: 326–329, 1969.
10. ELASHOFF, J. D., T. J. REEDY, AND J. H. MEYER. Analysis of gastric emptying data. *Gastroenterology* 83: 1306, 1982.
11. FALASCO, J. D., K. S. JOYNER, A. A. AVILON, J. GIBBS, AND G. P. SMITH. Secretin and bombesin decrease gastric emptying, but secretin does not decrease food intake (Abstract). *Proc. Annu. Meet. East. Psychol. Assoc.* 57: 11, 1986.
12. FISHER, R. S., W. LIPSHUTZ, AND S. COHEN. The hormonal regulation of pyloric sphincter function. *J. Clin. Invest.* 52: 1289–1296, 1973.
13. FORSYTH, P. A., H. P. WEINGARTEN, AND S. M. COLLINS. Role of oropharyngeal stimulation in cholecystokinin-induced satiety in the sham feeding rat. *Physiol. Behav.* 35: 539–543, 1985.
14. GERNER, T., AND J. F. W. HAFFNER. Pressure responses to CCK-PZ and gastrin in guinea-pig antrum and fundus in vitro. *Scand. J. Gastroenterol.* 13, *Suppl.* 49: 67, 1978.
15. GIBBS, J., AND G. P. SMITH. The neuroendocrinology of postprandial satiety. *Front. Neuroendocrinol.* 8: 223–245, 1984.

16. GIBBS, J., R. C. YOUNG, AND G. P. SMITH. Cholecystokinin decreases food intake in rats. *J. Comp. Physiol. Psychol.* 84: 488–495, 1973.
17. GUTIERREZ, J. G., W. Y. CHEY, AND V. P. DINOSO. Actions of cholecystokinin and secretin on the motor activity of the small intestine in man. *Gastroenterology* 67: 35–41, 1974.
18. KOULISHER, D., L. MORODER, AND M. DESCHODT-LANCKMAN. Degradation of cholecystokinin octapeptide, related fragments and analogs by human and rat plasma in vitro. *Regul. Pept.* 4: 127–139, 1982.
19. KULKOSKY, P. J., J. GIBBS, AND G. P. SMITH. Behavioral effects of bombesin administration in rats. *Physiol. Behav.* 28: 505–512, 1982.
20. LORENZ, D. N., G. KREIELSHEIMER, AND G. P. SMITH. Effect of cholecystokinin, gastrin, secretin and GIP on sham feeding in the rat. *Physiol. Behav.* 23: 1065–1072, 1979.
21. MCHUGH, P. R., AND T. H. MORAN. The stomach, cholecystokinin, and satiety. *Federation Proc.* 45: 1384–1390, 1986.
22. MORAN, T. H., AND P. R. MCHUGH. Gastric and nongastric mechanisms for satiety action of cholecystokinin. *Am. J. Physiol.* 254 (*Regulatory Integrative Comp. Physiol.* 23): R628–R632, 1988.
23. MORAN, T. H., AND P. R. MCHUGH. Cholecystokinin suppresses food intake by inhibiting gastric emptying. *Am. J. Physiol.* 242 (*Regulatory Integrative Comp. Physiol.* 11): R491–R497, 1982.
24. MORGAN, K. G., P. F. SCHMALZ, V. L. W. GO, AND J. H. SZURZEWSKI. Electrical and mechanical effects of molecular variants of CCK on antral smooth muscle. *Am. J. Physiol.* 235 (*Endocrinol. Metab.* 4): E324–E329, 1978.
25. PHAOSAWASDI, K., AND R. S. FISHER. Hormonal effects on the pylorus. *Am. J. Physiol.* 243 (*Gastrointest. Liver Physiol.* 6): G330–G335, 1982.
26. PORRECA, F., AND T. F. BURKS. Centrally administered bombesin affects gastric emptying and small and large bowel transit in the rat. *Gastroenterology* 85: 313–317, 1983.
27. RAMIREZ, M., AND J. T. FARRAR. The effect of secretin and cholecystokinin-pancreozymin on the intraluminal pressure of the jejunum in the unanesthetized dog. *Dig. Dis. Sci.* 15: 539–544, 1970.
28. ROBINSON, P. H., T. H. MORAN, M. GOLDRICH, AND P. R. MCHUGH. Development of cholecystokinin binding sites in rat upper gastrointestinal tract. *Am. J. Physiol.* 252 (*Gastrointest. Liver Physiol.* 15): G529–G534, 1987.
29. SCARPIGNATO, C., AND G. BERTACCINI. Bombesin delays gastric emptying in the rat. *Digestion* 21: 104–106, 1981.
30. SCARPIGNATO, C., B. MICALI, F. VITUALO, G. ZIMBARO, AND G. BERTACCINI. Inhibition of gastric emptying by bombesin in man. *Digestion* 23: 128–131, 1982.
31. SCHALLY, A. V., T. W. REDDING, AND H. W. LUCIEN. Enterogastrone inhibits eating by fasted mice. *Science Wash. DC* 157: 210–211, 1967.
32. SCHMALZ, P. F., AND G. W. BEELER. Motor responses of canine gastric antrum to pentagastrin and secretin. *Federation Proc.* 36: 557, 1977.
33. SMITH, G. P. Gut hormone hypothesis of postprandial satiety. In: *Eating and Its Disorders*, edited by A. J. Stunkard and E. Stellar. New York: Raven, 1984.
34. SMITH, G. P., AND J. GIBBS. Postprandial satiety. In: *Progress in Psychobiology and Physiological Psychology*, edited by J. M. Sprague and A. N. Epstein. New York: Academic, 8: 179–242, 1979.
35. STEWART, J. J., AND T. F. BURKS. Actions of pentagastrin on smooth muscle of isolated dog intestine. *Am. J. Physiol.* 239 (*Gastrointest. Liver Physiol.* 2): G295–G299, 1980.
36. SUGAWARA, K., J. ISAZA, J. CURT, AND E. R. WOODWARD. Effect of secretin and cholecystokinin on gastric motility. *Am. J. Physiol.* 217: 1633–1637, 1969.
37. TAYLOR, I. L., AND R. GARCIA. Effects of pancreatic polypeptide, caerulein, and bombesin on satiety in obese mice. *Am. J. Physiol.* 248: 277–280, 1985.
38. THOMPSON, J. C., H. R. FENDER, N. I. RAMUS, H. V. VILLAR, AND P. L. RAYFORD. Cholecystokinin metabolism in man and dogs. *Ann. Surg.* 182: 496–504, 1975.
39. VALENZUELA, J. E. Effect of intestinal hormones and peptides on intragastric pressure in dogs. *Gastroenterology* 71: 766–769, 1976.
40. VALENZUELA, J. E., AND C. DEFILIPPI. Inhibition of gastric emptying in humans by secretin, the octapeptide of cholecystokinin, and intraduodenal fat. *Gastroenterology* 81: 898–902, 1981.
41. WAGER-SRDAR, S. A., J. E. MORLEY, AND A. S. LEVINE. The effect of cholecystokinin, bombesin and calcitonin on food intake in virgin, lactating postweaning female rats. *Peptides* 7: 729–734, 1986.
42. WEINGARTEN, H. P., AND T. L. POWLEY. Gastric acid secretion of unanesthetized rats demonstrated with a new technique. *Lab Animal Sci.* 30: 673–680, 1980.
43. ZARBIN, M., J. WAMSLEY, R. INNIS, AND M. KUHAR. Cholecystokinin receptors: Presence and axonal flow in the rat vagus nerve. *Life Sci.* 29: 697–705, 1981.

Glossary

afferent - referring to peripheral neurons that carry signals to the CNS; as opposed to efferent neurons, which carry signals away from the CNS

antral muscle - a muscle of the stomach wall

bombesin - a gut peptide that produces a sustained suppression of eating at doses that produce only a transient inhibition of stomach emptying

catheter - a tube through which solutions can be drawn from the body

cholecystokinin (CCK) - a peptide that is released from the intestine and causes the pyloric sphincter to contract; CCK is also found in the brain

cholecystokinin octapeptide (CCK-8) - an 8-amino-acid form of the CCK molecule; because it is easier to synthesize than longer forms of the molecule, it is commonly used in CCK research

cimetidine - a drug that blocks gastric secretion

exogenous - from outside the body

gastric - pertaining to the stomach

gastric cannula - a tube implanted in the stomach

half life - the amount of time after an injection that it takes the level of a drug to fall to half of its peak level; the half life of a CCK injection in humans is only about 2 or 3 minutes

in vivo - in the living organism

intraluminal - inside the cavity or channel of a tubular organ such as the intestine

intraperitoneally - into the abdominal body cavity or peritoneum

kinetics - in this article, this term is used as a short form for pharmacokinetics, which refers to the time course of drug effects

pentagastrin - a gut peptide that at moderate doses influences neither stomach emptying nor eating

peptides - short chains of amino acids; many peptides such as CCK, pentagastrin, bombesin, and secretin are released from the gut; most peptides have also been found in the brain

pyloric sphincter - the ring of muscle that controls the flow of food from the stomach to the intestine; CCK causes the pyloric sphincter to constrict

pylorus - the passage from the stomach through which food exits to the intestine

saline lavage - a saline solution used for rinsing out a body cavity (e.g., the stomach)

satiety - the state that is presumed to motivate the cessation of eating; the opposite of hunger

secretin - a gut peptide that does not decrease food intake at a dose that inhibits stomach emptying

vagal - pertaining to the vagus nerve, a nerve that includes neurons that transmit signals from the gut to the brain and others that transmit signals from the brain to the gut

Essay Study Questions

1. What theory was tested by the experiments of Conover, Collins, and Weingarten?

2. Summarize the methods of Experiment 1.

3. Summarize the methods of Experiment 2.

4. What did Conover, Collins, and Weingarten find in their two experiments, and what did they conclude?

5. In view of the recent discovery that CCK can act directly on the brain, suggest an alternative interpretation of Conover et al.'s experiments.

Multiple-Choice Study Questions

1. Which of the following is not a peptide?
    a. secretin
    b. bombesin
    c. pentagastrin
    d. cimetidine
    e. cholecystokinin

2. CCK-8
    a. increases meal size.
    b. decreases meal size.
    c. increases the rate of gastric emptying.
    d. both a and c
    e. both b and c

3. CCK has a half-life in human plasma of about
    a. 30 minutes.
    b. 1 hour.
    c. 1.5 hours.
    d. 3 hours.
    e. 2 or 3 minutes.

4. Which of the following results of Conover, Collins, and Weingarten is consistent with the inhibition-of-gastric-emptying interpretation of CCK-produced satiety?
    a. CCK-8 reduced food consumption.
    b. CCK-8 reduced gastric emptying.
    c. Pentagastrin did not reduce food consumption or gastric emptying.
    d. all of the above
    e. none of the above

5. Which of the following results of Conover, Collins, and Weingarten was consistent with previous findings?
    a. their finding that CCK-8 reduces meal size
    b. their finding that CCK-8 retards gastric emptying
    c. their finding that bombesin reduces meal size
    d. their finding that secretin retards gastric emptying
    e. all of the above

6. The findings of Conover, Collins, and Weingarten strongly suggest that CCK-induced satiety is
    a. mediated by the effects of CCK on the CNS.
    b. not mediated totally by the effects of CCK on gastric emptying.
    c. mediated by the effects of CCK on afferent vagal fibers.
    d. both a and b
    e. both b and c

7. The results of Conover, Collins, and Weingarten prove that gastric emptying is
    a. not necessary for CCK-induced satiety.
    b. necessary for CCK-induced satiety.
    c. sufficient for CCK-induced satiety.
    d. both a and c
    e. both b and c

The answers to the preceding questions are on page 290.

Food-For-Thought Questions

1. How might CCK be used to augment a weight-control program?

2. As Experiment 2 turned out, the cimetidine conditions served little or no purpose. Why do you think that the authors included them?

3. We tend to think of the brain as the commander of the body. These two experiments illustrate the frequently overlooked point that signals originating in the gut can have a major impact on the brain and behavior. Discuss.

ARTICLE 10

Sleep on the Night Shift: 24-Hour EEG Monitoring of Spontaneous Sleep/Wake Behavior

L. Torsvall, T. Åkerstedt, K. Gillander, and A. Knutsson
Reprinted from Psychophysiology, 1989, volume 26, 352-357.

Shift work has been shown to disrupt sleep-wake cycles in a variety of ways, but the full impact of working a night shift has been difficult to assess because of the spontaneity and unpredictability of many of its consequences (e.g., on-the-job sleepiness and day-time naps). To more fully assess the consequences of shift work, Torsvall and his colleagues took continuous round-the-clock records of both scalp EEG activity and the EOG activity of 24 male papermill workers. These EEGs (electroencephalograms) and EOGs (electrooculograms) were recorded on portable medical tape-recorders, which the subjects carried with them for two 24-hour sleep-wake cycles, one during which they worked an afternoon shift and one during which they worked a night shift.

There were four major findings. First, the main period of morning sleep following a night shift was found, on the average, to be more than 2 hours shorter than the night sleep following an afternoon shift. Second, 7 of the 25 subjects supplemented their abbreviated sleep following a night shift with an afternoon nap; no subjects napped during the day following an afternoon shift. Third, the afternoon nap that followed night shifts contained a high proportion of slow-wave sleep. And fourth, 20% of the subjects fell asleep while working their night shift, during lulls in their work schedule. This 20% figure probably greatly underestimates the proportion of night shift workers who sleep on the job; the subjects in this study would have been much more likely than usual to resist the compulsion to sleep because they knew their activities were being monitored. In fact, 60% of the subjects admitted that they occasionally slept during their night shifts. This observation is worrying because many night-shift workers perform jobs that involve a great deal of responsibility for human life and well being.

These results indicate the advantages of round-the-clock recording in studies of sleep. They provide the first direct knowledge of the sleep patterns of people working night shifts.

Sleep on the Night Shift: 24-Hour EEG Monitoring of Spontaneous Sleep/Wake Behavior

LARS TORSVALL, TORBJÖRN ÅKERSTEDT,
KATJA GILLANDER, AND ANDERS KNUTSSON

Department of Stress Research, Karolinska Institute, Stockholm, Sweden

ABSTRACT

The present study sought to objectively describe the spontaneous sleep/wakefulness pattern of shift workers during a 24-hour period. Portable Medilog tape-recorders were used for ambulatory EEG monitoring of 25 male papermill workers (25–55 years) during days with night and afternoon work. The results showed that sleep after night work was two hours shorter than after afternoon work. The sleep reduction affected mainly Stage 2 and REM sleep while slow wave sleep was unchanged. In connection with night work 28% of the workers took a nap in the afternoon. These naps contained a large proportion of slow wave sleep and were, apparently, caused by the sleep deficit after the short main sleep period. The EEG recordings also revealed that 20% of the participants had sleep episodes *during* night work. These naps were as long as the afternoon naps, were experienced as "dozing offs" rather than naps, occurred at the time of the trough of the circadian wakefulness rhythm, and were concomitant with extreme subjective sleepiness and low rated work load. It was concluded that not only the sleep of shift workers was disturbed, but also the wakefulness—to the extent that sleepiness during night work sometimes reached a level where reasonable wakefulness could not be maintained. The latter observation is probably of special importance in work situations demanding a great responsibility for human lives or for great economic values.

DESCRIPTORS: Long-term ambulatory recording, EEG, Sleep/wakefulness pattern, Shift work, Naps, Subjective sleepiness.

Work schedules that include night work greatly interfere with the sleep/wakefulness pattern. Both survey and EEG studies demonstrate that morning sleep after a night shift is short and unrefreshing (Aanonsen, 1976; Knauth et al., 1980; Tepas, Walsh, & Armstrong, 1981; Torsvall, Åkerstedt, & Gillberg, 1981; Walsh, Tepas, & Moss, 1981; Tilley, Wilkinson, Warren, Watson, & Drud, 1982). A few survey studies have also revealed compensatory sleep, naps, during days with short main sleep periods (Tune, 1969; Evans, Cook, Cohen, Orne, & Orne, 1977; Åkerstedt & Torsvall, 1985). However, because of the spontaneous nature of these naps, they have never been documented in EEG studies. Furthermore, the major sleep episodes that have been studied have not been recorded under spontaneous conditions but have been more or less *scheduled* to occur at some "habitual" time of sleep.

This study was supported by grants from the Swedish Work Environment Fund.
Address requests for reprints to: L. Torsvall, Department of Stress Research, Box 60205, S-10401 Stockholm, Sweden.

Another major negative effect of night work is the increase in perceived sleepiness, particularly during work. This sleepiness, which is normally at its maximum during the last half of the night shift (Åkerstedt, Torsvall, & Gillberg, 1982), can lead to deterioration of performance (Browne, 1949; Bjerner, Holm, & Swensson, 1955; Hildebrandt, Rohmert, & Rutenfranz, 1974; Harris, 1977; Mitler et al., 1988), and also to falling asleep incidents during work (Prokop & Prokop, 1955; Kogi & Ohta, 1975; Åkerstedt, Torsvall, & Fröberg, 1983; Coleman & Dement, 1986).

The studies cited above were based on *reports* of sleepiness or sleep incidents. In a recent study, however, we made ambulatory EEG recordings of train drivers during work (Torsvall & Åkerstedt, 1987) and found sharply increased spectral density in the alpha (8–12 Hz) and theta (4–7.9 Hz) bands in the early morning. These changes covaried with increased subjective sleepiness, dozing off incidents, and performance errors. Sleep proper never developed, however, probably because of the drivers' awareness of the serious consequences of falling asleep, as well as because of the use of alertness maintenance devices. Presumably, sleep proper might be more likely to be observed in conventional industrial settings where the consequences of sleep are less immediate, less serious, and where wakefulness is less intensely enforced. The purpose of the present study was to investigate, with EEG methods, the spontaneous sleep/wake behavior over 24-hour periods in industrial shift workers.

Methods

Subjects

Twenty-five male papermill workers aged 25–55 years ($\overline{X}=38.6$) participated in the study. To encourage participation they received a modest economic compensation. The subjects were all free of alcohol and drugs before and during the recording days. The work involved handling of semi-automatic machinery and consisted of supervision (both directly and via visual display units, VDU) and minor repair work.

Procedure

The subjects worked a continuous 3-shift system (shift changes at 0600, 1400, and 2200 hours) with 2–4 days spent on each shift (Figure 1). During two 24-hour periods, one with night and one with afternoon work, the EEG (C_z–O_z) and EOG (oblique derivation) were recorded on a portable Medilog tape recorder. EMG recordings were not feasible because they interfered too much with normal functioning and social interaction. Except for 3 subjects, all recordings took place during the second night or afternoon shift. Half the subjects were recorded in the first half of the shift cycle and the rest in the second half. The electrodes were attached and the recording was started just prior to the work period. The subjects were encouraged to follow their normal habits of work, sleep, and leisure time activities. The equipment was removed prior to the next work period 24 hours later. The EEG recordings were scored by two of the authors (LT and TÅ—interrater reliability .79–.82) in accordance with the criteria of Rechtschaffen and Kales (1968).

The placement of the EEG electrodes was different from the recommended C_4–A_1. This compromise was necessary because only one EEG channel was available and both delta activity during sleep and alpha activity during waking needed to be optimized. However, C_z–O_z may actually be more informative than traditional derivations (Cooper, Osselton, & Shaw, 1980). The C_z–O_z derivation does, however, yield higher delta amplitudes (Dyson, Thornton, & Doré, 1984) and after extensive laboratory testing we decided to adjust amplitude criteria for delta activity to 100 μV because this yielded sleep stage data within a few percent of data obtained using the traditional derivation C_4–A_1.

During the two recording days the subjects provided subjective ratings of sleepiness at 3–5 hour intervals (7=very sleepy, 1=very alert). The first and second parts of the work period were also rated with respect to the percentage of time occupied by passive supervision. For a few days before and after the recording days a sleep diary was completed after awakening. Furthermore, the subjects filled out a small questionnaire about age and diurnal type (Torsvall & Åkerstedt, 1980), as well as two modified scales measuring rigidity of the sleeping habits and "vigor," i.e. the ability to overcome drowsiness (Folkard & Monk, 1979). The rigidity scale consisted of one item on the ease of sleeping at unusual times and another on the regularity of the sleeping habits (4=rigid, 1=flexible). Two items on the estimated effects of a night without sleep and the capacity to overcome drowsiness, respectively, made up the vigor scale (4=vigorous, 1=non-vigorous).

Differences between conditions were tested using paired t-tests. To adjust for the fact that a large number of tests were made, Bonferroni probabilities were used in interpreting all t-tests. As a result, conventional significance levels .05, .01, and .001 have been replaced by .0167, .0033, and .0003 respectively. The time of day pattern of sleepiness was tested using analysis of variance for repeated measurements (Winer, 1971). To correct for sphericity the p-values were based on degrees of freedom reduced by multiplication with the epsilon correction factor. However, for clarity, the original degrees of freedom are given in the text, together with the appropriate epsilon (ϵ) coefficient.

Results

The results from the sleep stage scoring of the EEG are shown in Table 1. The main sleep period after the night shift (day sleep) started around 06.40 ± 0.3 hours (mean ± 1SE), and after the afternoon shift (night sleep) around 23.27 ± 0.4 hours. The day sleep was reduced by more than two hours

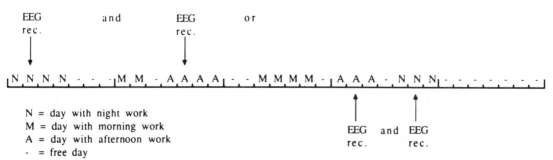

Figure 1. The shift schedule used at the company and the days on which EEG was recorded (indicated by arrows).

Table 1
Mean values for sleep variables during 24-hour periods in connection with night and afternoon shifts

	Mean Values (SDs in Parentheses)			
	Main Sleep During Days		Naps During Days With Night Shift	
Sleep Variables[a]	Afternoon Shift (n=25)	Night Shift (n=25)	Leisure Time (n=7)	Work Time (n=5)
TST, min	450 (15)	313 (17)***	44 (6)	43 (11)
sleep lat (stage 1), min	13 (3)	6 (1)*	7 (2)	—
lat SWS, min	25 (4)	27 (4)	12 (3)	28 (6)
lat REM, min	74 (7)	79 (4)	80 (0)	—
Wake, min	24 (5)	12 (2)*	6 (5)	6 (3)
Stage 1, min	52 (5)	26 (2)***	9 (5)	7 (2)
Stage 2, min	222 (8)	147 (9)***	17 (3)	20 (6)
SWS, min	83 (5)	77 (5)	17 (5)	12 (5)
REM, min	94 (7)	64 (7)**	1 (1)	—
Wake, %	5 (1)	7 (1)	8 (6)	10 (5)
Stage 1, %	11 (1)	8 (1)*	19 (9)	16 (3)
Stage 2, %	46 (1)	45 (2)	35 (5)	51 (7)
SWS, %	18 (1)	24 (1)***	36 (10)	23 (7)
REM, %	20 (1)	19 (2)	1 (1)	—
C1, SWS, min	36 (3)	37 (5)		
C2, SWS, min (n=21)	42 (3)	39 (4)		
C3, SWS, min (n=15)	18 (3)	24 (4)		
C1, REM, min	15 (2)	25 (3)*		
C2, REM, min (n=21)	27 (3)	25 (3)		
C3, REM, min (n=15)	21 (2)	20 (3)		

[a] TST = total sleep time, lat = latency, SWS = slow wave sleep (stages 3+4), REM = rapid eye movement sleep, C = sleep cycle.
*$p<.0167$, **$p<.0033$, ***$p<.0003$ (Bonferroni probabilities, paired t-tests, two-tailed), for differences between the main sleep periods of the two conditions.

compared with the night sleep. This affected particularly Stages 1, 2, and REM (rapid eye movement) sleep, and the time awake. The amount of slow wave sleep (Stages 3 and 4) was unchanged, although it made up a significantly greater proportion of the day sleep. For day sleep the sleep latency was only half as long as for night sleep, but there were no differences in latency to slow wave or REM sleep. When analysed by cycle it was found that day sleep contained significantly more REM sleep in cycle 1, whereas amounts of slow wave sleep did not differ.

During the day with afternoon work no naps were taken. In contrast, during the day with night work (Table 1) 7 subjects (28%) took a nap in the afternoon and 5 (20%) had sleep episodes *during* the night work period. One subject belonged to both nap groups. The leisure time nap started around 15.02 ± 0.5 hours, lasted for 44 minutes, and was dominated by Stage 2 and slow wave sleep. Only one napper reached the REM stage. The work time nap started around 02.50 ± 0.4 hours, lasted for 43 minutes, and was dominated by Stage 2 and slow wave sleep. None reached the REM stage. These naps ranged in length from 8–90 minutes. Figure 2

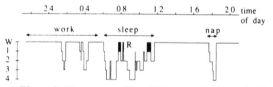

Figure 2. Hypnogram for a 24-hour ambulatory EEG recording of one worker during a day with night work.

shows the 24-hour pattern of sleep and wakefulness of one subject. It contains three short sleep episodes during night work, a short main sleep during the morning hours, and a leisure time nap around 1800 hours.

The seven leisure time nappers did not differ from the non-nappers (all subjects who did not take a leisure time nap, i.e. including work time nappers, n=18) with respect to age, diurnal type or vigor, but they scored higher on the rigidity scale (2.5 ± 0.3 vs. 1.8 ± 0.1; $t(23)=3.06$, $p<.05$) (t-tests with Bonferroni adjustments). These nappers had a significantly shorter prior main sleep after the night shift (249 ± 26 vs. 339 ± 18 min; $t(23)=2.81$, $p<.05$), but there were no significant differences in amounts of different sleep stages. There were no differences be-

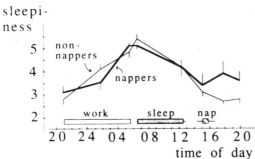

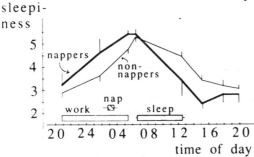

Figure 3. Mean (±1 SE) subjective ratings of sleepiness during a day with night work. Above: Comparison between leisure time nappers and non-nappers. Below: comparison between work time nappers and non-nappers.

tween the groups in rated amount of passive supervision during the night shift.

ANOVA for repeated measurements of sleepiness ratings (Figure 3) showed no overall difference between the leisure time nappers and the non-nappers ($F(1/23)=0.85$), nor was there any significant interaction ($F(7/161)=1.80$, $\epsilon=.63$). The overall time of day effect was, however, highly significant ($F(7/161)=25.62$, $p<.001$, $\epsilon=.63$).

Work time nappers did not differ significantly from non-nappers (all subjects who did not take a work time nap, i.e. including leisure time nappers, n=20) with respect to age, diurnal type, rigidity, or vigor. Neither were there any differences in sleep length or amounts of different sleep stages during the main sleep period after night work. Because the preceding 24-hour period was not recorded, no prior EEG sleep data were available. However, the sleep diaries showed that the sleep length of the nappers did not differ significantly from the non-nappers during the night before (305±35 vs. 341±31 min; $t(23)=0.72$) the recorded 24-hour night shift period. Furthermore, the work time nappers rated their night work as consisting of more passive supervision, especially during the first half of the night, although the difference did not reach significance (72±13% vs. 40±6% of the time; $t(23)=2.35$).

ANOVA for repeated measurements (Figure 3) showed no significant difference between work time nappers and non-nappers in rated sleepiness ($F(1/23)=0.05$), but a highly significant effect of time of day ($F(7/161)=24.78$, $p<.001$, $\epsilon=.63$), and also a significant interaction effect ($F(7/161)=3.01$, $p<.05$, $\epsilon=.63$). The latter was evidenced in greater sleepiness in the nappers during work hours and lower sleepiness during leisure time.

After the data were collected the subjects were interviewed about their experience of sleep during work. Sixty percent had experienced sleep during night shifts and 88% of these subjects estimated the length of such episodes to be less than 5 minutes. All of them reported that sleep occurred during periods when the work load was low and while sitting. They all also reported a struggle against sleep before finally succumbing to it.

Discussion

In the present study of spontaneous sleep/wake behavior the amount and composition of the main sleep during the days with afternoon and night work confirm the results of previous studies (Torsvall et al., 1981; Walsh et al., 1981; Tilley et al., 1982). That is, sleep after the night shift was reduced by more than two hours, the loss affecting mainly Stage 2 and REM sleep, but not slow wave sleep. Because slow wave sleep is influenced mainly by prior sleep loss/time awake (Webb & Agnew, 1972; Borbély, Baumann, Brandeis, Strauch, & Lehmann, 1981; Åkerstedt & Gillberg, 1986), a certain increase in slow wave sleep during the day might have been expected. Presumably, this increase was prevented by the early curtailment of the sleep episode. Interestingly, REM amounts during day sleep were increased during cycle 1, but were significantly decreased when the entire sleep episode was considered. These effects were, very likely, due to the well-known morning enhancement of REM sleep (Czeisler, Weitzman, Moore-Ede, Zimmerman, & Knauer, 1980) together with the similarly well-known dependence of REM (and Stage 2) on the length of the sleep episode (Agnew & Webb, 1973).

Twenty-eight percent of the participants added a spontaneous afternoon nap to the main sleep after night work. This agrees with earlier data on self-reported napping prevalence in shift work (Tune, 1969; Åkerstedt & Torsvall, 1985). The reason for the nap appears to be the shorter prior main sleep length. Indeed, if the nap length is added to the length of the main sleep the significant difference between non-nappers and leisure time nappers disappears (although total sleep time is still much shorter after the night shift than after the afternoon shift for both groups). This compensatory view of napping is also supported by the fact that the afternoon naps contained large amounts of slow wave sleep. The reason for the short main sleep length is not clear, although the higher ratings of sleep rigidity suggest a greater need for stable sleep habits in these subjects.

Another influence on napping is the circasemidian rhythmicity of alertness that frequently causes a temporary increase in sleepiness in the afternoon (Broughton, 1975; Richardson, Carskadon, Orav, & Dement, 1982; Dinges, in press). The ratings suggest that the nap group experienced a more pronounced alertness dip than the non-nappers. This does not necessarily mean, however, that their circasemidian rhythm is more pronounced. The alertness dip may merely have been enhanced by the sleep deficit (Mitler et al., 1988).

Moreover, Broughton (1975) has proposed that the afternoon temporary increase in sleepiness represents a pressure for slow wave sleep and that the afternoon nap is a 12-hour harmonic of the circadian nocturnal slow wave sleep activity. Sleep debt would increase this intensity. Support for this theory has come from several studies (Weitzman, Czeisler, Zimmerman, & Ronda, 1980; Gagnon, De Koninck, & Broughton, 1985; Zulley & Campbell, 1985), and the large amount of slow wave sleep in the afternoon naps in the present study could possibly be an expression of the same phenomenon. On the other hand, Dijk, Beersma, and Daan (1987), in a recent study of napping and homeostatic sleep regulation, failed to support such a notion.

The most striking finding about napping, however, was that 20% of the workers showed incidents of proper sleep *during* night work. Actually, this may be an underestimation of the actual figures because the awareness of the tape recorder may have influenced the results. It should be emphasized that sleeping on the job was not condoned by the company, and could theoretically be a reason for dismissal. Interestingly, the frequency of napping in the present group is close to the proportion of train drivers reporting dozing off during night driving (Torsvall & Åkerstedt, 1987).

The work time naps occurred during the second half of the night shift when the sleepiness rhythm was approaching its peak. The nappers also showed a phase advanced sleepiness rhythm. The reason for this earlier sleepiness may have been of circadian origin, although the diurnal type test does not support this. Neither does prior sleep loss appear to have been involved, although this observation is based only on ratings and not on EEG data. An alternative explanation (although in the present study just short of significance) may be understimulation through a slackening of work load. This should, of course, be seen against the background of the more general night shift sleepiness due to sleep loss and circadian influences.

Also, the interviews indicated that sleep events practically always occurred during periods of low work load, while the subject was sitting. Interestingly, the sleep episodes were reported to be always preceded by a struggle against severe sleepiness, i.e., most workers were aware of their state before falling asleep. This is supported by the associated high sleepiness ratings. Thus, the workers seemed to be aware that they were going to fall asleep. The latter also suggests that the sleep episodes may have been more or less intentional. On the other hand, these states of severe sleepiness could last for extended periods of time and could equally well terminate without any sleep occurring. Thus, sleep was a very uncertain outcome of sleepiness. Similarly, questionnaire data in another study have shown that reports of extreme sleepiness were twice as frequent as reports of falling asleep (Åkerstedt et al., 1983).

Furthermore, in the present study the length of sleep incidents was grossly underestimated by the subjects, which suggests a reduced awareness during sleepiness. Impaired awareness during sleepiness has been described by several authors, particularly in relation to narcolepsy (Aguirre, Broughton, & Stuss, 1985; Guilleminault & Dement, 1977) or night driving (O'Hanlon & Kelley, 1977; Lisper, Laurell, & van Loon, 1986). We have, ourselves, observed consistent underestimation of dozing off episodes in laboratory studies (Torsvall & Åkerstedt, 1988). One mechanism behind these effects may be that sleep must be established for several minutes before it may be perceived as sleep (Sewitch, 1984). Taken together, the observations suggest that, although severe sleepiness is associated with a high risk of falling asleep, it is not an absolute indication of imminent sleep—particularly not in individuals who, because of night work, are compelled to operate for long periods of time at, or close to, this sleepiness level.

In summary, the present data have demonstrated that sleep in connection with night work occurs, not only immediately *after* the night shift, but also in the afternoon during leisure time and, importantly, also *during work* itself. It appears that circadian rhythmicity, sleep loss, and possibly passive work tasks, induce sleepiness during work to the extent that it reaches levels at which reasonable wakefulness is difficult to maintain and sleep ensues. This observation is probably of particular significance in work situations that require responsibility for human lives or great economic values.

REFERENCES

Aanonsen, A. (1976). *Shift work and health.* Oslo: Universitetsforlaget.

Agnew, H.W., & Webb, W.G. (1973). The influence of time course variables on REM sleep. *Bulletin of the Psychonomic Society, 2,* 131-133.

Aguirre, M., Broughton, R., & Stuss, D. (1985). Does memory dysfunction exist in narcolepsy-cataplexy? *Journal of Clinical and Experimental Neuropsychology, 7,* 14-24.

Åkerstedt, T., & Gillberg, M. (1986). A dose-response study of sleep loss and spontaneous sleep termination. *Psychophysiology, 23,* 293-297.

Åkerstedt, T., & Torsvall, L. (1985). Napping in shift work. *Sleep, 8,* 105-109.

Åkerstedt, T., Torsvall, L., & Fröberg, J. (1983). A questionnaire study of sleep/wake disturbances and irregular work hours. *Sleep Research, 12,* 358.

Åkerstedt, T., Torsvall, L., & Gillberg, M. (1982). Sleepiness and shift work: Field studies. *Sleep, 5*(Suppl. 2), 95-106.

Bjerner, B., Holm, Å., & Swensson, Å. (1955). Diurnal variation of mental performance: A study of three-shift workers. *British Journal of Industrial Medicine, 12,* 103-110.

Borbély, A.A., Baumann, F., Brandeis, D., Strauch, I., & Lehmann, D. (1981). Sleep deprivation: Effects on sleep stages and EEG power density in man. *Electroencephalography & Clinical Neurophysiology, 51,* 483-493.

Broughton, R. (1975). Biorhythmic variations in consciousness and psychological functions. *Canadian Psychological Review, 16,* 217-239.

Browne, R.C. (1949). The day and night performance of teleprinter switchboard operators. *Occupational Psychology, 23,* 121-126.

Coleman, R.M., & Dement, W.C. (1986). Falling asleep at work: A problem for continuous operations. *Sleep Research, 15,* 265.

Cooper, R., Osselton, J.W., & Shaw, J.C. (1980). *EEG technology* (p. 159). London: Butterworths.

Czeisler, C.A., Weitzman, E.D., Moore-Ede, M.C., Zimmerman, J.C., & Knauer, R.S. (1980). Human sleep: Its duration and organization depend on its circadian phase. *Science, 210,* 1264-1267.

Dijk, D.J., Beersma, D.G.M., & Daan, S. (1987). EEG power density during nap sleep: Reflection of an hourglass measuring the duration of prior wakefulness. *Journal of Biological Rhythms, 2,* 207-219.

Dinges, D.F. (in press). The nature of sleepiness: Causes, contexts and consequences. In A. Baum & A.S. Stunkard (Eds.), *Eating, sleeping and sex: Perspectives on behavioral medicine.* Hillsdale, NJ: Lawrence Erlbaum.

Dyson, R.J., Thornton, C., & Doré, C.J. (1984). EEG electrode positions outside the hairline to monitor sleep in man. *Sleep, 7,* 180-188.

Evans, F., Cook, M., Cohen, H., Orne, E., & Orne, M. (1977). Appetitive and replacement naps: EEG and behavior. *Science, 197,* 687-689.

Folkard, S., & Monk, T.H. (1979). Towards a predictive test of adjustment to shift work. *Ergonomics, 22,* 79-91.

Gagnon, P., De Koninck, J., & Broughton, R. (1985). Reappearance of electroencephalogram slow waves in extended sleep with delayed bedtime. *Sleep, 8,* 118-128.

Guilleminault, C., & Dement, W.C. (1977). Amnesia and disorders of excessive sleepiness. In R.R. Drucker-Colin & J.L. McGaugh (Eds.), *Neurobiology of sleep and memory* (pp. 439-456). New York: Academic Press.

Harris, W. (1977). Fatigue, circadian rhythms and truck accidents. In R.R. Mackie (Ed.), *Vigilance* (pp. 133-146). New York: Plenum Press.

Hildebrandt, G., Rohmert, W., & Rutenfranz, J. (1974). 12 and 24 hour rhythms in error frequency of locomotive drivers and the influence of tiredness. *International Journal of Chronobiology, 2,* 175-180.

Knauth, P., Landau, K., Dröge, C., Schwitteck, M., Widynski, M., & Rutenfranz, J. (1980). Duration of sleep depending on the type of shift work. *International Archives of Occupational and Environmental Health, 46,* 167-177.

Kogi, K., & Ohta, T. (1975). Incidence of near accidental drowsing in locomotive driving during a period of rotation. *Journal of Human Ergology, 4,* 65-76.

Lisper, H.O., Laurell, H., & van Loon, J. (1986). Relation between time to falling asleep behind the wheel on a closed truck and changes in subsidiary reaction time during prolonged driving on a motorway. *Ergonomics, 29,* 445-453.

Mitler, M.M., Carskadon, M.A., Czeisler, C.A., Dement, W.C., Dinges, D.F., & Graeber, R.C. (1988). Catastrophes, sleep, and public policy: Consensus report. *Sleep, 11,* 100-109.

O'Hanlon, J.F., & Kelley, G.R. (1977). Comparison of performance and physiological changes between drivers who perform well and poorly during prolonged vehicular operation. In R.R. Mackie (Ed.), *Vigilance* (pp. 189-202). New York: Plenum Press.

Prokop, O., & Prokop, L. (1955). Ermüdung und Einschlafen am Steuer. *Zentralblatt für Verkehrs-Medizin, Verkehrs-Psychologie und Angrenzende Gebiete, 1,* 19-30.

Rechtschaffen, A., & Kales, A. (1968). *A manual of standardized terminology, techniques, and scoring system for sleep stages of human subjects.* Los Angeles: University of California, Brain Information Service/Brain Research Institute.

Richardson, G.S., Carskadon, M.A., Orav, E.J., & Dement, W.C. (1982). Circadian variation of sleep tendency in elderly and young adult subjects. *Sleep, 5,* S82-S94.

Sewitch, D.E. (1984). NREM sleep continuity and the sense of having slept in normal sleepers. *Sleep, 7,* 147-154.

Tepas, D.I., Walsh, J.K., & Armstrong, D.R. (1981). Comprehensive study of the sleep of shift workers. In L.C. Johnson, D.I. Tepas, W.P. Colquhoun, & M.J. Colligan (Eds.), *Biological rhythms, sleep and shift work* (pp. 347-356). New York: SP Medical & Scientific Books.

Tilley, A.J., Wilkinson, R.T., Warren, P.S.G., Watson, W.G., & Drud, M. (1982). The sleep and performance of shift workers. *Human Factors, 24,* 624-641.

Torsvall, L., & Åkerstedt, T. (1980). A diurnal type scale: Construction, consistency, and validation in shift work. *Scandinavian Journal of Work Environment and Health, 6,* 283-290.

Torsvall, L., & Åkerstedt, T. (1987). Sleepiness on the job: Continuously measured EEG changes in train drivers. *Electroencephalography & Clinical Neurophysiology, 66,* 502-511.

Torsvall, L., & Åkerstedt, T. (1988). Extreme sleepiness: Quantification of EOG and spectral EEG parameters. *International Journal of Neuroscience, 38,* 435-441.

Torsvall, L., Åkerstedt, T., & Gillberg, M. (1981). Age, sleep and irregular work hours: A field study with EEG recording, catecholamine excretion, and self-ratings. *Scandinavian Journal of Work Environment and Health, 7,* 196-283.

Tune, G.S. (1969). Sleep and wakefulness in a group of shift workers. *British Journal of Industrial Medicine, 26,* 54-58.

Walsh, J.K., Tepas, D.I., & Moss, P.D. (1981). The EEG sleep of night and rotating shift workers. In L.C. Johnson, D.I. Tepas, W.P. Colquhoun, & M.J. Colligan (Eds.), *Biological rhythms, sleep and shift work* (pp. 371-381). New York: SP Medical & Scientific Books.

Webb, W.B., & Agnew, H.W. (1972). Stage 4 sleep: Influences of time course variables. *Science, 174,* 1354-1356.

Weitzman, E.D., Czeisler, C.A., Zimmerman, J.C., & Ronda, J.M. (1980). Timing of REM and stage 3 and 4 during temporal isolation in man. *Sleep, 2,* 391-408.

Winer, B.J. (1971). *Statistical principles in experimental design.* New York: McGraw-Hill.

Zulley, J., & Campbell, S.S. (1985). Napping behavior during "spontaneous internal desynchronization": Sleep remains in synchrony with body temperature. *Human Neurobiology, 4,* 123-126.

(Manuscript received July 1, 1987; accepted for publication September 15, 1988)

Glossary

alpha activity - rhythmic 8-to-12-Hz EEG waves; alpha waves can often be recorded from the scalp during relaxed wakefulness

ambulatory - walking about

circadian - about one day

circasemidian - about half a day

C_z, O_z, C_4, and A_1 - these are all conventional sites for the placement of EEG electrodes; for example, C_z is on the midline at the very top of the head; certain types of scalp EEG waves tend to have a greater amplitude at some sites than at others

delta activity - the largest, slowest EEG waves; delta waves are defined as those between 0.5 and 3.5 Hz; delta waves occur during stage 3 and stage 4 sleep

diurnal - about one day; circadian

EEG - electroencephalogram

EOG - electrooculogram; a psychophysiological measure of eye movement; the EOG is recorded through electrodes taped to the skin near the eye

hypnogram - a graphic representation of the various stages (wakefulness, 1, 2, 3, and 4) of a single night's sleep of one subject

Hz - herz; cycles per second

narcolepsy - a type of sleep disorder; narcoleptics experience sudden uncontrollable attacks of sleepiness, and thus they frequently fall asleep at inopportune times during the day

REM sleep - REM sleep (i.e, rapid-eye-movement sleep) is a period of sleep during which there are rapid eye movements under the closed eyelids; REM sleep is associated with Stage 1 EEG

slow-wave sleep - refers to both Stage 3 sleep EEG, which has some slow EEG waves, and Stage 4 sleep EEG, which has a predominance of them

Stage 1 sleep - any period of sleep characterized by low-amplitude, high-frequency activity, which is similar to that of wakefulness; typically, all but the first period of Stage 1 EEG during a night's sleep are accompanied by rapid eye movements

theta - moderately slow EEG waves, between 4 and 7 Hz; scalp theta waves occur in a variety of circumstances, but they are particularly prominent during drowsiness

Essay Study Questions

1. In what way was the study of Torsvall and his collaborators an advance over previous research on the sleep patterns of night-shift workers?

2. What measures were recorded in addition to the EEG and EOG?

3. Describe the work schedule of the subjects. During which 24-hour periods were the subjects studied?

4. What effect did working the night shift have on the ensuing main sleep of the subjects?

5. What did this study suggest about the incidence of sleeping on the night shift?

Multiple-Choice Study Questions

1. In a recent study, Torsvall and Åkerstedt found that the waking scalp EEG recordings of night-shift train drivers had
    a. more alpha and theta waves in the early morning.
    b. fewer alpha and theta waves in the early morning.
    c. more alpha and theta waves in the late afternoon.
    d. fewer alpha and theta waves in the late afternoon.
    e. both b and c

2. Following a night shift, the main sleep of the subjects
    a. was longer than usual.
    b. was shorter than usual.
    c. had a greater proportion of slow-wave sleep.
    d. both a and c
    e. both b and c

3. Both the leisure-time and on-the-job naps of the night-shift workers contained little or no
    a. Stage 3 sleep.
    b. Stage 4 sleep.
    c. REM sleep.
    d. Stage 2 sleep.
    e. both a and b

4. There was a tendency, albeit not a statistically significant one, for night shift workers who slept on the job to
    a. have jobs that were more demanding.
    b. have jobs that were more passive.
    c. be older.
    d. be more sleepy during the day.
    e. have had substantially less sleep during the previous night.

5. Perhaps the most important practical observation made by Torsvall and his collaborators was that
    a. nappers were more effective workers than non-nappers
    b. non-nappers were more effective workers than nappers.
    c. night-shift workers sleep more than day-shift workers.
    d. night-shift workers tend to sleep on the job, even when they know that their activities are being monitored.
    e. the on-the-job night-shift naps typically lasted only a few seconds, but the subjects thought that they lasted several minutes.

6. The average duration of each of the five on-the-job naps observed by Torsvall and his collaborators was about
    a. 30 seconds.
    b. 2 minutes.
    c. 4 minutes.
    d. 9 minutes.
    e. 43 minutes.

The answers to the preceding questions are on page 290.

Food-For-Thought Questions

1. This study illustrates how portable recording devices can be used to study every-day biopsychological phenomena that cannot be studied in the laboratory. Because this is one of the first efforts to directly measure the effects of night work on subsequent spontaneous sleep patterns, it raises several interesting questions. Suggest a study that might be done to answer one of these questions.

2. It is not clear to what extent the sleep patterns of the subjects were attributable to their preceding night shift as opposed to the fact that their shifts were changing every few days. Design a study to differentiate between these two possibilities.

ARTICLE 11

Transplanted Suprachiasmatic Nucleus Determines Circadian Period

M.R. Ralph, R.G. Foster, F.C. Davis, and M. Menaker

Reprinted from Science, 1990, volume 247, 975-978

Many aspects of behavior and of nervous system activity follow a circadian cycle, that is, they display a pattern of change that repeats itself about every 24 hours (e.g., the sleep-wake cycle). Under natural conditions, the timing of circadian cycles is controlled by regular cues from the environment, most notably by the daily cycle of light and dark. However, the fact that circadian rhythms occur even in animals that have lived from birth in laboratory environments that have been kept constant (e.g., under continuous illumination) indicates that the brain has its own circadian timing mechanism that can function in the absence of external temporal cues. This neural circadian timing mechanism is thought to be located in the suprachiasmatic nucleus (SCN), which is located near the midline of the brain in the floor of the third ventricle, just above the optic chiasm.

Ralph and his colleagues ablated (lesioned) the SCNs of two kinds of hamsters: (1) normal hamsters, which typically display circadian activity rhythms of about 24 hours when living under constant laboratory conditions, and (2) mutant hamsters, which display substantially shorter circadian activity rhythms when living under constant laboratory conditions. These ablations abolished the circadian activity rhythms of both kinds of hamsters. Then 3 or 4 weeks later, Ralph and his colleagues surgically removed the SCNs and surrounding tissue from hamster fetuses (the donors) and injected them into the third ventricle of the hamsters with SCN lesions (the hosts) just adjacent to the lesion site. In some cases, transplants were made from normal hamsters to mutants, and in others, they were made from mutants to normals. These neurotransplants restored the circadian activity rhythms of the hosts, and amazingly, hamsters receiving an SCN transplant from a normal hamster fetus displayed 24-hour activity cycles, whereas those receiving transplants from a mutant fetus displayed shorter cycles.

This experiment provides convincing evidence that the timing of the circadian activity cycle is controlled by tissue in the SCNs or in their immediate vicinity. It also provides an illustration of the great potential of neurotransplantation as a medical and scientific tool. Another recent study illustrates this latter point in a particularly interesting way; Balaban, Teillet, and LeDouarin (Science, 1989, 241, 1339-1342), removed a section of the brainstem of a chick fetus and replaced it with the corresponding segment of the quail fetus brainstem. When the chicks were born, they crowed like quails.

SCIENCE

Reprint Series
23 February 1990, Volume 247, pp. 975–978

Copyright © 1989 by the American Association for the Advancement of Science

Transplanted Suprachiasmatic Nucleus Determines Circadian Period

MARTIN R. RALPH,* RUSSELL G. FOSTER, FRED C. DAVIS, MICHAEL MENAKER

The pacemaker role of the suprachiasmatic nucleus in a mammalian circadian system was tested by neural transplantation by using a mutant strain of hamster that shows a short circadian period. Small neural grafts from the suprachiasmatic region restored circadian rhythms to arrhythmic animals whose own nucleus had been ablated. The restored rhythms always exhibited the period of the donor genotype regardless of the direction of the transplant or genotype of the host. The basic period of the overt circadian rhythm therefore is determined by cells of the suprachiasmatic region.

THERE IS CONSIDERABLE EVIDENCE to suggest that the suprachiasmatic nucleus (SCN) of the hypothalamus is the site of circadian pacemaker cells that generate overt circadian rhythms in mammals. The evidence that supports this view is diverse. (i) The SCN is the target of direct and indirect retinal projections required for entrainment of circadian rhythms to environmental cycles (1, 2). (ii) The SCN exhibits strong circadian rhythms of glucose utilization in vivo (3). (iii) Ablation of the SCN or its surgical isolation within the brain eliminates overt behavioral rhythmicity (4–6) and rhythmic electrical activity in the brain (7). (iv) Tissue explants containing the SCN continue to express circadian rhythms in electrical activity (8, 9) and vasopressin release (10) in vitro. (v) Circadian rhythmicity can be restored to SCN-lesioned arrhythmic hosts by implantation of fetal brain tissue containing SCN cells (11–14).

Despite this evidence, however, the pacemaker role of the SCN circadian oscillator has not been confirmed. In addition, the role of the nucleus has come into question because methamphetamine given on a longterm basis to arrhythmic, SCN-lesioned rats will restore circadian rhythmicity (15). Moreover, in the rat (16) and in lower vertebrates (17), structures outside the SCN are able to generate circadian rhythms.

Although in the aggregate the evidence is compelling, final proof that the SCN is the site of a central driving oscillator for mammalian circadian systems requires that characteristics of the overt rhythm such as phase and period be unambiguously attributable to the activity of SCN cells. The discovery of the τ mutation in hamsters provided the opportunity to test directly the pacemaker role of the SCN by tissue transplantation. The mutation has the primary behavioral effect of reducing the period of the circadian rhythm from 24 hours to about 22 hours in heterozygotes and to about 20 hours in homozygotes (18). If the SCN drives overt behavioral rhythmicity in hamsters, then the period of the rhythm that is restored by SCN transplantation should reflect the genotype of the donor tissue and not that of the lesioned host.

All animals used in these experiments were raised in our colony, and only male animals were used as hosts. These were placed in running wheel cages for activity recording after reaching 8 weeks of age and were kept in constant dim light or constant dark for the duration of the experiment. After the period of the host rhythms had been established (7 to 21 days), animals were anesthetized and placed in a Kopf model 900 stereotaxic instrument for SCN ablation. Lesions were made by current in-

---

M. R. Ralph, R. G. Foster, M. Menaker, Department of Biology, University of Virginia, Charlottesville, VA 22903.
F. C. Davis, Department of Biology, Northeastern University, Boston, MA 02115.

*To whom correspondence should be addressed at Department of Psychology, University of Toronto, Toronto, Canada M5S 1A1.

jection (4 mA for 10 s) through a platinum/iridium (90/10) wire electrode insulated except for 0.3 mm at the tip (electrode placement: anteroposterior, +0.6 mm from bregma; dorsoventral, 8.4 mm from the top of the skull; tooth bar, −2 mm). Functional ablation of the SCN was determined by visual analysis of the subsequent locomotor activity record. As is common with this type of lesion, ultradian rhythmicity persisted in many records after SCN ablation. Lesions were considered to be functionally complete if 24-hour periodicity was not visible in the activity record. The location and extent of the lesion was determined later during histological examination.

In most cases, transplants were performed between 3 and 4 weeks after SCN ablation. When transplants involved hosts and donor tissue with the same genotype, the operation was performed at least 3 weeks after ablation so that there would be ample opportunity to identify rhythmicity that persisted in incompletely lesioned animals. In four cases, in which transplants involved donor and host of different genotype, the implantation was performed 1 week after SCN ablation to make a preliminary assessment of whether the timing of the two operations influenced the success rate of the transplantation procedure.

Fetal tissue was obtained on embryonic day 13.5 (day 1 is the first 24 hours after conception) from donors of known circadian genotype. Pregnant females were anesthetized with sodium pentobarbital, and fetuses were removed and decapitated immediately. The brains were removed quickly and placed in culture medium [60% Eagle's minimal essential medium, with Earle's salts; 40% Hanks balanced salt solution (MEM-BSS)] maintained at 36°C. After all of the brains had been collected, blocks of tissue containing either SCN or control tissue (cortex) were excised and placed in separate dishes (19). For implantation, the SCN from two donors were drawn into a 5-μl Wiretrol micropipette (Drummond Scientific, Broomall, Pennsylvania) for stereotaxic placement in the third ventricle of the host near the site of the SCN lesion. This site was chosen to be consistent with tissue placement in other SCN transplant studies. The total volume of material implanted was about 1 μl (20).

Circadian rhythmicity was restored unambiguously in about 80% of the arrhythmic hosts that received SCN implants. Only rhythmicity that was visible to naïve observers of the raw activity data is presented in this report. Rhythmicity was not restored in animals that received cortical tissue implants ($n = 4$), although apparently healthy implants were found later during immunocytochemical analysis. Time series (fast Fourier) analysis was used to confirm the presence of rhythmicity restored by SCN transplantation and to confirm the absence of donor rhythmicity in the activity of control animals. This analysis indicated the presence of residual rhythmicity in some animals after SCN lesions. Because of the difference in period length between host and donor phenotype, this residual rhythmicity did not interfere with the interpretation of the rhythmicity restored by transplantation.

The period of restored rhythms always matched that predicted by the genotype of the donor tissue. Examples of rhythmicity restored by transplantation are shown in Fig. 1. When wild-type tissue was used, the period of the restored rhythm was always about 24 hours; heterozygous donor tissue always produced rhythms with periods close to 22 hours; and homozygous mutant tissue always produced periods close to 20 hours. This result was obtained regardless of the genotype of the host. No influence of the host genotype on the period of rhythms could be detected.

Our data for SCN transplants between different genotypes are summarized in Fig. 2. For all combinations of donor and host, the restored period was significantly different from that of the host ($P < 0.01$), and for each genotype of donor tissue the periods of the restored rhythms always fell within the range shown by intact adults of that genotype.

Perhaps the most parsimonious interpretation of our results, when taken together with the work of others (1–14), is that the SCN contains cells that are not only able to generate oscillations with a circadian period, but are also responsible for the generation of overt behavioral circadian rhythms in intact mammals. Al-

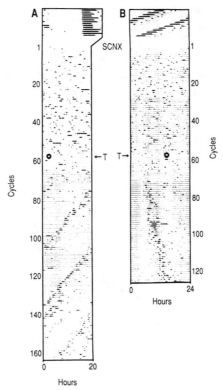

Fig. 1. (**A**) Expression of homozygous mutant rhythmicity in a wild-type host. The endogenous rhythm of the intact host is shown at the top of the activity record (period, 24.05 hours). SCNX is SCN ablation; at this point, the plotting interval was changed from 24 hours to 20 hours to help visualize rhythmicity that was restored later by neural transplant. Implantation of fetal SCN tissue was performed on the day indicated by a T at the time indicated by the circle. The period of the restored rhythm in this example was 19.5 hours. (**B**) Expression of wild-type rhythmicity in a heterozygous mutant host. The period of the host rhythm was 21.7 hours. Transplantation resulted in the restoration of a 24.2-hour rhythm seen in the lower third of the record.

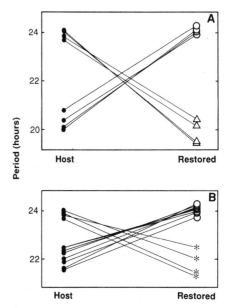

Fig. 2. Reciprocal transplantation of SCN tissue between wild-type and mutant animals. Periodicity was determined from eye-fit lines drawn through activity onsets on at least 20 consecutive days of data and was confirmed later with time series analysis. To reduce measurement error, data were plotted at intervals close to their free-running period before the final period determination was made. This procedure reduced the inherent variability of the measurement to less than 1% for a single determination of period. For each host, the endogenous rhythm (left) was eliminated by SCN ablation and restored by SCN implants (right). The range of period of the intact adult population for each genotype is indicated by vertical shaded bars (right axis). (**A**) Reciprocal transplants between wild-type and homozygous mutants. (**B**) Reciprocal transplants between wild-type and heterozygous mutants. Symbols represent the following: (●) host; (○) SCN tissue from wild-type donor; (*) SCN tissue from heterozygous donor; and (△) SCN tissue from homozygous mutant donor.

though behavioral patterns can be modified by transplantation of large brain regions early in the development of some vertebrates (21), our work shows that a discrete behavioral pattern can be transplanted with a neural gift of limited size and well-defined function.

After behavioral observation we attempted to determine the extent of SCN lesions, the amount of extra-SCN donor tissue within the implant, and whether neural connections had been established between the host brain and implant. For each animal examined (22) (n = 16) in which rhythmicity had been restored, the SCN lesion appeared complete [no evidence of the host SCN when antisera directed against vasoactive polypeptide (VIP), vasopressin, or neuropeptide Y (NPY) were used], and a plug of donor tissue was found within the third ventricle. These plugs were always found in close apposition to the ependymal wall. VIP-positive perikarya and fibers were always identified within these implants, and in most cases cells and fibers formed a discrete "ball" (Fig. 3, A and B) reminiscent of the organization of VIP within the SCN. We could never clearly trace VIP fibers extending from the graft and crossing the host-graft border, although this was strongly suggested in some sections (Fig. 3C). Vasopressin-positive perikarya that resemble the vasopressin immunoreactive perikarya within the SCN were also consistently found in the implant (Fig. 3, E to G). These perikarya were often difficult to identify because of their small size (long axis around 10 μm) and weak immunostaining. In these cells, much of the soma was occupied by the nucleus (Fig. 3E). Vasopressin immunoreactive perikarya were often associated with a fine plexus of varicose fibers showing weak vasopressin immunostaining. These SCN-like vasopressin perikarya and fibers contrast with vasopressin cells of the magnocellular system, which show large, strongly immunoreactive perikarya (long axis around 25 μm) and fibers. Such cells were also identified within some of the implants (Fig. 3D), suggesting that we had occasionally transplanted part of the magnocellular system. NPY-positive fibers were always found crossing the host-graft border (Fig. 3, H and I), but cell bodies were rarely found within the implant. As NPY perikarya have not been identified within the SCN and were not identified within the graft, we assume that the majority of NPY fibers within the graft came from the host. In four unsuccessful SCN implants, we identified weakly stained VIP cells and fibers within the graft, but found no evidence of vasopressin immunoreactive

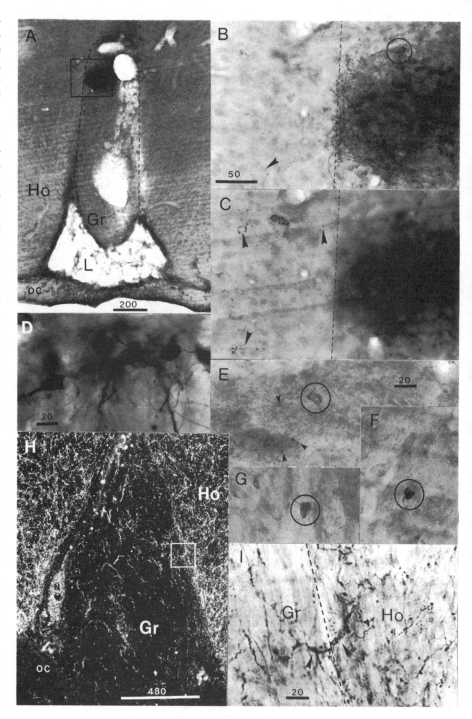

Fig. 3. (**A**) Photomicrograph of a graft placed within the third ventricle that restored a 22-hour rhythm. The approximate borders between the host and graft tissue have been indicated with a dotted line. This section has been immunostained with an antibody to VIP. The area within the square has been enlarged in (B) and (C). (**B**) A "ball" of VIP perikarya and fibers (right side), reminiscent of the organization of VIP within the SCN. (**C**) The same area as (B) but at a different plane of focus. At this plane of focus immunoreactive fibers [indicated with arrowheads in (B) and (C)] are identified that appear to have arisen from the graft and cross the host-graft border (dotted line). (**D**) Magnocellular-like vasopressin cells were occasionally found within the graft. (**E, F,** and **G**) Vasopressin immunoreactive perikarya (circles) and fibers (arrowheads) within the graft that resemble the vasopressin cells of the SCN. Perikarya are faintly stained and small and were often associated with a fine plexus of varicose fibers. These cells contrast with the vasopressin cells of the magnocellular system, which have large densely stained perikarya associated with heavily stained fibers (D). (**H**) Dark-field photomicrograph of a successful graft placed within the third ventricle. This section has been immunostained with an antibody to NPY. The area within the white square has been enlarged in (**I**) to show immunoreactive NPY fibers (arrowheads) crossing the host-graft boundary (dotted line). As no perikarya were identified within the graft we assume that these NPY fibers came from the host tissues. Abbreviations: Gr, graft; Ho, host tissue; N, necrotic tissue; OC, optic chiasma; and L, lesion site and base of third ventricle. Scale bars: (A) 200 μm; (B and C) 50 μm; (D) 20 μm; (E, F, and G) 20 μm; (H) 480 μm; and (I) 20 μm.

perikarya or fibers within the graft or evidence that NPY fibers were entering the graft from the host. In contrast to implants that contained the SCN, cortical implants never restored rhythmicity to the host. Cortical implants always contained a few NPY perikarya and many fibers. In cortical implants, NPY fibers were always seen to cross the host-graft border, and most crossing fibers seemed to originate from the host.

Although most of our implants contained some portion of extra-SCN tissues (Fig. 3, A and D), the immunocytochemical analysis showed that grafts that restored rhythmicity always contained cells with SCN characteristics (VIP and vasopressin). Therefore, the period of the overt rhythm is determined by cells within, or very close to the SCN. This observation is in agreement with reports showing that the SCN is required for successful restoration of rhythmicity (11–14, 23).

In most of our locomotor data, rhythmicity was visually apparent within 6 to 7 days after transplantation. Although surprisingly short, this latency does not preclude the possibility that neural reconnections drive the behavior since dense neural outgrowth has been reported from other transplanted tissue with a similar time course (24). Immunocytochemical analysis indicates that neural connections have been made between graft and host brain; however, it was not possible to determine the source of fibers crossing the graft boundary.

The fact that the genotype of the host does not appear to affect significantly the expression of the transplanted rhythm is somewhat surprising, especially in view of evidence for the existence of oscillators outside the SCN in the mammalian brain (15, 16). We interpret the absence of a host contribution to the circadian period to mean that either the SCN is essentially autonomous in determining the primary characteristics of rhythmicity in hamsters or that the host brain fails to make the connections with the tissue graft that are required for the brain to influence this period. In either case, our results strengthen the view that the SCN occupies a position at the top of the circadian hierarchy in mammals.

REFERENCES AND NOTES

1. R. Y. Moore, in *Biological Rhythms and Their Central Mechanisms*, M. Suda, O. Hayaishi, H. Hakagawa, Eds. (North-Holland, Amsterdam, 1979), pp. 343–354.
2. B. Rusak and Z. Boulos, *Photochem. Photobiol.* **34**, 267 (1981).
3. W. J. Schwartz and H. Gainer, *Science* **197**, 1089 (1977).
4. R. Y. Moore and V. B. Eichler, *Brain Res.* **42**, 201 (1972).
5. F. K. Stephan and I. Zucker, *Proc. Natl. Acad. Sci. U.S.A.* **69**, 1583 (1972).
6. B. Rusak and I. Zucker, *Physiol. Rev.* **59**, 449 (1979).
7. S. T. Inouye and H. Kawamura, *Proc. Natl. Acad. Sci. U.S.A.* **76**, 5962 (1979).
8. D. J. Green and M. U. Gillette, *Brain Res.* **245**, 198 (1982).
9. G. Groos and J. Hendricks, *Neurosci. Lett.* **34**, 283 (1982).
10. D. J. Earnest and C. D. Sladek, *Brain Res.* **382**, 129 (1986).
11. Y. Sawaki et al., *Neurosci. Res.* **1**, 67 (1984).
12. R. Drucker-Colin et al., *Brain Res.* **311**, 353 (1984).
13. M. N. Lehman et al., *J. Neurosci.* **7**, 1626 (1987).
14. P. J. DeCoursey and J. Buggy, *Soc. Neurosci. Abstr.* **12**, 210 (1986).
15. K.-I. Honma, S. Honma, T. Hiroshige, *Physiol. Behav.* **40**, 767 (1987).
16. K. Abe, J. Kroning, M. A. Greer, V. Critchlow, *Neuroendocrinology* **29**, 119 (1979); F. K. Stephan, J. M. Swann, C. L. Sisk, *Behav. Neural. Biol.* **25**, 346 (1979).
17. In the avian pineal [S. A. Binkley, J. B. Riebman, K. B. Reilly, *Science* **202**, 1198 (1978); N. H. Zimmerman and M. Menaker, *Proc. Natl. Acad. Sci. U.S.A.* **76**, 999 (1979)], in the lizard pineal [M. Menaker and S. Wisner, *ibid.* **80**, 6119 (1983)], and in the amphibian retina [J. C. Besharse and P. M. Iuvone, *Nature* **305**, 133 (1983)].
18. M. R. Ralph and M. Menaker, *Science* **241**, 1225 (1988).
19. Tissue for implantation was obtained in the following manner. Pregnant females were anesthetized and prepared for surgery on day 13.5 of gestation. Fetuses were located and decapitated in utero, and the heads were removed to a sterile petri dish containing BSS. The entire litter was collected at one time. Fetal brains were then removed under dissecting microscope and placed in a second dish. After all of the brains had been collected, each was oriented with the ventral surface visible under the microscope. All dissections were performed with the tissue immersed in MEM-BSS. This procedure allowed the ends of the developing optic nerves to float in the media so that the optic chiasm could be located easily. A coronal incision was made where the two nerves fused, and a second, parallel cut was made 1 to 1.5 mm caudal to this. Two parasagittal cuts were then made about 1.5 mm on either side of midline, with the scissors held at a 45° angle so that these cuts passed into the third ventricle. This resulted in the excision of two small blocks of tissue connected by the optic chiasm. Neural tissue for implantation was then teased away from the chiasm and pia mater.
20. Tissue blocks for implantation were placed in a group at the tip of a Wiretrol micropipette. The pipette was graduated in increments of 1.0 µl so that the total volume to be injected could be estimated. The tissue occupied about 1 µl of the 1.5 to 2 µl volume that was injected into the host brain.
21. E. Balaban, M.-A. Teillet, N. Le Douarin, *Science* **241**, 1339 (1988).
22. Animals were perfused intracardially with 350 ml of 4% paraformaldehyde in 0.01M phosphate buffer containing 15% picric acid; the brain was removed and placed in the same fixative for an additional 24 to 48 hours. Frontal sections (80 µm) were cut on a vibratome and processed for immunocytochemistry as described [R. G. Foster, G. Plowman, A. Goldsmith, B. Follett, *J. Endocrinol.* **115**, 211 (1987)] with antibodies directed against NPY (1:1000; Peninsula Laboratories, Belmont, CA), vasopressin (1:500; Incstar, Stillwater, MN), and VIP (1:1000; Peninsula Laboratories). Controls were performed by incubating sections in absorbed primary antisera [40 nm of peptide (Peninsula Laboratories) added to 1 ml of diluted primary antisera for 24 hours at 4°C] or omitting the primary antisera. In both controls, immunostaining was abolished in the graft and host brain, except for faint staining within necrotic tissue and astrocytes around the lesion site, suggesting artifactual staining within these tissues. Necrotic tissue and astrocytes around the lesion site also showed immunostaining with VIP, NPY, and vasopressin antibodies (Fig. 3).
23. R. Aguilar-Roblero, L. P. Morin, R. Y. Moore, *Soc. Neurosci. Abstr.* **14**, 49 (1988).
24. M. K. Floeter and E. G. Jones, *Dev. Brain Res.* **22**, 19 (1985).
25. Supported by PHS grants MH09483 to M.R.R., HD13162 to M.M., and HD18686 to F.C.D.

28 June 1989; accepted 28 November 1989

Glossary

ablation - the surgical destruction of tissue; a lesion

bregma - a point on the skull where two major bone sutures (seams) intersect; this intersection is often used as a reference point for stereotaxic surgery

circadian period - the duration of each cycle of a circadian rhythm when a subject is living in a constant environment (i.e., in an environment without temporal cues); the circadian periods of each individual animal are regular, but they often differ somewhat from 24 hours

circadian rhythm - any bodily change that displays regularly recurring cycles of change that each last about 24 hours (e.g., the sleep-wake cycle)

donor - an animal from which tissue has been removed for transplantation

entrainment - the control of the timing of circadian cycles by external temporal stimuli such as those provided by the light-dark cycle

explant - a piece of tissue that is removed from a living animal and is maintained alive in a suitable external medium

genotype - the entire genetic constitution of an individual

glucose - a simple sugar that is the primary source of energy for neurons and other cells

heterozygote - two genes control the expression of each trait; in heterozygotic individuals, the two genes controlling the trait in question are different

homozygote - two genes control the expression of each trait; in homozygotic individuals, the two genes controlling the trait in question are identical

host - an animal that receives a transplant

in vitro - in a tissue culture; outside the living organism

methamphetamine - a stimulant drug

micropipette - a fine glass tube drawn to a hollow microscopic point

mutant - any organism that carries a genetic trait that differentiates it from other members of its species; members of mutant strains pass their unique characteristics on to their offspring

pacemaker - any source of a signal that controls the timing of a recurring biological cycle (e.g., a cardiac pacemaker)

suprachiasmatic nucleus (SCN) - a neural structure that is thought to contain the circadian pacemaker circuit; there are two suprachiasmatic muclei, a left one and a right one; they are located near the midline in the floor of the third ventricle, just above the optic chiasm

ultradian - biological cycles that are shorter than 24 hours (e.g., cycles of hunger)

vasopressin - a hormone that acts to conserve bodily fluids; some cells of the SCN contain vasopressin

vasoactive peptide (VIP) - a peptide hormone; some cells of the SCN contain VIP

Essay Study Questions

1. Prior to the experiment of Ralph, Foster, Davis, and Menaker, what evidence indicated that the suprachiasmatic nuclei contain circuits that can control the timing of circadian rhythms?

2. What evidence suggests that neural timing mechanisms capable of controlling circadian rhythms may exist in parts of the brain other than the suprachiasmatic nuclei?

3. Describe the procedure used by Ralph and his colleagues to transplant suprachiasmatic nuclei.

4. Which circadian cycle did Ralph and his colleagues study, and how did they measure it?

5. Why were the SCN transplants positioned in the third ventricle?

6. As a control procedure, cortical tissue was implanted in some hosts. What did this control procedure prove?

7. Why was it surprising that the genotype of the host had no effect?

8. How did the effects of transplants from homozygous and heterozygous mutant donors differ.

Multiple-Choice Study Questions

1. Which of the following is evidence that the suprachiasmatic nuclei contain circadian pacemaker cells?
 a. The SCNs receive direct connections from the biological clock.
 b. Explants of SCN tissue display no rhythmicity.
 c. Explants of SCN tissue display no circadian rhythmicity.
 d. Ablation of the SCNs usually eliminates circadian rhythmicity.
 e. Methamphetamine injections can induce circadian rhythms.

2. Prior to the neural transplantation experiment of Ralph and his colleagues, it had been shown by several researchers that
 a. circadian rhythmicity can be restored to SCN-lesioned hosts by implantation of fetal SCN cells.
 b. methamphetamine destroys SCN cells.
 c. methamphetamine inhibits the rejection of SCN cells.
 d. tissue explants display 20-hour circadian sleep-wake cycles.
 e. transplanted SCN tissue survives for only 7 days.

3. Why did Ralph and his colleagues wait 3 or 4 weeks after making the lesions in the hosts to replace the lesioned tissue with fetal transplants?
 a. to increase the probability of rejection
 b. to decrease the probability of rejection
 c. to provide ample opportunity to determine whether any rhythmicity survived in the hosts following the lesions
 d. to allow time for degeneration
 e. to adequately prepare the implantation site

4. The measurements used to describe the stereotaxic placement of lesioning electrodes are often given in millimeters from
 a. smegma.
 b. bregma.
 c. lambda.
 d. the tip of the nose.
 e. the electrode.

5. Circadian rhythmicity was restored in
 a. 20% of the donors.
 b. 70% of the donors.
 c. 50% of the donors.
 d. 95% of the donors.
 e. none of the above

6. Circadian rhythmicity was not restored in any of the rats that received
    a. SCN transplants.
    b. cortical transplants.
    c. transplants from heterozygous mutants.
    d. SCN lesions.
    e. transplants from homozygous mutants.

7. How long did it take for the SCN transplants to restore the circadian rhythmicity of the activity of the SCN-lesioned hosts?
    a. about 24 hours
    b. about 48 hours
    c. about 6 or 7 days
    d. about 3 or 4 weeks
    e. about 3 or 4 months

The answers to the preceding questions are on page 290.

Food-For-Thought Questions

1. Suggest how the neurotransplantation procedure might be used to study the neural basis of memory. Do you think that it will ever prove possible to transfer memories from one animal to another?

2. Following the removal of a small tumor from the vicinity of the SCN, an infant displayed no circadian cycles. Is the lack of a circadian rhythm a problem? Would you recommend neurotransplantation as a treatment—why or why not?

ARTICLE 12

Female Visual Displays Affect the Development of Male Song in the Cowbird

M.J. West and A.P. King
Reprinted from Nature, 1988, volume 334, 244-246.

How do birds come to sing the songs characteristic of their species? The birds of many species do not develop normal species-specific songs unless they have had the opportunity as juveniles to interact with adult conspecifics (members of the same species); this suggests that they learn to sing by imitating the songs of mature birds. Such imitation learning undoubtedly plays an important role in the development of species-specific bird song, but the following experiment by West and King shows that it is not the entire story—at least not for the cowbird. In an earlier experiment, West and King had shown that exposure of juvenile male cowbirds to mature female cowbirds influences the development of the males' songs despite the fact that female cowbirds do not sing. West and King report in the following article how such nonacoustic bird song learning takes place.

In the first stage of their study, West and King recorded the interactions of juvenile male cowbirds and adult female cowbirds during the fall (i.e., outside the breeding season), and special attention was paid to the females' behavior during the 18,837 recorded male songs. During 1.1% of these immature songs, the females made a particular wing stroke that is a component of the copulatory posture, which is normally elicited by the songs of a mature male. Prior to a wing stroke the males tended to sing a variety of songs, whereas after the wing stroke they tended to restrict their songs to the type that had produced the wing stroke.

In the second stage of the study, the songs of the juvenile males that were recorded during the first stage of the study were played to mature females during their breeding period. Songs that had elicited wing strokes during the first phase of the study elicited a full copulatory posture in the females 89% of the time. In comparison, songs that had preceded wing-stroke-eliciting songs elicited the full copulatory posture 57% of the time, and songs that had followed wing-stroke-eliciting songs elicited the full copulatory posture 72% of the time.

These results illustrate a remarkable form of social learning. Outside of the mating season, the songs of the mature male cowbird elicit components of the copulatory posture in mature females. When juvenile male cowbirds sing in the presence of a mature female, those songs that are similar to those of the mature male cowbird tend to be reinforced by the resulting postural changes of the female.

Female visual displays affect the development of male song in the cowbird

Meredith J. West* & Andrew P. King†

* Department of Psychology, University of North Carolina, Chapel Hill, North Carolina 27599, USA
† Department of Psychology, Duke University, Durham, North Carolina 27706, USA

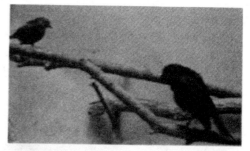

The role of social stimulation in avian vocal learning is well documented[1]. The separate contribution of social, as opposed to vocal, stimulation has been difficult to address, however, because in almost all cases both tutor and pupil sing. The opportunity to isolate such effects arose in cowbirds (*Molothrus ater ater*) after discovering that males housed with non-singing female cowbirds made vocal changes which related directly to the female preferences for native song[2-4]. Here we report how females communicate with males about songs. We describe a visual display by females, a wing stroke, that is elicited by specific vocalizations. The songs that trigger wing strokes are in turn highly effective releasers of copulatory postures, and thus this previously unnoticed female display has biological significance. The data not only provide the first evidence of the tutorial role of male–female interactions during song ontogeny, they also clearly implicate visual stimulation in song learning, a process that has until now been assumed to be affected only by auditory information[5].

The study was conducted in two parts. First, interactions between captive male-female pairs were videotaped in March and April, the time when cowbirds are found on their breeding habitat in the wild. Second, playbacks of songs that had elicited specific reactions from females during the videotaped sessions were presented one year later to females in breeding condition, thus testing the songs' effectiveness as releasers of copulatory postures.

We examined social interactions in eight male-female pairs in the first part of the study. The males were wild-caught juveniles captured when under 50 days of age; the females were adults captured the year before. Individual pairs were housed in sound-attenuating chambers and maintained according to previously described procedures[6]. The males were exposed to tape recordings of male songs for three months in the fall (autumn). Synchronized audio and video recordings were made in March and early April, at which point song types could be identified. These were stereotyped acoustic patterns differing from one another in the frequency contours of individual notes, note syntax, and the modulation of terminal whistles. Ten hours of singing were videotaped for four pairs and 12 hours for four other pairs. Audio taping continued into May to record the males' final repertoires as the pairs came into breeding condition.

The videotapes were analysed to mark the occurrence of every song and to record reactions by the female during each song. Only behaviours that began after the song's onset and before its finish, an interval of less than one second, were coded. In 88 hours of recording, 18,837 songs (range per male 1,320-4,614) were recorded. On average, 94% of the songs (range 86-98%) produced no visible change in the females' behaviour. When females did respond however, a striking display was a wing stroke, that is, rapid lateral movements of one or both wings away from the body (Fig. 1). It was called to our attention in part by the alert behaviour of the males, who frequently interrupted the pace of their singing after a wing stroke to approach and inspect the subsequent behaviour of the female (Fig. 2).

Wing strokes occurred during 237 of the 18,837 songs, ranging in frequency from 2 to 70 across the eight pairs, yielding a mean of 29.6 per female and a ratio of 1.1 wing strokes to every 100 songs (wing strokes h^{-1}/songs h^{-1}). Two features of the wing stroke suggest it may be a precursor to a full copulatory posture, a display seen only in the breeding season. First, the wing stroke resembles the initial wing movement of a copulatory posture. Second, and most striking, the wing stroke occurs extremely rapidly while the male's one-second song is still in progress, a defining feature of the release of the copulatory posture, in this species.

The second part of the study, the playback tests, was carried out one year later. The subjects were six of the females studied in part one and two additional females which had been housed identically with males. In the time between coding of wing strokes and conclusion of playback testing, the females resided in same-sex pairs in sound-attenuating chambers.

Five song series were obtained from five of the eight videotaped male-female pairs. Each song series was composed

Fig. 1 A wing stroke is shown in the middle two panels. The male began his song 400 ms before the wing stroke. The upper panel shows the female's behaviour at song onset and the bottom panel shows her behaviour 900 ms later, just before the finish of the song. The photographs were made from the videotaped sequence used in experiment 4.

Fig. 2 The male's orientation and approach towards the female after producing a song that had elicited a wing stroke. The song had begun at 8:09:45:6 and ended at 8:09:46:1. The wing stroke began at 8:09:46:0 and ended at 8:09:46:2, shown in the upper panel. The female is now stretching her wing. This sequence is part of the playback series comprising experiment 3.

of 16 songs, the eight songs that had preceded a song eliciting a wing stroke, the wing-stroke song itself, and the seven songs following it. The major criterion for selecting sequences was audio recording quality. Each song series was run as a separate experiment. In each, the 16 songs were presented one at a time in six daily trials, separated in time by a minimum of 90 minutes. Each experiment took about two weeks to complete and the songs were played back an average of five times (range 4-8). The songs were not presented in their original order (hence the wing-stroke song was not necessarily the ninth song) but were ordered differently each day with the constraints that no song should occur twice in one day and that all songs should be played an equal number of times at different times of day.

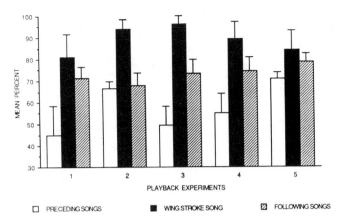

Fig. 3 The mean per cent and standard error of copulatory responses by the eight females for each experiment. Differences in potency among the 16 songs were tested by Friedman analyses of variance of ranks. Significant differences were obtained for each. The $xr2$ values were as follows (df = 15): Exp. 1, 78, $P < 0.001$; Exp. 2, 31.2, $P < 0.01$; Exp. 3, 36.0, $P < 0.01$; Exp. 4, 38, $P < 0.01$; and Exp. 5, 25, $P < 0.05$. In each experiment, the wing-stroke song, song nine, received the highest mean rank. Wilcoxon signed rank tests were used to test for two differences: differences in potency between the wing-stroke song and the mean potency of songs 1-8 and differences between the mean potency of songs 1-8 and songs 10-16. For the first comparison, the Wilcoxon T value was 0, $P < 0.01$ for experiments 1-4 and $T = 2$, $P < 0.02$ for experiment 5. For the second comparison, T was 1 or 0 for experiments 1, 3, and 4, $P < 0.02$ and the T value was >6 (n.s.), for experiments 2 and 5.

Playback testing began when the females came into breeding condition as judged by egg-laying. Song potency was defined as the percentage of trials in which a song elicited a copulatory posture, that is, the female lowered and spread her wings, arched her neck, and separated the feathers around the cloacal area.

The females responded differently to the 16 songs. In each experiment, the wing-stroke song yielded the highest mean potency, 89% (range 81-96%; Fig. 3). In contrast, the mean percent of responses to the preceding eight songs was 57% (range 0-100%) whereas the seven songs following the wing stroke elicited a mean percentage of responses of 72% (range 0-100%), producing statistically significant differences in each experiment (Fig. 3). Thus, although males sang other effective songs, only the wing-stroke song produced consistently high levels of copulatory responses.

The song sequences were examined with respect to acoustic content and organization. The five males sang four to seven different song types before the wing-stroke song. In four of the sequences (experiments 1-4, see Fig. 3), no new song types occurred after the wing-stroke song. The song type eliciting the wing stroke accounted for an average of 19% (range 0-38%) of the singing during the first eight songs in contrast to 63% (range 43-86%) of all singing after the wing stroke, a non-overlapping difference in each experiment. The song type eliciting the wing stroke was also sung more repetitively after the wing stroke: the mean number of immediate repetitions before the wing stroke was 1 (range 0-2) compared to 3.75 (range 2-6) after it. These changes, in addition to the higher potencies of songs following wing strokes, highlight the conspicuous nature of the wing stroke to males who immediately attempted to make use of the information contained in the female's display.

Thus, the present study underscores the multi-modal nature of the stimulation available to males as they learn to sing, and adds credence to a multi-phasic view of song learning whereby it is influenced by both natal and juvenile experiences[7]. Most importantly, the data bring into focus visual dimensions of song learning that have not been described before. Males must not only listen to the nature of the sounds they produce, but also look at the reactions their sounds provoke.

Financial assistance was provided by the NSF and the National Institute of Neurological and Communicative Disorders and Strokes.

Received 11 March; accepted 31 May 1988.

1. Petrinovich, L. in *Social Learning: Psychological and Biological Perspectives* (eds Zentall, T. & Galef, B. G.) 255-278 (Lawrence Erlbaum, New York, 1988).
2. King, A. P. & West, M. J. *Nature* **305**, 704-706 (1983).
3. West, M. J. & King, A. P. *Ethology* **70**, 225-235 (1985).
4. King, A. P. & West, M. J. *Anim. Behav.* (in the press).
5. Kroodsma, D. E. in *Acoustic Communication in Birds* Vol. 2 (eds Kroodsma, D. E. & Miller, E. H.) 1-28 (Academic Press, New York, 1982).
6. West, M. J. & King, A. P. *J. comp. Psychol.* **100**, 296-303 (1986).
7. McGregor, P. K. & Krebs, J. P. *Behaviour* **79**, 127-147 (1982).

Glossary

avian - pertaining to birds

cloaca - a passage that is used for urinary and fecal discharge and for reproduction; birds have a cloaca, but mammals do not

copulatory - pertaining to the act of sexual intercourse

cowbird - a small North American blackbird that frequently associates with cattle; it builds no nest and lays its eggs in the nests of other species

natal - pertaining to birth or to the period around birth

ontogeny - the development of the individual organism through its life; as opposed to phylogeny, which refers to the development (evolution) of the species through generations

releasers - particular simple stimuli that elicit complex but inflexible patterns of behavior; the song of the mature male cowbird releases the full female copulatory posture during the breeding season

syntax - the principles of the relation between the elements of a series, such as between the words of a sentence or notes of a song

wing stroke - one component of the female cowbird copulatory posture

Essay Study Questions

1. Why has the role of visual factors in the learning of bird songs not been previously recognized?

2. In most biopsychological research articles, the time of year that the experiment was performed is not reported. In contrast, West and King describe the time of year during which all phases of their study were conducted. Why?

3. What two lines of evidence suggested that the wing strokes elicited in female cowbirds by some of the songs of juvenile male cowbirds are copulatory in nature?

4. What evidence is there that wing strokes are potent positive reinforcers for juvenile males?

5. What effect did the taperecorded juvenile male songs have on the behavior of females that were in breeding condition?

6. Compare the acoustic properties of the juvenile songs recorded immediately preceding a wing stroke with those immediately following a wing stroke.

Multiple-Choice Questions

1. What percentage of juvenile male cowbird songs elicited wing strokes in females that were not in breeding condition?
    a. 1.1%
    b. 10.1%
    c. 30.1%
    d. 50.1%
    e. 70.1%

2. What percentage of juvenile male songs that had elicited wing strokes in females that were not in breeding condition released a full copulatory posture in females that were in breeding condition?
    a. 1.9%
    b. 19%
    c. 29%
    d. 69%
    e. 89%

3. What percentage of the eight juvenile songs preceding a wing stroke and the seven following the wing stroke elicited a copulatory posture in females that were in breeding condition?
    a. 99% and 1.9%, respectively
    b. 57% and 72%, respectively
    c. 1.9% and 89%, respectively
    d. 72% and 1.9%, respectively
    e. 89% and 57%, respectively

4. In contrast to previous research on the development of bird song, the article of West and King emphasizes the important role of
    a. acoustic stimuli.
    b. taperecorded songs.
    c. juvenile experience.
    d. learning.
    e. imitation.

5. A juvenile song that elicited a female wing stroke was
    a. not repeated for the remainder of that test session.
    b. sung more frequently after the wing stroke than before.
    c. always followed by three provocative trills.
    d. then repeated by the female to encourage the male.
    e. usually accompanied by a male wing stroke the next time that it was sung.

6. This study emphasizes two features of bird song development that have not received wide recognition. It emphasizes that bird song development
    a. occurs at more than one stage of development and is influenced by visual stimuli.
    b. is both multiphasic and natal.
    c. is influenced by both visual and acoustic stimuli.
    d. is influenced by both natal and juvenile visual stimuli.
    e. is both seasonal and acoustic.

The answers to the preceding questions are on page 290.

Food-For-Thought Questions

1. Cowbirds do not build nests; the females lay their fertilized eggs in the nests of other birds. How might the factors that influence song development differ between cowbirds and nest-building species?

2. Sometimes studying the behavior of other species can suggest interesting hypotheses about human behavior. Do you think that subtle social reinforcers of a sexual nature might have a lasting effect on the development of human behavior? For example, are not some of the adult male behaviors that are appropriately derided by many adult females the very same behaviors that were reinforced by adolescent female titters in the school yard? Can you think of anecdotal examples of such influences? How might this hypothesis be tested?

ARTICLE 13

Caenorhabditis Elegans: A New Model System for the Study of Learning and Memory

C.H. Rankin, C.D.O. Beck, and C.M. Chiba

Reprinted from Behavioral Brain Research, 1990, volume 37, 89-92.

One method of trying to identify the neural bases of learning and memory is to study very simple forms of learning in subjects with very simple nervous systems. Although the nervous systems of the invertebrates that have commonly served as subjects in simple-system studies of learning and memory are far more simple than those of mammals, they are still very complex. For example, the widely studied marine mollusk Aplysia has a nervous system that is composed of tens of thousands of neurons. In contrast, the nematode worm C. elegans (Caenorhabditis elegans) has a nervous system that is composed of only 302 neurons. Accordingly, if C. elegans could be shown to learn, it would provide biopsychologists with an excellent preparation for the study of the neural bases of the learning. In the following article, Rankin, Beck, and Chiba reported that C. elegans are capable of three forms of non-associative learning: habituation, dishabituation, and sensitization.

Each C. elegans was tested on a Petri plate under a stereomicroscope, which fed its image into a videorecording system. In each experiment, a mechanical tapper delivered controlled taps to the side of the Petri plate. Each tap normally caused the C. elegans to reverse its direction of movement and momentarily travel backwards before resuming its forward motion. This reversal reflex response was the focus of Rankin, Beck, and Chiba's experiments.

To demonstrate habituation of the reversal reflex response, the Petri plate was tapped 40 times, once every 10 seconds. The amplitude of the reversal reflex decreased progressively, and by the last two taps little or no reversal was elicited. Then, disinhibition was demonstrated by shocking the habituated subjects and administering a further 10 taps to assess the disinhibitory effect of the shocks. Following the dishabituating shock, the reversal reflex recovered to about half of its original amplitude. In the sensitization experiment, C. elegans were shocked and then exposed to a series of Petri-plate taps. The shocks sensitized the reversal reflex responses; they increased the amplitude of the responses to over twice that observed in a group of unshocked control worms. Finally, Rankin, Beck, and Chiba demonstrated that the habituation produced by 50 Petri-plate taps can be retained for 24 hours.

With their demonstration that C. elegans is capable of three kinds of non-associative learning, Rankin, Beck, and Chiba have introduced a potentially valuable simple system for the study of the neural bases of learning and memory.

Short Communication

Caenorhabditis elegans: a new model system for the study of learning and memory

Catherine H. Rankin, Christine D.O. Beck and Catherine M. Chiba

Department of Psychology, University of British Columbia, Vancouver, B.C. (Canada)

(Received 16 July 1989)
(Revised version received 26 July 1989)
(Accepted 12 September 1989)

Key words: Non-associative learning; Habituation; Dishabituation; Sensitization; Long-term memory; *Caenorhabditis elegans;* Model system

The extensive information on the neuroanatomy, development and genetics of *Caenorhabditis* (*C.*) *elegans* make it an ideal candidate model system for the analysis of the mechanisms underlying learning and memory. A first step in this analysis is the demonstration of the capacity of *C. elegans* to learn. In these experiments non-associative learning in *C. elegans* was investigated by observing changes in reversal reflex response amplitude to a mechanical vibratory stimulus. The results from these studies of non-associative learning show that *C. elegans* is capable of short-term habituation, dishabituation and sensitization, as well as long-term retention of habituation training lasting for at least 24 h. These findings set the stage for detailed developmental, genetic and physiological analyses of learning and memory.

Simple-system approaches using invertebrate species have led to considerable progress in our understanding of mechanisms underlying learning and memory[2]. Although the invertebrate species used as model systems to date are simple when compared with vertebrates, they are still very complex, consisting of tens of thousands of neurons, with many unknown connections and interactions. With this level of complexity it would be almost impossible to determine all of the cells that might influence a reflex circuit, or play a role in learning. *Caenorhabditis* (*C.*) *elegans* is a simpler system than those previously studied, and offers the possibility that all of the neurons involved in a given behavior can be analyzed. We propose that *C. elegans* has a number of characteristics that make it an excellent candidate model system for studies of learning and memory.

As a simple system, *C. elegans* has already proven useful for studies of anatomy, development and genetics. The anatomy of the nervous system has been mapped at the electron-microscope level, including a wiring diagram of the worm's 302 neurons showing putative electrical and chemical synapses[20,21]. Through the use of genetic analysis and laser microsurgery, the function and neurotransmitters of many of the neurons are known[6]. Although the neurons of

---

Correspondence: C.H. Rankin, Department of Psychology, University of British Columbia, Vancouver, B.C., V6T 1Y7 Canada.

C. elegans are too small and inaccessible for current physiological recording techniques, the electrophysiology of the neurons of a larger, related species of nematode (*Ascaris*) with a similar neuroanatomy can be studied[10,11,16]. In addition, the complete cell lineages of the 1000 somatic cells in *C. elegans* are now available[18], as are a nearly complete genomic map and a large library of mutant strains[9]. Thus, if *C. elegans* can learn, it offers the promise of behavioral, genetic and physiological analyses of learning and memory in a very simple nervous system. Our results show for the first time that *C. elegans* is capable of a variety of forms of non-associative learning as well as both short- and long-term memory.

Since *C. elegans* had been anecdotally reported to show response decrement to repeated touches[4,8], we began our analysis of learning and memory with studies of non-associative learning (habituation, dishabituation and sensitization). In these studies individual adult hermaphrodites (N2) were observed on Nematode Growth Medium[1] agar plates through a stereomicroscope with attached videorecording equipment. A mechanical tapper[14] mounted on a micromanipulator delivered single taps or trains of taps to the side of the Petri plate containing an individual animal. A tap to the plate caused worms to swim backwards for some distance before changing direction and moving forward. The distance the worms traveled backwards in response to the tap was the dependent measure in these studies. Using stop-frame video analysis, the path of the reversal was traced onto acetate sheets. The length of the path was then quantified using a microcomputer (Macintosh SE) with attached digitizing tablet (Summagraphics Bit Pad Plus) and measurement software (MacMeasure).

Habituation is defined as a decrease in response amplitude due to repeated stimulation, while dishabituation is the facilitation of that decremented response by a novel or noxious stimulus[12,15]. In order to study these processes, 40 trains of taps at 10-s interstimulus interval were delivered followed by a brief, electric shock, followed by 10 additional trains of taps to assess the effects of the shock. A train of taps consisted of 6 taps at 8.5 pps for 600 ms (controlled by the output of a Grass S88 stimulator). Trains of electrical shocks were delivered using a hand-held spanning electrode with wires placed in the agar on either side of the worm; a train of shocks consisted of 10 ms shocks of 60 V at 10 pps for 600 ms. To test for habituation and dishabituation a repeated measures ANOVA and Fisher's planned comparisons were used to compare the initial response with the mean of the last two responses of the habituation series and the mean of the habituation series with the mean of the first two responses following shock. There was significant habituation as evidenced by a decrease in response amplitude following 40 stimuli (initial response mean = 58.95 ± 7.27, habituated response mean = 1.54 ± 1.03; $F = 28.23$, df = 2.18, $P = 0.0001$), and significant dishabituation as evidenced by an increase in response amplitude following shock (dishabituated response mean = 18.32 ± 4.8). Thus, *C. elegans* is capable of both habituation and dishabituation (summarized in Fig. 1).

Sensitization is defined as the facilitation of non-decremented or baseline responses by a strong stimulus[12,15]. Worms in the experimental group ($n = 10$) were given two baseline single taps two minutes apart; two minutes later they were given a stronger stimulus of a train of taps (see above), followed after two minutes by two test taps two minutes apart. Worms in the control group ($n = 10$) received 5 single taps separated by two-minute intervals. For each worm the baseline responses were averaged to obtain a mean baseline response. Within each group the mean baseline responses and each test measurement (tests 1 and 2) were used in calculating a repeated measures ANOVA and Fisher's planned comparisons. Between-group t-tests were calculated for the mean baseline responses and for test measurements for control and experimental groups. Within-group comparisons indicated that the experimental group showed a significant increase in response amplitude to test tap 1 two minutes following the train (baseline mean = 27.92 ± 4.41, test 1 mean = 50.66 ± 9.15; $F = 4.503$, df = 2,18, $P = 0.026$), but not to the test tap 2 two minutes later (test 2 mean = 26.23 ± 4.29), while in contrast test taps 1

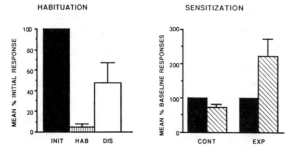

Fig. 1. *C. elegans* demonstrates significant habituation, dishabituation and sensitization. For graphic representation the data for habituation and dishabituation were standardized by setting the initial response for each animal at 100% and calculating the habituation and dishabituation results as a percent of the initial response. The standardized initial response and the mean ($\pm$ S.E.M.) standardized habituation and dishabitiation responses are plotted. INIT, initial response of habituation series; HAB, mean of the last two responses of habituation series (responses to stimuli 39 and 40); DIS, mean of the first two responses following shock. Significant habituation to 4.5% of the initial response was observed following 40 stimuli. Significant dishabituation to 48% of initial response was observed following shock. For graphic representation of the sensitization data the mean baseline response for each worm was set at 100% and the test 1 response was standardized to the mean baseline response. The baseline response (black bars) and mean ($\pm$ S.E.M.) of the standardized test 1 response (striped bars) were plotted for the control (CONT; $n = 10$) and the experimental (EXP; $n = 10$) groups. The mean of the test 1 response for the control worms was 73% of baseline showing significant habituation. The mean of the test 1 response for the experimental worms was 223% of baseline, showing significant sensitization.

and 2 for the control group showed a significant decrease from mean baseline response levels (baseline mean = 31.78 $\pm$ 3.37, test 1 mean = 22.7 $\pm$ 2.84; test 2 mean = 23.45 $\pm$ 2.51; $F = 4.96$, df = 2.18, $P = 0.019$). Between-group comparisons showed that there were no significant differences between control and experimental groups in the mean baseline scores ($t = 0.694$, df = 18, n.s.), and that the baseline-to-test 1 difference was significantly greater for the experimental group than for the control group ($t = 3.022$, df = 18, $P = 0.0037$). Thus, the train of taps produced a short duration sensitization in *C. elegans* (summarized in Fig. 1).

These data are all from experiments that measure short-term memory, that is, tests within minutes of the stimuli. To investigate whether *C. elegans* was capable of long-term memory (24-h retention) a protocol for long-term habituation from experiments with *Aplysia*[3] was modified. On Day 1, all worms received a baseline block of 20 trains of taps at a 10-s interstimulus interval. The experimental group then received 4 additional training blocks separated by one hour. On Day 2, all worms were given a single test block. A mean reversal amplitude for Day 1 for each experimental and each control animal was calculated by averaging the lengths of the 20 responses in the baseline block habituation series, and for Day 2 by averaging the the lengths of the 20 responses in the test block habituation series. These individual means were then used to calculate group mean reversal amplitude scores for the Day 1 baseline block and the Day 2 test block. These data are shown in Fig. 2. Comparing the baseline and test-block means within-group analyses showed that the experimental group had significantly smaller responses on Day 2 than on Day 1 (Day 1 mean = 24.74 $\pm$ 4.59, Day 2 mean = 14.36 $\pm$ 2.21; $t = 2.49$, df = 9, $P = 0.017$), while there was no difference between Days 1 and 2 for the control group (Day 1 mean = 22.15 $\pm$ 2.51, Day 2 mean = 24.25 $\pm$ 4.58; $t = 0.352$, df = 9, n.s.). In addition, between group comparisons showed that there were no significant differences between experimental and control groups on Day 1 ($t = 0.31$, df = 18, n.s.), while on Day 2 the experimental group had significantly lower responses than the control group ($t = 1.94$, df = 18, $P = 0.034$). Thus *C. elegans* is

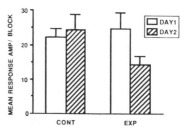

Fig. 2. *C. elegans* is capable of long-term habituation. The data are presented as the mean of the mean response amplitudes for the Day 1 (white bars) baseline block of 20 stimuli and the Day 2 test block of 20 stimuli (striped bars) for control (CONT; $n = 10$) and experimental (EXP; $n = 10$) groups. There were no differences between control and experimental groups on Day 1. The experimental group showed significantly smaller reversals on Day 2 than on Day 1 and significantly smaller reversals on Day 2 than did the control group.

capable of 24-h retention of habituation training.

These data show for the first time that the simple nematode *C. elegans* has a rich repertoire of non-associative learning abilities and is capable of both short- and long-term memory. *C. elegans* can now serve as a model system for the analysis of the mechanisms underlying these learning processes. The 3 areas where *C. elegans* has been studied in the greatest detail (anatomy, development and genetics) can serve as the bases for these explorations of the mechanisms of plasticity.

The reversal response to taps appears to be mediated at least in part by the well-studied touch circuit[4,5,7], since mutants (*mec-3* and *mec*-4) with defective touch receptors[5] showed no response to the tap stimulus (Rankin, unpublished observations). The known anatomy of the touch and locomotion circuits[4,5,7] can now be probed for possible sites of plasticity using genetic manipulation and laser ablation techniques. In addition the development of these circuits has been studied[19], and can be combined with developmental analyses of learning to further elucidate the functions of elements of the circuits. Finally, using genetic manipulations the roles of various neurotransmitters, proteins and enzymes can be investigated.

The authors thank Dr. Ann Rose for supplying worms, culturing supplies, and advice, Dr. Ken Lukowiak for the loan of the tapper, M. Chalfie for mutants mec-3 and mec-4, as well as helpful discussions, Donald Law for data collection, and M. Mana and Drs. P. Graf, A.G. Phillips, P.B. Reiner, and J. Steeves for their comments on an earlier version of this manuscript.

1 Brenner, S., The genetics of *Caenorhabditis elegans*, *Genetics*, 110 (1974) 71–94.
2 Carew, T.J. and Sahley, C.L., Invertebrate learning and memory: from behavior to molecules, *Annu. Rev. Neurosci.*, 9 (1986) 435–487.
3 Carew, T.J., Pinsker, H.M. and Kandel, E.R., Long-term habituation of a defensive withdrawal reflex in *Aplysia*, *Science*, 175 (1972) 451–454.
4 Chalfie, M. and Sulston, J.E., Developmental genetics of the mechanosensory neurons of *Caenorhabditis elegans*, *Dev. Biol*, 82 (1981) 278–370.
5 Chalfie, M. and Au, M., Genetic control of differentiation of the *Caenorhabditis elegans* touch receptor neurons, *Science*, 243 (1989) 1027.
6 Chalfie, M. and White, J.G., The nervous system, in W.B. Wood (Ed.) *The Nematode Caenorhabditis elegans* Cold Spring Harbor Lab. Pub., Cold Spring Harbor, New York, 1988, pp. 81–122.
7 Chalfie, M., Sulston, J.E., White, J.G., Southgate, E., Thomson, J.N. and Brenner, S., The neural circuit for touch sensitivity in *Caenorhabditis elegans*, *J. Neurosci.*, 5 (1985) 956–964.
8 Croll, N.A., Components and patterns in the behavior in the nemotode *Caenorhabditis elegans*, *J. Zool. Lond.* 176 (1975) 159–176.
9 Coulson, A., Sulston, J.E., Brenner, S. and Karn, J., Towards the physical map of the genome of the nematode *Caenorhabditis elegans*, *Proc. Natl. Acad. Sci. U.S.A.*, 83 (1986) 7821–7825.
10 Davis, R.E. and Stretton, A.O.W., Passive membrane properties of motorneurons and their role in long-distance signalling in the nematode *Ascaris*, *J. Neurosci*, 9 (1989) 403–414, ibid., p. 415.
11 Davis, R.E. and Stretton, A.O.W., Signalling properties of *Ascaris* motor neurons: graded active responses, graded synaptic transmission and tonic transmitter release, *J. Neurosci.*, 9 (1989) 415–425.
12 Groves, P.M. and Thompson, R.F., Habituation: a dual-process theory, *Psychol. Rev.*, 77, (1970) 419–450.
13 Hodgkin, J., Edgley, M., Riddle, D.L. and Albertson, D.G., Genetics. In W.B. Wood (Ed.), *The Nematode Caenorhabditis Elegans* Cold Spring Harbor Lab., Cold Spring Harbor, New York, 1988, pp. 491–586.
14 Lukowiak, K. and Peretz, B., Age-dependent CNS control of the habituating gill withdrawal reflex and of correlated activity in identified neurons in *Aplysia*, *J. Comp. Physiol.*, 103 (1975) 1–17.
15 Marcus, E.A., Nolen, T.G., Rankin, C.H. and Carew, T.J., Behavioral dissociation of dishabituation, sensitization and inhibition of *Aplysia, Science*, 241 (1988) 210–213.
16 Stretton, A.O.W. Fishpool, R.M., Southgate, E., Donmoyer, J.E., Walrond, J.P., Moses, J.E.R. and Kass, I.S., Structure and physiological activity of the motoneurons of the nematode *Ascaris*, *Proc. Natl. Acad. Sci. U.S.A.*, 75 (1978) 3493–3497.
17 Sulston, J.E., Schierenberg, E., White, J.G. and Thomson, J.N., The embryonic cell lineage of the nematode *Caenorhabditis elegans*, *Dev. Biol.*, 100 (1983) 64–119.
18 Sulston, J.E., Horvitz H.R., and Kimble, J. Cell Lineage. In W.B. Wood (Ed.), *The Nematode Caenorhabditis elegans*, Cold Spring Harbor Lab. Pub., Cold Spring Harbor, New York, 1988, pp. 457–490.
19 Walthall, W.W. and Chalfie, M., Cell–cell interactions in the guidance of late-developing neurons in *Caenorhabditis elegans, Science*, 239 (1988) 643–645.
20 White, J.G., Southgate, E. and Durbin, R. *Neuroanatomy*. In W.B. Wood (Ed.), *The Nematode Carnorhabditis elegans*, Cold Spring Harbor Lab. Pub., Cold Spring Harbor, New York, 1988, pp. 433–456.
21 White, J.G., Southgate, E., Thomson, J.N. and Brenner, S., The structure of the nervous system of *Caenorhabditis elegans*, *Phil. Trans R. Soc. Lond. Biol. Sci.*, 314 (1986) 1–340.

Glossary

agar - a nutritive, jelly-like medium; microscopic laboratory organisms are commonly reared in agar in a covered Petri plate

Caenorhabditis elegans - a simple microscopic nematode worm with a 302-neuron nervous system

dishabituation - a stimulus-produced release from inhibition; dishabituation is usually produced by noxious or otherwise novel stimuli; for example, dishabituation of the C. elegans reversal reflex response was produced by a train of electric shocks

enzymes - chemicals that influence the occurrence and rate of chemical reactions without actually participating in them

genomic map - map of the position of various genes on the chromosomes of a particular species

habituation - a decrease in the amplitude of an elicited response caused by repeated administration of the eliciting stimulus; for example, the amplitude of the C. elegans reversal reflex response declined with repeated Petri-plate taps

learning - the process by which organisms acquire the capacity to change their behavior as the result of experience

micromanipulator - an electromechanical device for generating precise movements

nematodes - a class of tapered cylindrical worms; many species of nematodes are parasitic, but others, like C. elegans, live in the soil

non-associative learning - any change in the capacity for behavior that results from the single or repeated experience of the same stimulus or from the experience of two or more unrelated stimuli

Petri plate - a small glass dish, which is used for a variety of purposes in biology laboratories

putative synapses - suspected synapses

reversal reflex response - the response of C. elegans and other nematodes to physical vibration; the worm momentarily reverses its direction of movement

sensitization - an increase above baseline in the strength of the response that is elicited by one stimulus as a result of the noncontingent (unassociated) administration of another stimulus; for example, sensitization of the reversal reflex response that is produced in C. elegans by a Petri-dish tap was sensitized by a preceding train of electric shocks

stereomicroscope - a microscope with two eye pieces, which permits the viewing of microscopic material in three dimensions

Essay Study Questions

1. Pavlovian and operant conditioning are examples of associative learning. What is the fundamental difference between these forms of associative learning and forms of non-associative learning such as habituation, dishabituation, and sensitization?

2. The C. elegans nervous system is a potentially valuable preparation for the study of learning and memory. Why?

3. Why do you think that Rankin, Beck, and Chiba focused all of their studies on the reversal reflex response?

4. How did Rankin, Beck, and Chiba demonstrate habituation, dishabituation, and sensitization?

5. Dishabituation and sensitization are similar forms of learning. Compare them.

Multiple-Choice Study Questions

1. The C. elegans nervous system comprises
    a. tens of thousands of neurons.
    b. thousands of neurons.
    c. 8 neurons.
    d. 42 neurons.
    e. none of the above

2. C. elegans is known to have
    a. 8 synapses.
    b. 302 synapses.
    c. 1,012 synapses.
    d. 604 synapses.
    e. none of the above

3. Following 40 Petri-dish taps, administered one every 10 seconds, the amplitude of the C. elegans reversal reflex response was reduced to about
    a. 4% of its original level.
    b. 12% of its original level.
    c. 24% of its original level.
    d. 48% of its original level.
    e. 62% of its original level.

4. Why have there been so few electrophysiological studies of the C. elegans nervous system?
    a. because it has so few neurons
    b. because its neurons are very small and not readily accessible
    c. because its nervous system is enclosed within a protective mantle
    d. because a C. elegans stereotaxic atlas has not yet been prepared
    e. because C. elegans is a parasite and cannot be readily handled

5. Before it is possible to demonstrate _____, it is necessary to first demonstrate _____.
    a. conditioning; learning
    b. nonassociative learning; associative learning
    c. habituation; sensitization
    d. habituation; dishabituation
    e. dishabituation; habituation

6. Rankin, Beck, and Chiba arbitrarily defined long-term memory as memory
    a. for associations.
    b. that lasts for at least 24 hours.
    c. that lasts for more than 10 seconds.
    d. that is more than 50% of baseline.
    e. that involves a synaptic change.

The answers to the preceding questions are on page 290.

Food-For-Thought Questions

1. Many people who would object to an electric shock being delivered to a monkey, a dog, or a cat have no qualms about shocks being delivered to nematode worms. Discuss.

2. Do you think that research on learning in C. elegans can reveal anything about the mechanisms of learning in mammals? Discuss.

ARTICLE 14

Social Influences on the Selection of a Protein-Sufficient Diet by Norway Rats (*Rattus norvegicus*)

M. Beck and B.G. Galef, Jr.
Reprinted from Journal of Comparative Psychology, 1989, volume 103, 132-139.

Laboratory experiments have identified two fundamentally different processes by which animals may come to select nutritionally complete diets. When animals become deficient in a particular nutrient (1) they may automatically develop a preference for its taste, or (2) they may learn to prefer foods that contain the missing nutrient by experiencing the health-degrading effects of their current diet and the health-promoting effects of foods that are rich in the missing nutrient. However, although these two processes work well in contrived laboratory situations, they cannot convincingly account for the uncanny ability of dietary generalists, such as rats, to select complete diets in their natural environments, where only a small proportion of the nutrients that they need for normal development can be tasted, and where there are hundreds of potential foods. Accordingly, Beck and Galef hypothesized that rats must have other means of selecting nutritionally complete diets.

Beck and Galef hypothesized that protein-deficient juvenile rats living in an environment in which they were unable to learn to choose a nutritionally adequate diet by themselves might be able to learn to do so through their interactions with experienced adult rats. In their experiments, each juvenile rat subject was given access to one protein-rich diet and three-protein deficient diets for one week—the protein was put in the least palatable of the diets to make the task of learning to choose the protein diet more difficult. Juvenile rats that spent the week living with three adult rats that had been trained to eat the protein diet consumed more of the protein diet (Experiment 2) and gained more weight (Experiment 1) than did control rats that spent the week alone. Beck and Galef found that the juvenile rats' choice of food was more influenced by the particular food that the demonstrator rats ate than it was by the site at which they ate it.

The ability of generalist feeders such as rats to select nutritionally adequate diets in their natural environment has been attributed entirely to an interaction between species-typical compensatory changes in taste preference and specific appetites and aversions learned by experiencing the consequences of eating various foods. However, the observations of Beck and Galef show that in an environment in which naive rats are unable to learn by themselves to consume a nutritionally adequate diet, they are able to learn what to eat from the feeding patterns of experienced rats.

Social Influences on the Selection of a Protein-Sufficient Diet by Norway Rats (*Rattus norvegicus*)

Matthew Beck and Bennett G. Galef, Jr.
McMaster University, Hamilton, Ontario, Canada

> Investigated effects of interactions between naive and knowledgeable rats (*Rattus norvegicus*) on selection of a nutritionally adequate diet by the naive. We found that during a 7-day test, isolated rats choosing among 4 foods, 3 of which were protein-deficient and 1 of which was protein-rich, failed to learn to prefer the protein-rich diet and lost weight. Conversely, those rats that interacted with conspecifics trained to eat the protein-rich diet developed a strong preference for that diet and thrived. The authors also found that Ss were more strongly influenced in their diet selection by the flavor of the foods eaten by conspecifics than by the locations where conspecifics fed. The results suggest that social influence may be important in development of adaptive patterns of diet choice by rats (or other dietary generalists) that need to find nutritionally adequate diets in demanding environments.

Students of dietary self-selection have described two complementary processes that may lead animals foraging in the world outside the laboratory to select nutritionally adequate foods from among the myriad potentially ingestible, beneficial, toxic, and useless substances found in natural habitat. First, animals have the ability to detect the presence of some nutrients in complex foods and can use tastes associated with those nutrients to select valuable foods to eat (Richter, 1943; Rozin, 1976). Second, animals can associate the postingestional consequences of a food with its taste and can, therefore, learn to select foods that provide valuable nutrients (Booth, 1985; Harris, Clay, Hargreaves, & Ward, 1933; Rozin, 1976). In the literature, these two abilities—the ability to detect directly some nutrients in foods and the ability to learn about the nutritional value of foods from the consequences of their ingestion—have been treated as sufficient to explain the development of adaptive patterns of food choice by generalists in natural environments (Rozin, 1976).

One can, however, imagine circumstances in which neither direct detection of the taste of nutrients in foods, individual learning about the consequences of eating various foods, nor both processes considered together would be adequate to explain the selection of a nutritionally adequate diet by animals. Dietary generalists such as rats can detect directly the sensory qualities correlated with only a handful of nutrients (e.g., sugars, salts, fat, and water) from among dozens of substances required for growth, self-maintenance, and reproduction. The ability of rats to learn about the nutritional value of ingested substances is also limited and need not result in adaptive food choices (see Westoby, 1974, for review). As the number of potential foods available for an animal to sample and the time to onset of the rewarding consequences of eating a nutritionally valuable food increase (Harriman, 1955; Harris et al., 1933; Rozin, 1969), and as the relative palatability of a food containing a necessary nutrient decreases (Kon, 1931; Scott & Quint, 1946; see Epstein, 1967, for review), the ability of an animal to acquire a preference for a nutritionally valuable food rapidly declines. Thus, although in benign environments the sensory–affective systems (Young, 1959) and learning abilities of individuals often prove adequate to the task of diet selection, in more challenging situations (where an animal may find many useless, potential foods to sample, where some needed nutrients are available only in unpalatable foods, and where foods have long-delayed, postingestional consequences), the probability that an isolated individual will find an adequate diet before exhausting its internal reserves can be quite low.

Many vertebrate generalists live throughout their lives as members of social groups. All mammals and most birds spend their first weeks or months of life in intimate association with a parent or parents who, by virtue of their very reproductive success, demonstrate the nutritional adequacy of the foods they have been eating. A naive juvenile, maturing in a demanding environment and unable to learn to identify an adequate diet by itself, may prosper simply by allowing its sampling of foods to be guided by the food choices of either its parents or other adults that it observes feeding. Socially acquired information concerning the foods others eat (Galef, 1986, in press) may provide a complement or supplement to individual learning that facilitates identification of valuable foods.

This series of experiments was undertaken to examine the possibility that naive, juvenile rats living in circumstances in which they were unable to select a nutritionally adequate diet

---

This article is based on part of a dissertation (Beck, 1989) submitted to the Graduate School of the Faculty of Science of McMaster University in partial fulfillment of the requirements of the doctoral degree. Portions of this work were presented at the meetings of the Eastern Psychological Association, Buffalo, New York, April 1988.

Financial support was provided by grants from the Natural Sciences and Engineering Research Council of Canada and McMaster University Research Board to Bennett G. Galef, Jr. Matthew Beck received support as an Ontario Graduate Scholar.

We thank Harvey Weingarten, Mertice Clark, Jim Kalat, and two anonymous reviewers for their helpful comments on earlier drafts of the article.

Correspondence concerning this article should be addressed to Bennett G. Galef, Jr., Department of Psychology, McMaster University, Hamilton, Ontario L8S 4K1, Canada.

for themselves could learn to choose a nutritionally adequate diet if allowed to interact with adult conspecifics that were doing so.

Experiment 1

Free-living, adult omnivores may be challenged occasionally by a failure of one or another of the foods on which they have come to depend. To survive, every juvenile must develop de novo a nutritionally adequate diet composed of solid foods. Because in natural circumstances it is juvenile animals that must most frequently solve the problem of diet selection, the study of how generalists develop nutritionally adequate diets ought to focus on the young (Galef & Beck, in press).

In this experiment we determined whether both weanling and adolescent rats, either in isolation or in a social context, could develop a preference for a single, protein-rich diet when it was presented together with a number of more palatable, protein-poor diets. We anticipated that in the test situation we used, isolated animals would have difficulty identifying the nutritionally adequate diet. Rats cannot directly detect the presence of protein in food, so detection of the nutritionally adequate food on the basis of its sensory qualities was not possible in our experimental situation. Furthermore, to make individual learning about the consequences of eating the protein-rich diet difficult, we both provided several alternative foods for our subjects to choose among (Harris et al., 1933; Rozin, 1969) and placed the needed protein in the least palatable of these foods (Kon, 1931; Scott & Quint, 1946).

Method

Subjects

Thirteen weanling (70 to 90 g) and 26 adolescent (150 to 175 g), male, Long-Evans rats (*Rattus norvegicus*) served as subjects in this experiment. All were born and reared in the McMaster University vivarium and all were descended from breeding stock acquired from Charles River Canada (St. Constant, Quebec). An additional 47 (175 to 200 g) male rats from the McMaster University colony served as demonstrators.

Diets

During the experiment each subject was presented with four diets. Three of these diets were both relatively palatable and relatively poor in protein (4.4% protein by weight); one was both relatively unpalatable and relatively rich in protein (17.5% protein by weight). (A 12%-protein diet is considered adequate for young rats; *Guide to the Care and Use,* 1980). Each of the three protein-poor diets was composed of 80% by weight protein-free, basal mix (Teklad Diets, Madison, Wisconsin, Catalogue No. TD 86146; in g/kg, 808.5 g corn starch, 108.1 g vegetable oil, 7.0 g cod liver oil, 54.1 g mineral mix, and 2.7 g vitamin mix), 10% corn starch, 5% granulated sugar, and 5% high-protein casein (Teklad Diets, Catalogue No. 160030). The three, different, protein-poor diets were flavored with, respectively, 1% by weight McCormick's fancy ground cinnamon (Diet Cin), 2% by weight Hershey's pure cocoa (Diet Coc), or 1% by weight Club House ground thyme (Diet Thy).

The single protein-rich diet was composed of 80%-by-weight protein-free, basal mix and of 20% high-protein casein. One-percent-by-weight Club House ground nutmeg was added to the high-protein diet (Diet HP–Nut) to give it a distinctive flavor and smell. Diet HP–Nut, lacking any palatable sugar, loaded with unpalatable casein (Kon, 1931), and flavored with nutmeg (which previous experiments have suggested was the least preferred of the four flavors we used), proved, as we had intended, to be the least preferred of the four diets when all were offered to hungry young rats in a 2-hr simultaneous choice.

Apparatus

The feeding behavior of subjects and demonstrators was observed in 1 m wide × 0.3 m high × 1 m deep cages that were constructed of angle iron and hardware cloth, floored with galvanized sheet metal, and carpeted to a depth of 2 to 3 cm with woodchip bedding. Each cage (see left panel of Figure 1) contained a single 30 × 30 cm, wooden, nest box with two 5 × 5 cm entrances and a watering station. The diets were presented in round, 10-cm diameter, Pyrex bowls placed in the positions indicated in Figure 1a.

Procedure

The experiment was performed in three stages:

Demonstrator training. Demonstrators were housed individually and trained with a taste-aversion conditioning procedure, to avoid

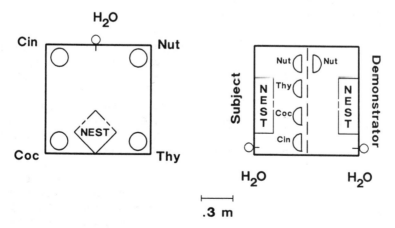

Figure 1. Overhead schematic of apparatuses used in Experiment 1 (left) and Experiments 2 and 3 (right).

eating each of the three, protein-poor diets (Diets Cin, Coc, and Thy). Each demonstrator was first food-deprived for 12 hr and was then given access for 2 hr to a weighed food bowl containing Diet Cin. Following the 2-hr period of access to Diet Cin, each demonstrator was injected intraperitoneally with 10-g/kg, 2% LiCl solution. The demonstrators were next given 34 hr to recover from the effects of toxicosis while maintained ad lib on powdered Purina Laboratory Rodent Chow. After recovery from toxicosis, each demonstrator was trained first to avoid Diet Coc, then to avoid Diet Thy by the same procedure that had been used to condition an aversion to Diet Cin. One hour after pairing of LiCl with Diet Thy, each demonstrator was given ad lib access to high-protein, nutmeg-flavored diet (Diet HP-Nut) for 24 hr.

Given the rationale for the present study, it would have been best to train demonstrators to eat Diet HP-Nut by allowing them to learn that each available alternative diet was inadequate and that eating any one exclusively or all three in combination caused illness. Unfortunately, the older demonstrators became, the more likely they were to attack subjects placed in their cages. If we had not used artificial training techniques to speed the demonstrators' learning of aversions to Diets Coc, Cin, and Thy, they would have become too old during training to allow subjects to be placed safely with them. However, even using the less-than-ideal methods that practical considerations required, we were able to examine the effect that a demonstrator's eating nutritious Diet HP-Nut had on the acquisition of a preference for that diet by naive subjects.

Habituation. Subjects within each age-weight class were randomly assigned to groups and, depending on the group to which a given subject was assigned, placed either alone or with demonstrators in an apparatus. Subjects assigned to the Weanling-No Demonstrator Group ($n = 7$) and to the Adolescent-No Demonstrator Group ($n = 11$) were each placed alone in an apparatus. Subjects assigned to the Weanling-Three Demonstrator Group ($n = 6$), Adolescent-One Demonstrator Group ($n = 8$), and Adolescent-Three Demonstrator Group ($n = 7$) were each placed in an apparatus with either one or three demonstrators, as appropriate.

A bowl containing powdered Purina Laboratory Rodent Chow (Diet P) was placed in the middle of each apparatus, and all subjects and demonstrators were left undisturbed for 24 hr. This 24-hr period of habituation of subjects and demonstrators, both to one another and to the apparatus, had been found in earlier studies to reduce variance in the amount eaten by subjects during the test phase of the experiment described in the next section.

Testing. Immediately after habituation, the bowl containing Diet P was removed from each apparatus and four, 10-cm diameter, round, Pyrex food bowls containing, respectively, Diets Cin, Coc, Thy, and HP-Nut were placed in the corners of the apparatus in the locations indicated in the left panel of Figure 1. For the next week the subjects and demonstrators were left undisturbed, except for daily weighing of subjects and refilling of foodbowls.

Data Analysis

To determine the efficiency of subjects' diet selections in this experiment, we examined the cumulative change in body weight of subjects as a percentage of their respective body weights at the start of the test phase of the experiment. To make these percentage scores suitable for parametric statistical analyses, they were arcsine transformed.

Results and Discussion

The main results of Experiment 1 are presented in Figure 2, which shows the mean cumulative percent changes in body

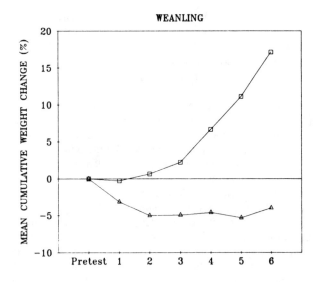

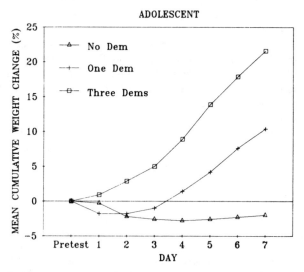

Figure 2. Mean cumulative percentage weight change of weanling (top) and adolescent (bottom) subjects during the test phase of Experiment 1. (Dem = demonstrator).

weight of, respectively, weanling and adolescent subjects that were caged either alone or with trained demonstrators during the test phase of the experiment. As can be seen in the upper panel of Figure 2, weanling subjects with three demonstrators gained a significantly greater percentage of their starting weight during the experiment than did weanling subjects feeding in isolation, Student's $t(11) = 3.20$, $p < .004$.

Similarly, as can be seen in the lower panel of Figure 2, adolescent subjects also benefited appreciably from the presence in their respective enclosures of demonstrators trained to eat the nutritionally adequate Diet HP-Nut. During the 7 days of the experiment, those adolescents interacting with demonstrators gained a significantly greater percentage of their starting body weight than did those adolescents choosing among foods in isolation, $F(2, 23) = 19.74$, $p < .001$.

Post hoc analyses of the weight gains of adolescent subjects revealed (a) that adolescent subjects sharing their enclosure with a single demonstrator gained a significantly greater percentage of body weight than did isolated, adolescent subjects (Tukey's test, $q = 12.4$, $p < .01$) and (b) that adolescent subjects interacting with three demonstrators gained a significantly greater percentage of body weight than did adolescent subjects interacting with but a single demonstrator (Tukey's test, $q = 11.2$, $p < .05$).

The results of Experiment 1 thus demonstrated that both weanling and adolescent rats that lived in an environment where they failed to select a nutritionally adequate diet for themselves chose an adequate diet if they had the opportunity to interact with conspecifics eating that diet. The results of Experiment 1 also revealed certain problems that both guided and restricted the design of later experiments. First, those weanling rats tested in isolation did so poorly in the diet-selection task that many became seriously debilitated during the 6 days of testing. For ethical reasons, it was decided not to continue to use weanlings in this series of studies.

Second, although the data of Experiment 1 provided evidence consistent with the view that interaction with knowledgeable conspecifics facilitated selection of adequate diets by naive, young rats, that evidence was indirect. In three of the five groups, demonstrators and subjects were eating from the same food bowls, making it impossible to determine the actual pattern of diet selection exhibited by subjects. Perhaps subjects with and without demonstrators ate similar proportions of Diets HP-Nut, Thy, Coc, and Cin. The presence of a demonstrator or demonstrators may have either increased the amount of all four diets eaten by subjects or reduced subjects' energy expenditures for thermoregulation, thus enhancing their weight gain by means other than directing feeding to the protein-rich Diet HP-Nut.

Experiment 2

In this experiment the procedures used in Experiment 1 were modified to permit the food intake of adolescent subjects to be measured directly.

Method

Subjects

Twelve, experimentally naive, male rats, weighing 150 to 175 g at the beginning of the experiment, served as subjects. An additional 18, 175- to 200-g, male rats served as demonstrators.

Apparatus

Experiment 2 was conducted in 1 × 1 × .3-m cages, each divided in half by a screen (1-cm grid) that separated each subject from its respective demonstrator (see Figure 1, right panel). Each of the two compartments in each cage (referred to herein as, respectively, the subject's and demonstrator's compartments) contained both a watering station and a 30 × 15 × 15-cm wooden nest box with a single 5 × 5 cm entrance.

Procedure

Experiment 2 was conducted in two stages:

Habituation. Subjects were randomly assigned to No Demonstrator ($n = 6$) and Three Demonstrator ($n = 6$) Groups, and each subject was placed alone in the subject's compartment of an apparatus. The demonstrator's compartment of each apparatus that contained a subject assigned to the No Demonstrator Group was left empty, whereas three demonstrators were placed in the demonstrator's compartment of each apparatus containing a subject assigned to the Three Demonstrator Group.

After the subjects and demonstrators were appropriately distributed, all were given ad lib access for 24 hr to a bowl containing Diet P placed in the middle of their respective compartments. The subjects and demonstrators were then left undisturbed for 24 hr to become habituated to the experimental situation.

Testing. At the end of the 24-hr period of habituation, the food bowls containing Diet P were removed from both demonstrators' and subjects' compartments, and each subject was presented with four, 10-cm diameter, semicircular food cups, each containing a different diet. The four food cups were attached, in the positions indicated in Figure 1, right panel, to the screen partition separating each subject's compartment from its demonstrator's compartment. As also indicated in the right panel of Figure 1, a food cup containing Diet HP-Nut was placed in each demonstrator's compartment directly across the screen partition from each subject's food cup containing Diet HP-Nut. For the following week, the subjects and demonstrators were left undisturbed except for daily weighings of all food cups and subjects.

Results and Discussion

The main results of Experiment 2 are presented in Figure 3, which shows the mean amounts of Diet HP-Nut eaten by subjects in Three Demonstrator and No Demonstrator Groups as a percentage of the total amount that subjects ate daily during testing.

Analyses of food choices of subjects revealed that subjects in the Three Demonstrator Group ate both more Diet HP-

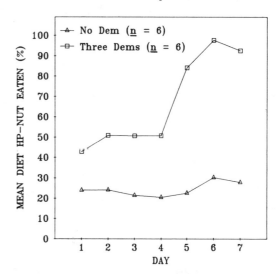

Figure 3. Mean amount of high-protein, nutmeg-flavored diet (Diet HP-Nut) ingested as a percentage of total amount eaten daily by isolated subjects (No Dem) and subjects choosing diets in the presence of demonstrators (Three Dems) during testing in Experiment 2.

Nut, Student's $t(10) = 2.44$, $p = .04$, and a greater percentage of Diet HP–Nut, Student $t(10) = 3.26$, $p < .01$, throughout the 7 days of testing than did subjects in the No Demonstrator Group. As one might expect, the correlation between the total amounts of Diet HP–Nut eaten by individual subjects during the 7 days of the experiment and their total percentage of weight gain during the same period was significantly positive (Pearson's $r = .87$, $p < .001$).

The data clearly indicate that the presence of demonstrators eating Diet HP–Nut significantly increased the intake of that diet by subjects and this increased intake of Diet HP–Nut was highly correlated with increased gains in body weight.

Experiment 3

Taken together, the results of Experiments 1 and 2 show that the success of naive rats in selecting an adequate diet can be influenced by interaction with conspecifics eating that nutritionally adequate diet. However, the results of these experiments provided no information as to how social influence was exerted on subjects by their demonstrators.

Previous studies of social influences on diet choice by rats have shown (a) that both adult and juvenile rats prefer diets that they have smelled on other rats (Galef, Kennett, & Wigmore, 1984; Galef & Stein, 1985; Galef & Wigmore, 1983; Posadas-Andrews & Roper, 1983; Strupp & Levitsky, 1984) and (b) that juvenile rats, but not adult rats, prefer to eat in locations where other rats are feeding (Galef, 1977b; Galef & Clark, 1971). Thus, both local enhancement effects (Thorpe, 1963) and the influence of olfactory cues on food preference (see Galef, 1986, for review) might have been responsible for the observed effects of demonstrators on diet selections by subjects in Experiments 1 and 2. Experiment 3 was undertaken to determine whether (a) the smell of Diet HP–Nut on a demonstrator, (b) the physical presence of a demonstrator in the vicinity of a food bowl containing Diet HP–Nut, or (c) both of these would influence the diet choices of naive, adolescent subjects.

Method

Subjects

Eighteen, experimentally naive, male, 150- to 175-g, Long-Evans rats served as subjects and an additional 18 of their conspecifics, 175 to 200 g in weight, served as demonstrators.

Apparatus

The present experiment was conducted in the apparatus illustrated in the right panel of Figure 1 and described in the *Apparatus* section of Experiment 2.

Procedure

The procedure of this experiment was identical to that of Experiment 2, except in (a) the number of demonstrators with which each subject interacted, (b) the foods fed to demonstrators, and (c) the locations where demonstrators were fed. In this experiment each subject was placed across a screen partition from a single demonstrator. After habituation each subject was assigned to one of three groups.

The subjects assigned to the Same Food–Same Place Group ($n = 6$) each shared an apparatus with a demonstrator that had access to a single food cup containing Diet HP–Nut. Each demonstrator's food cup containing Diet HP–Nut was placed directly across the screen partition from each subject's food cup containing Diet HP–Nut (see Figure 4, Panel A). Thus, subjects assigned to the Same Food–Same Place Group were treated identically to subjects assigned to the Three Demonstrator Group of Experiment 2, except that each subject in the Same Food–Same Place Group of this experiment interacted through the screen partition with a single demonstrator rather than with three demonstrators.

Each subject assigned to the Same Food–Different Place Group ($n = 6$) interacted with a demonstrator eating Diet HP–Nut from a food cup located directly across the screen partition from each subject's food cup containing Diet Cin (see Figure 4, Panel B). As the name of the Same Food–Different Place Group implies, the subjects in this group were exposed to demonstrators eating the high-protein, nutmeg-flavored diet in a location distant from the location where each subject's own bowl of high-protein, nutmeg-flavored diet was placed.

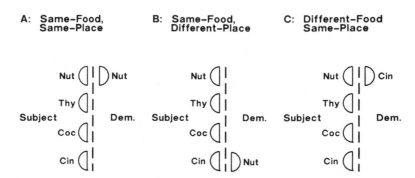

Figure 4. Overhead schematic of the positions of food cups presented to subjects and demonstrators in the Same Food–Same Place (Panel A), Same Food–Different Place (Panel B), Different Food–Same Place (Panel C) Groups in Experiment 3. (Nut = nutmeg-flavored diet; Cin = cinnamon-flavored diet; Thy = thyme-flavored diet; Coc = cocoa-flavored diet; and Dem. = demonstrator).

Last, those subjects assigned to the Different Food–Same Place Group ($n = 6$) each shared an apparatus with a demonstrator eating Diet Cin directly across the screen partition from the subject's food cup containing Diet HP–Nut (see Figure 4c).

For the week of the experiment, the subjects and demonstrators were left undisturbed except for daily weighing of food cups and subjects.

Results

The main results of Experiment 3 are presented in Figure 5, which shows, respectively, the mean amount of Diet HP–Nut eaten by subjects in each of the three groups as a percentage of the total amount eaten daily by subjects during testing (Figure 5, top panel) and the mean percent cumulative weight change exhibited by subjects in each of the three groups (Figure 5, bottom panel).

Because there was extreme heterogeneity of variance across groups in the amount of Diet HP–Nut that they ate ($F_{max} = 20.3$), statistical analyses of diet intake were carried out on log-transformed data. We found a significant effect of treatment on the total amount of Diet HP–Nut eaten by subjects during the 7 days of the test phase of the experiment, $F(2, 15) = 5.61, p < .025$. Furthermore, protected t tests (Wike, 1985) revealed that subjects in the Different Food–Same Place Group ate significantly less Diet HP–Nut during the 7 days of the experiment than did subjects in the Same Food–Same Place Groups (least significant difference [LSD] = .42, $p < .01$) and that subjects in the Same Food–Same Place Group did not differ from subjects in the Same Food–Different Place Group in the percentage of Diet HP–Nut that they ate.

As would be expected, given the observed, positive correlation between the amount of Diet HP–Nut eaten by subjects and their percentage weight gain in the test situation (Pearson's $r = .750, p < .001$), there was also a significant effect of treatment on the percentage weight gain shown by subjects assigned to the three groups, $F(2, 15) = 3.85, p < .05$. Protected t tests revealed that subjects in the Different Food–Same Place Group gained a significantly smaller percent body weight than did subjects in each of the other two groups (LSD = .67, both $ps < .01$)."

Subjects in the Different Food–Same Place Group were exposed to a demonstrator eating Diet Cin in the vicinity of each subject's food cup containing Diet HP–Nut. If the food eaten by demonstrators influenced diet selection by subjects, then one would expect subjects in the Different Food–Same Place Group (i.e., subjects whose demonstrators were eating Diet Cin), to eat a greater percentage of Diet Cin than subjects in either of the other two groups, whose demonstrators were eating Diet HP–Nut.

As can be seen in Figure 6, which shows the mean amount of Diet Cin, as a percentage of the total amount of low-protein diets (Diets Cin, Coc, and Thy) eaten by subjects in each of the three groups during the 7 days of testing, there was a significant effect of treatment on intake of Diet Cin, $F(2, 15) = 5.51, p < .025$. Protected t tests revealed that subjects in the Different Food–Same Place Group ate significantly more Diet Cin than did subjects in either the Same Food–Same Place or Same Food–Different Place Groups (LSD = 29.3, both $ps < .01$).

Discussion

All results of Experiment 3 consistent with the hypothesis that the food choices of adolescent subjects were more strongly influenced by the foods that other rats were eating than by the locations where those rats were feeding. These findings ought not, of course, be used to infer that the location where adult rats eat cannot be an important influence on the feeding behavior of young rats. As mentioned in the introduction to this experiment, in other situations, particularly those where

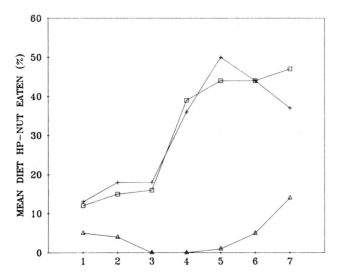
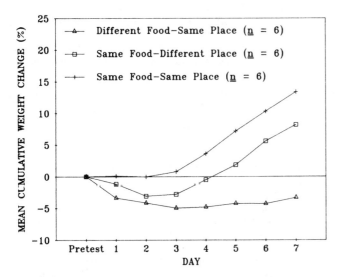

Figure 5. Mean amount of high-protein, nutmeg-flavored diet (Diet HP–Nut) ingested as a percentage of total amount eaten daily by subjects during testing (top) and mean cumulative percentage weight change of subjects (bottom) in the three groups of Experiment 3.

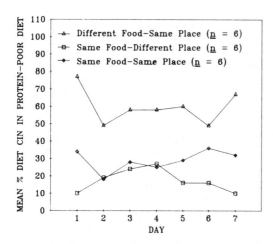

Figure 6. Mean amount of cinnamon-flavored diet (Diet Cin) ingested as a percentage of total amount of protein-poor, cinnamon-, cocoa-, and thyme-flavored diets eaten daily by subjects in the three groups of Experiment 3 during testing.

the same food is available at several locations, the feeding site selection of young rats is profoundly influenced by where adult or juvenile conspecifics are feeding (Galef, 1977a; Galef & Clark, 1971; Strupp & Levitsky, 1984).

General Discussion

In the attempt to explain how generalist feeders, living in the world outside the laboratory, come to select nutritionally adequate diets, attention has focused on the specific hungers, learned aversions, and learned appetites exhibited by individual animals. Little attention has been paid to the fact that during at least the early part of their lives, many generalists live as members of social groups and that naive, group-living animals can exploit more informed others as sources of information about what foods to eat and what foods to avoid eating.

A series of articles from our laboratory have demonstrated profound effects of social interaction on the foraging patterns and diet choices of rats (reviewed in Galef, 1977a, 1986, in press). The results of the present series of experiments extend these earlier findings by showing that in environments where individual rats were unable to learn to select a nutritionally adequate diet from among an array of alternatives, social interaction of naive rats with successful conspecifics could facilitate the acquisition of adaptive diet choices by the naive. Our data suggest that socially acquired information can enable young rats to survive, even to thrive, in environments where many would succumb if they had to depend on their individual abilities to select nutritionally adequate diets.

When a new area is colonized by members of a generalist species, most immigrants may fail to find foods that meet their dietary needs. Many may die. However, those individuals that, for whatever reason, find an adequate diet in circumstances where others fail can serve as models both for less successful contemporaries and for members of future generations. Presence of successful models could make benign otherwise undesirable habitat (Galef & Beck, in press). The ability of rats and, presumably, of other vertebrate generalists to be guided in their choice of foods by the food choices of conspecifics ought not to be ignored in future discussions of the development of adaptive patterns of diet choice by free-living animals.

References

Beck, M. (1989). *The role of social information in the development of preferences for nutritious foods by Long-Evans rats.* Unpublished doctoral dissertation, McMaster University, Hamilton, Ontario, Canada.

Booth, D. A. (1985). Food-conditioned eating preferences and aversions with interoceptive elements: Conditioned appetites and satieties. *Annals of the New York Academy of Sciences, 443,* 22–41.

Epstein, A. N. (1967). Oropharyngeal factors in feeding and drinking. In C. F. Code (Ed.), *Handbook of physiology: Vol. 1. Alimentary canal* (pp. 197–218). Washington, DC: American Physiological Society.

Galef, B. G., Jr. (1977a). Mechanisms for the social transmission of food preferences from adult to weanling rats. In L. M. Barker, M. Best, & M. Domjan (Eds.), *Learning mechanisms in food selection* (pp. 123–150). Waco, TX: Baylor University Press.

Galef, B. G., Jr. (1977b). Social transmission of food preferences: An adaptation for weaning in rats. *Journal of Comparative and Physiological Psychology, 91,* 1136–1140.

Galef, B. G., Jr. (1986). Olfactory communication among rats of information concerning distant diets. In D. Duvall, D. Müller-Schwarze, & R. M. Silverstein (Eds.), *Chemical signals in vertebrates: Vol. 4. Ecology, evolution, and comparative biology* (pp. 487–505). New York: Plenum Press.

Galef, B. G., Jr. (in press). An adaptationist perspective on social learning, social feeding and social foraging in Norway rats. In D. A. Dewsbury (Ed.), *Contemporary issues in comparative psychology.* Sunderland, MA: Sinauer.

Galef, B. G., Jr., & Beck, M. (in press). Diet selection and poison avoidance by mammals individually and in social groups. In E. M. Stricker (Ed.), *Handbook of neurobiology* (Vol. 11). New York: Plenum Press.

Galef, B. G., Jr., & Clark, M. M. (1971). Social factors in the poison avoidance and feeding behavior of wild and domesticated rat pups. *Journal of Comparative and Physiological Psychology, 75,* 341–357.

Galef, B. G., Jr., Kennett, D. J., & Wigmore, S. W. (1984). Transfer of information concerning distant foods in rats: A robust phenomenon. *Animal Learning & Behavior, 12,* 292–296.

Galef, B. G., Jr., & Stein, M. (1985). Demonstrator influence on observer diet preference: Analyses of critical social interactions and olfactory signals. *Animal Learning & Behavior, 13,* 31–38.

Galef, B. G., Jr., & Wigmore, S. W. (1983). Transfer of information concerning distant foods: A laboratory investigation of the "information-centre" hypothesis. *Animal Behaviour, 31,* 748–758.

Guide to the care and use of experimental animals (Vol. 1). (1980). Ottawa: Canadian Council on Animal Care.

Harriman, A. (1955). Provitamin A selection by Vitamin A depleted rats. *Journal of Physiology, 86,* 45–50.

Harris, L., Clay, J., Hargreaves, F., & Ward, A. (1933). The ability of Vitamin B deficient rats to discriminate between diets containing and lacking the vitamin. *Proceedings of the Royal Society, Section B, 113,* 161–190.

Kon, S. (1931). The self-selection of food constituents by the rat. *Biochemical Journal, 25,* 473–481.

Posadas-Andrews, A., & Roper, T. J. (1983). Social transmission of food preferences in adult rats. *Animal Behaviour, 31,* 265–271.

Richter, C. P. (1943). Total self regulatory functions in animals and human beings. *Harvey Lecture Series, 38,* 63–103.

Rozin, P. (1969). Adaptive food sampling patterns in vitamin deficient rats. *Journal of Comparative and Physiological Psychology, 69,* 126–132.

Rozin, P. (1976). The selection of foods by rats, humans and other animals. In J. S. Rosenblatt, R. A. Hinde, E. Shaw, & C. Beer (Eds.), *Advances in the study of behavior* (Vol. 6, pp. 21–76). New York: Academic Press.

Scott, E. M., & Quint, E. (1946). Self-selection of diet IV: Appetite for protein. *Journal of Nutrition, 32,* 293–301.

Strupp, B. J., & Levitsky, D. A. (1984). Social transmission of food preference in adult hooded rats (*Rattus norvegicus*). *Journal of Comparative Psychology, 98,* 257–266.

Thorpe, W. H. (1963). *Learning and instinct in animals* (2nd ed.). London: Methuen.

Westoby, M. (1974). An analysis of diet selection by large generalist herbivores. *American Naturalist, 108,* 290–304.

Wike, E. L. (1985). *Numbers: A primer of data analysis.* Columbus, OH: Charles E. Merrill.

Young, P. T. (1959). The role of affective processes in learning and motivation. *Psychological Review, 66,* 104–125.

Received June 13, 1988
Revision received September 6, 1988
Accepted September 8, 1988 ■

Glossary

conspecifics - members of the same species; a human is the conspecific of a human

dietary generalists - animals that consume a wide variety of foods; for example, rats and humans

foraging - searching the environment for provisions; for example, foraging for food

habituation - in the general sense in which the term <u>habituation</u> is used in this experiment, it refers to getting the subjects accustomed to the test situation

lithium chloride (LiCl) - a toxic substance that is commonly used in taste-aversion conditioning experiments to induce gastrointestinal upset

nutrients - substances in food that are necessary for an organism to develop normally

omnivores - animals that eat both plants and animals

taste-aversion conditioning - a procedure whereby rats learn to avoid particular tastes; this is typically accomplished in the laboratory by allowing subjects to eat a novel-tasting food and then injecting them with a substance, such as lithium chloride (LiCl), that induces gastrointestinal upset; the animals subsequently have an aversion for the novel taste

thermoregulation - the regulation of body temperature

toxic - poisonous

weanling - a young mammal that has recently been weaned, that is, a young mammal that has recently stopped suckling

vivarium - a facility designed for the housing of animals

Essay Study Questions

1. Prior to Beck and Galef's experiments, two complementary processes were thought to account for the ability of animals to select nutritionally adequate diets. What were they?

2. Why are the two processes by which individual animals had been shown to select adequate diets in laboratory experiments insufficient to explain how generalist feeders such as rats can accomplish this task in the real world?

3. Why were the demonstrators in Beck and Galef's experiments first subjected to taste-aversion conditioning? Describe the taste-aversion conditioning procedure.

4. The methods of Experiment 2 were based on shortcomings of Experiment 1. Explain.

5. What did the results of Experiment 3 indicate about the effect of adult feeding sites on the food preferences of young rats?

6. Briefly summarize the results and conclusions of each of the three experiments.

Multiple-Choice Questions

1. There is evidence that wild rats usually select diets rich in protein because
    a. they have an innate preference for the taste of protein.
    b. they develop a preference for the taste of protein if they become protein deficient.
    c. when they are juveniles, they learn from adults to prefer the taste of protein.
    d. both b and c
    e. none of the above; rats cannot taste the protein in their food

2. There is now evidence that rats can learn to select diets rich in protein by
    a. developing learned aversions for a protein-deficient diet that makes them ill.
    b. developing preferences for protein-rich food by experiencing its health-promoting effects.
    c. by watching the feeding patterns of adult rats.
    d. all of the above
    e. both a and c

3. In each of their three experiments, Beck and Galef
    a. tested both weanling and adolescent rats.
    b. compared the effects of one and three demonstrators.
    c. used demonstrators that had been subjected to taste-aversion conditioning.
    d. housed the subjects and demonstrators in the same chamber of the test apparatus.
    e. all of the above

4. Beck and Galef stopped testing weanling rats after Experiment 1
    a. for ethical reasons.
    b. because the weanling rats that were housed with demonstrators became seriously debilitated.
    c. because a few of them continued to suckle.
    d. to increase the statistical significance of their effects.
    e. both a and b

5. Why did Beck and Galef put the protein in the nutmeg-flavored diet rather than in one of the three other diets?
   a. because nutmeg was the least preferred flavor
   b. because nutmeg was the most preferred flavor
   c. to make it difficult for the subjects to learn to select the protein diet
   d. none of the above
   e. both a and c

6. On the basis of their experiments, Beck and Galef concluded that
   a. the places where adult rats eat cannot influence the food choices of adolescents.
   b. the places where adult rats do influence the food choices of adolescents.
   c. the particular foods that adult rats eat influence the food choices of adolescents.
   d. both a and c
   e. both b and c

The answers to the preceding questions are on page 290.

Food-For-Thought Questions

1. Of what relevance are Beck and Galef's results to understanding human food preferences? Can you think of some examples of comparable observational effects on dietary selection in humans?

2. The tendency of animals to learn about dietary selection from "successful" conspecifics is undoubtedly adaptive in most situations for most species. But most species do not have advertising agencies that take advantage of this tendency to sell nonnutritive foods for profit. Discuss.

3. It is not unreasonable to assume that humans have the same abilities to select healthy diets as do rats. Then, why do so many people in wealthy industrialized countries suffer from a combination of obesity and malnutrition?

ARTICLE 15

Human Amnesia and Animal Models of Amnesia: Performance of Amnesic Patients on Tests Designed for the Monkey

L.R. Squire, S. Zola-Morgan, and K.S. Chen
Reprinted from Behavioral Neuroscience, 1988, volume 102, 210-221.

Humans with bilateral damage to the medial temporal lobes (e.g., H.M.) or to the medial diencephalon (e.g., patients with Korsakoff's syndrome) typically suffer from anterograde and retrograde amnesia. A number of laboratory tests have been developed to study the behavioral deficits of monkeys with similarly placed lesions. With the following innovative experiments, Squire, Zola-Morgan, and Chen completed the circle by showing that human amnesics have difficulty performing five memory tasks that were developed to study brain-damage-produced amnesia in monkeys.

The amnesic subjects in these experiments included people with Korsakoff's syndrome and people with brain lesions produced by anoxia (an interruption of oxygen supply) or ischemia (an interruption of blood flow). The control subjects were nonamnesic alcoholics. The subjects were tested on five different tests of monkey memory. Monkeys with medial diencephalic or medial temporal lobe damage had been shown to perform four of these memory tests poorly: (1) the delayed nonmatching-to-sample test, (2) the object-reward-association test, (3) the 8-pair concurrent-discrimination test, and (4) the object-discrimination test. The fifth test was a memory test that monkeys with medial temporal lobe lesions had been shown to perform well: the 24-hour concurrent-discrimination test. Human amnesics are capable of retaining certain kinds of material (e.g., motor skills and multiple-trial discrimination learning), and the 24-hour concurrent-discrimination test was assumed to provide a measure of these surviving kinds of human memory.

The amnesic patients were impaired on all five monkey tests. Accordingly, the performance of the amnesic patients was similar to that of monkeys with medial temporal lobe and medial diencephalic lesions on the delayed nonmatching-to-sample, the object-reward-association, the 8-pair concurrent-discrimination, and the object-discrimination tests; but it was different on the 24-hour concurrent-discrimination test. Squire, Zola-Morgan, and Chen argued that the lack of correspondence between the performance of the amnesic humans and the lesioned monkeys on the 24-hour concurrent-discrimination test suggests that humans and monkeys use a different strategy to perform this test.

It is common for biopsychologists to adapt tests of human brain dysfunction to the study the abilities of other species; it is rare for human abilities to be assessed with tests specifically designed for the study of other species. With this paper, Squire, Zola-Morgan, and Chen lend strong support to the enterprise of modeling human amnesia in laboratory animals.

Human Amnesia and Animal Models of Amnesia: Performance of Amnesic Patients on Tests Designed for the Monkey

Larry R. Squire and Stuart Zola-Morgan
Veterans Administration Medical Center, San Diego
and Department of Psychiatry, School of Medicine,
University of California, San Diego

Karen S. Chen
Department of Neurosciences, School of Medicine,
University of California, San Diego

The performance of amnesic patients was assessed on five tasks, which have figured prominently in the development of animal models of human amnesia in the monkey. The amnesic patients were impaired on four of these tasks (delayed nonmatching to sample, object-reward association, 8-pair concurrent discrimination learning, and an object discrimination task), in correspondence with previous findings for monkeys with bilateral medial temporal or diencephalic lesions. Moreover, performance of the amnesic patients correlated with the ability to verbalize the principle underlying the tasks and with the ability to describe and recognize the stimulus materials. These tasks therefore seem to be sensitive to the memory functions that are affected in human amnesia, and they can provide valid measures of memory impairment in studies with monkeys. For the fifth task (24-hour concurrent discrimination learning), the findings for the amnesic patients did not correspond to previous findings for operated monkeys. Whereas monkeys with medial temporal lesions reportedly learn this task at a normal rate, the amnesic patients were markedly impaired. Monkeys may learn this task differently than humans.

Damage to the medial temporal region or the midline diencephalic region of the human brain causes amnesia in the absence of other intellectual impairment (Scoville & Milner, 1957; Squire, 1986; Victor, Adams, & Collins, 1971; Zola-Morgan, Squire, & Amaral, 1986). Recent studies have succeeded in establishing an animal model of amnesia in the monkey (for reviews, see Mishkin, 1982; Squire & Zola-Morgan, 1983); for example, bilateral lesions of the medial temporal region in monkey produced severe anterograde and retrograde amnesia on several memory tasks (Mishkin, 1978; Salmon, Zola-Morgan, & Squire, 1987; Zola-Morgan & Squire, 1985a). The lesions, which included hippocampal formation, entorhinal cortex, parahippocampal gyrus, and amygdala, were intended to reproduce the surgical removal sustained by the well-studied amnesic patient H. M. (Corkin, 1984; Scoville & Milner, 1957). Bilateral lesions of medial thalamic structures, including the mediodorsal thalamic nucleus, also produced memory impairment (Aggleton & Mishkin, 1983a, 1983b; Zola-Morgan & Squire, 1985b). As in human amnesia (Cohen, 1984; Squire & Cohen, 1984), the memory impairment in monkeys was selective in that the ability to acquire new skills was spared (Zola-Morgan & Squire, 1984).

The development of the animal model has depended on behavioral tasks designed for the monkey, which are analogous to ones failed by human amnesic patients; for example, some tasks require monkeys to retain information across a delay or to learn multiple associations concurrently. Other tasks are designed to be analogous to ones that human amnesic patients can perform normally, for example, tasks of motor skill learning or pattern discrimination tasks, which resemble skill learning in that they involve repetition and incremental improvement over many trials. However, little information is available about how human amnesic patients would perform if they were given precisely the same tasks that have been used with monkeys (but see Aggleton, Nicol, Huston, & Fairbairn, in press; Oscar-Berman & Zola-Morgan, 1980a, 1980b, 1982); and no information is available at all in the case of delayed nonmatching to sample, the task most widely used in studies of memory impairment in the monkey.

Would the tasks failed by monkeys with medial temporal or diencephalic lesions also be failed by amnesic patients? Would they be so easy for human subjects that only the most severely amnesic patients would fail them? Are the tasks that operated monkeys succeed at also ones that amnesic patients could perform normally? Whereas it is clear that motor-skill learning is intact both in monkeys with medial temporal or diencephalic lesions (Zola-Morgan & Squire, 1984) and in amnesic patients (Brooks & Baddeley, 1976; Cohen, 1981; Corkin, 1968), certain other tasks that can be learned well by

---

This work was supported by the Medical Research Service of the Veterans Administration, by National Institute of Mental Health Grant MH 24600, and by the Office of Naval Research.

We thank Joyce Zouzounis, Armand Bernheim, Patty Feldstein, Jody Lee, Brian Leonard, and Kim Rivero-Frink for research assistance, and Art Shimamura for his comments on this article.

Correspondence concerning this article should be addressed to Larry R. Squire, Veterans Administration Medical Center (V116A), 3350 La Jolla Village Drive, San Diego, California 92161.

operated monkeys have not been given to amnesic patients. Specifically, monkeys with medial temporal lesions can learn a 24-hour concurrent discrimination task (Malamut, Saunders, & Mishkin, 1984), in which 20 object pairs are presented once daily and the same member of each pair is always rewarded.

The present study assessed the ability of amnesic patients to perform five different tasks that have figured prominently in the development of animal models of human amnesia in the monkey. Four of the tasks (delayed nonmatching to sample, object-reward association, 8-pair concurrent discrimination learning, and object discrimination) are failed by monkeys with bilateral medial temporal or diencephalic lesions (Aggleton & Mishkin, 1983b; Mahut, Zola-Morgan, & Moss, 1982; Mishkin, 1978; Moss, Mahut, & Zola-Morgan, 1981; Phillips & Mishkin, 1984; Zola-Morgan & Squire, 1985a). The fifth task (24-hour concurrent discrimination learning) can be successfully learned by monkeys with medial temporal lesions and has been considered to provide a measure of the kind of learning ability that is spared in amnesia (Malamut et al., 1984).

Experiments 1A and 1B: Delayed Nonmatching to Sample

Method

Subjects

Patients with Korsakoff's syndrome. This group consisted of 3 men and 2 women (Patients K1, K3, K4, K5, and K6 in Squire & Shimamura, 1986) living in supervised facilities in San Diego County. They averaged 50.4 years of age during the final year of the study, had 12.2 years of education, and had an average full-scale Wechsler Adult Intelligence Scale (WAIS) score of 101.6 (WAIS-R score = 94.2). Their average Wechsler Memory Scale (WMS) score was 77.4, and the average WAIS − WMS difference score was 24.2 (range = 16 to 38). On the Wechsler Memory Scale-Revised (WMS-R), the average index scores were as follows: Attention and Concentration, 83.8; Verbal Memory, 70.8; Visual Memory, 69.2; General Memory, 63.4; Delayed Memory, 53.6.

Descriptions of the following tests as well as scores for two control groups are presented in Squire and Shimamura (1986). Scores for one of the control groups in that study (alcoholic subjects) are also included here to facilitate interpretation of the scores obtained by the amnesic patients.

For immediate and delayed (12 min) story recall, the patients averaged 4.0 and 0.0 segments, respectively (maximum score = 21 segments; alcoholics = 6.7 and 4.9). For copy and delayed (12 min) reconstruction of the Rey-Osterreith figure, they averaged 26.6 and 1.4, respectively (maximum score = 36; alcoholics = 28.9 and 16.4). For paired-associate learning (10 word pairs; see Jones, 1974), the average total score after three study/test trials was 3.0 (maximum score = 30; alcoholics = 21.9). On the Rey Auditory Verbal Learning Test (15 words), free recall on five successive study/test trials was 3.2, 4.0, 4.4, 4.8, and 4.8. For yes/no recognition of 15 old words and 15 new words, the average scores on five successive study/test trials were 18.2, 23.0, 23.8, 25.2, and 26.0. For free recall, alcoholics averaged 6.1, 8.0, 9.1, 10.6, and 12.4; for recognition, they averaged 26.2, 29.4, 29.4, 29.8, and 29.9. The average Dementia Rating Scale score (Coblentz, Mattis, Zingesser, Kasoff, Wiesniewski, & Katzman, 1973) was 128.0 (maximum score = 144; alcoholics = 139.1), and the average score excluding the memory subscale score was 110.2 (maximum score = 119; alcoholics = 114.1). A high score on this test signifies good performance.

Additional amnesic cases. This group of 3 males (Patients AB, GD, and LM from Squire & Shimamura, 1986) had memory impairment resulting from either an anoxic or ischemic episode. They averaged 50.3 years of age, had 15.7 years of education, and had an average WAIS score of 116.3 (WAIS-R score = 107.3). Their WMS score averaged 92.3. The average WAIS − WMS difference score was 24. On the WMS-R, the average index scores were as follows: Attention and Concentration, 109.3; Verbal Memory, 78.3; Visual Memory, 85.3; General Memory, 76.3; Delayed Memory, 58.3.

For immediate and delayed story recall, these 3 patients averaged 7.0 and 0.0, respectively. For copy and delayed reconstruction of the Rey-Osterreith figure, they averaged 31.3 and 7.2, respectively. For paired-associate learning, the average total score for three trials was 4.3. On the Rey Auditory Verbal Learning Test, free recall on five successive study/test trials was 4.0, 5.7, 6.7, 6.0, and 5.7. For yes/no recognition of 15 old words and 15 new words, the average score on five successive study/test trials was 26.0, 25.7, 25.7, 25.3, and 27.3. The average Dementia Rating Scale score was 135.0; the average score excluding the memory subscale was 116.3.

This detailed description of the amnesic patients is intended to facilitate comparisons between this study and other studies that involve different patients or different etiologies of amnesia. However, our interest in the present study is in the overall performance of amnesic patients, not in possible differences between groups of patients. Accordingly, our data analysis presents all the amnesic patients together. The patients were tested on the tasks described in Experiments 1 to 5 during a 4-year period, 1982 to 1986.

Alcoholic control group. Two separate groups of alcoholic subjects were tested, one in Experiment 1A, and the other one in Experiment 1B. All were current or former participants in alcoholic treatment programs in San Diego County. The first group (7 men and 1 woman) had an average drinking history of 19 years but had abstained from alcohol an average of 9 weeks prior to participating in the study. There was no history of liver disease or severe head injury (i.e., no episode of unconsciousness lasting more than 1 hour). They averaged 53 years of age, had 13.1 years of education, and had WAIS-R subtest scores of 18.6 for Information (18.9 for the amnesic patients) and 42.4 for Vocabulary (49.4 for the amnesic patients). On a test of story recall, they scored 5.9 for immediate recall and 4.9 after a 12-min delay.

The second group of alcoholic subjects (4 men and 2 women) had an average drinking history of 14 years, with an average of 1.3 years of abstinence prior to the present study. They averaged 55 years of age, had 13 years of education, and had WAIS-R subtest scores of 19.0 for Information and 43.2 for Vocabulary. On immediate and delayed story recall, they scored 6.3 and 6.0, respectively.

Materials

A table-top version of the Wisconsin General Testing Apparatus (WGTA) was constructed for use with human subjects (Oscar-Berman & Zola-Morgan, 1980a). The WGTA was placed on a table between the experimenter and the subject. An opaque sliding screen could be raised by the experimenter to reveal a stimulus tray (53 × 28 cm), which could then be slid forward within reach of the subject. The stimulus tray consisted of three reinforcement wells equidistant from each other. The stimuli consisted of easily discriminable junk objects (e.g., a toy gun, a yogurt container, and a plastic block).

Procedure

Experiment 1A. Subjects were first trained during a single session on the basic version of the delayed-nonmatching-to-sample task. Subjects were instructed that each time the screen was raised, a penny would always be under one of the objects, and that they should try to obtain the penny every time. Each trial consisted of two parts. First, a single object was placed over the central well covering a penny reward. The screen was then raised, and the subject was able to displace the object and retrieve the penny. After a delay of 5 s with the screen down, the original object and a second, novel object were presented, each covering a lateral well of the stimulus tray. The penny was always under the novel object. A new pair of objects was used for every trial (intertrial interval = 5 to 10 s), and the position of the novel object (left or right) varied according to a pseudorandom schedule (Gellermann, 1933). The two objects for each trial were drawn randomly from a pool of 100 junk objects. Trials continued until a subject achieved a learning criterion of 9 correct out of 10 consecutive trials. At that point, they were asked to state in their own words the principle that determined where the penny was hidden (e.g., "The penny is always under the *new* object.").

On the following day, subjects were retrained on the basic task to a criterion of 9 correct out of 10 trials. At that point, they were again asked to state the principle that determined the placement of the penny. Subjects were then given 50 additional trials, which required that they remember the sample stimulus across delays of up to 60 s. The 50 trials consisted of 10 trials with a 5-s delay, 20 trials with a 15-s delay, and 20 trials with a 60-s delay. Half of the 15-s and 60-s delays were filled with a distraction task, which involved placing groups of random numbers into correct numerical order. The five kinds of trials (5-s delay, 15-s unfilled delay, 15-s filled delay, 60-s unfilled delay, and 60-s filled delay) were presented in a mixed fashion, such that each kind of trial was distributed evenly through the 50 trials. After all 50 trials were completed, subjects were asked a final time to state the principle that determined the placement of the penny.

Experiment 1B. This experiment was designed to assess how well patients could remember the sample object across a delay, independently of their ability to remember the nonmatching principle. To accomplish this objective, a card stating the principle was given to the subjects during testing. In addition, nonmatching-to-sample performance was tested on 2 consecutive days: first without a distraction task and then with the same distraction task used in Experiment 1A. Thus, on the first day, no distraction task was used and all the delays were unfilled. On the second day, just as in Experiment 1A, half of the 15-s and 60-s delays were filled with a distraction task. On average, 40.6 months (range = 8 to 49) elapsed between Experiments 1A and 1B. During this interval, the amnesic patients had received additional testing on delayed nonmatching to sample, using various experimental designs, such that when Experiment 1B began, they had received an average of 473 trials on the task (range = 157 to 1,149).

Subjects were first trained on the basic task as before, with a delay of 5 s. At that point, they were given a card stating the principle, and the card remained in view throughout the remainder of testing. The card read, "When there are two objects, the penny will always be under the novel object, *not* the one you just saw." With the card in place, subjects were given 50 additional trials in mixed order: 10 trials with a 5-s delay, 20 trials with a 15-s delay, and 20 trials with a 60-s delay. No distraction tasks were used on this test day.

On the following day, subjects were tested again in the identical way, except that now some of the delays were filled with a distraction task. Thus subjects were trained to criterion, given the card, and then given 50 additional trials in mixed order: 10 trials with a 5-s delay, 10 trials with a 15-s unfilled delay, 10 trials with a 15-s filled delay, 10 trials with a 60-s unfilled delay, and 10 trials with a 60-s filled delay. (The distraction task in this case involved sorting junk objects rather than ordering random numbers).

A final test was carried out with the same amnesic patients to determine whether the information provided on the card was essential to sustain performance, or whether (with all the practice they had received) patients could perform reliably based on the nonmatching principle. This final test was given 6.3 months later (range = 3 to 8), and patients were tested in the same manner as before, except that now the card stating the nonmatching principle was not provided. At this time, the amnesic patients had received an average of 603 total trials on delayed nonmatching to sample (range = 277 to 1,274).

Patients were first trained on the basic task to a criterion of 9 out of 10 correct trials and then asked to state the principle that determined the location of the penny. On the second day, they were retrained on the basic task to criterion performance and again asked to state the nonmatching principle. They were then given 50 additional trials in mixed order: 10 trials with a 5-s delay, 20 unfilled trials with a 15-s delay, and 20 unfilled trials with a 60-s delay. No distraction tasks were used on this test day.

On the following day, patients were tested again in an identical way, except that some of the delays were now filled with a distraction task. Thus, they were trained to criterion, asked to state the principle, and then given 50 additional trials in a mixed order: 10 trials with a 5-s delay, 10 trials with a 15-s unfilled delay, 10 trials with a 15-s delay (with the distraction task described in Experiment 1A), 10 trials with a 60-s unfilled delay, and 10 trials with a 60-s delay (and distraction task). At the completion of these 50 trials, subjects were asked to state the principle a final time.

Results

Experiment 1A

All subjects reached the learning criterion for the basic task on the first test day, with a delay of 5 s between presentation of the sample and the choice. The alcoholic subjects took a median of 1 trial to reach criterion (not including the run of 9 out of 10 correct criterion trials); the amnesic patients required 29 trials ($p < .05$). On the next day, the alcoholic subjects required 0 trials to reach criterion; the amnesic patients required 9 trials ($p = .05$). When the delay between presentation of the sample and the choice was then varied between 5, 15, and 60 s, the amnesic patients were markedly impaired (Figure 1). A 2 × 3 analysis of variance (ANOVA; Groups × Unfilled Delays) revealed a significant effect of group, $F(1, 14) = 19.1$, $p < .001$, but no effect of delay and no interaction, $ps > .1$. These findings show that the amnesic patients were impaired to an equivalent extent across all the delay intervals. They did achieve 90% correct performance on 2 consecutive days during training on the basic task. However, they were unable to sustain this level of performance during 50 succeeding trials, even on those trials when the delay interval was the same (5 s) as had been used during initial training.

A 2 × 2 × 2 ANOVA (Groups × Delays × Conditions; distraction task vs. no distraction task) confirmed the difference between the two groups, $F(1, 14) = 31.0$, $p < .001$, but revealed no other main effects and no interactions, $ps > .1$. In particular, performance was no different across unfilled delays than across delays that were filled by the distraction task.

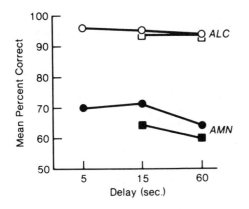

Figure 1. Delayed nonmatching to sample. (After reaching criterion on the basic nonmatching task [using a 5-s delay], subjects were tested at three different delays, with [square symbols] or without [circles] a distraction task interposed during the delay. ALC = alcoholics, $n = 8$; AMN = amnesics, $n = 8$.)

Experiment 1B

The basic task was learned rapidly by both groups (alcoholics: median = 0 trials; amnesics: median = 0.5 trials), and from this point on, no group required a median of more than one trial to reach criterion. Figure 2A shows performance when no distraction task was given, once when the nonmatching principle was provided on a card during testing (solid lines) and once 6 months later when the principle was not provided (dashed lines). The figure shows that the amnesic patients performed almost identically whether or not the card was provided ($p > .1$). For the amnesic patients, a 2 × 3 repeated measures ANOVA (Testing Occasions × Unfilled Delays) revealed a significant effect of delay on nonmatching performance, $F(2, 14) = 5.8, p < .05$. Performance was normal at the 5-s delay and poorer at the longer delays. In addition, on each of the two testing occasions, the amnesic patients and the alcoholic subjects performed differently across the delays. The Group × Delay interaction was significant for the first set of scores (solid lines, nonmatching principle available on a card) obtained by the amnesic patients, $F(2, 24) = 4.3, p < .05$, and just short of significance for the second set of scores (dashed lines, nonmatching principle not available) obtained by the amnesic patients, $F(2, 24) = 3.0, p < .07$.

Figure 2B shows the results on the following day when a distraction task was given during half of the 15-s and 60-s delay intervals. For the amnesic patients, a 2 × 3 repeated measures ANOVA (Testing Occasions × Unfilled Delays) revealed a significant effect of delay, $F(2, 14) = 4.0, p < .05$. Performance was good at the 5-s delay and poorer at the longer delays. In addition, the amnesic patients performed more poorly overall than the control subjects. This difference was significant in the case of the first set of scores (solid lines) obtained by the amnesic patients (2 groups × 3 unfilled delays), $F(1, 12) = 10.3, p < .01$. The same comparison fell just short of significance in the case of the second set of scores (dashed lines) obtained by the amnesic patients, $F(1, 12) = 3.6, p < .08$. The Group × Delay interactions did not reach significance, $Fs < 2.0, ps > 0.1$.

Effect of the Distraction Task

Figure 2B shows that just as in Experiment 1A, performance on filled and unfilled trials was about the same. Thus, in one sense, the distraction task had no measurable effect on nonmatching performance. However, the distraction task did have an overall disruptive effect on performance. That is, simply presenting the distraction task during some of the delay intervals impaired performance across all the delay intervals. This point follows from a comparison of Figures 2A and 2B. Amnesic patients performed better (average of 3 delays = 91.1%, Figure 2A) when no distraction tasks were presented during the test day than they did when distraction tasks were included (average of 3 delays = 82.1%, Figure 2B).

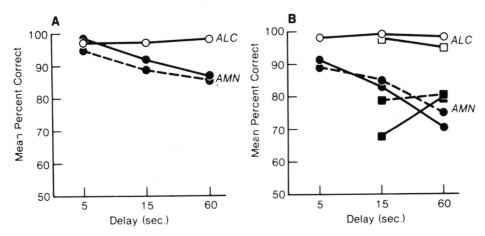

Figure 2. Delayed nonmatching to sample 41 months after the test in Figure 1. (A. Performance across three delays. Solid lines: Subjects were given a card stating the nonmatching principle. Dashed lines: On a second test, 6 months later, the card was not provided. B: Performance across three delays, with [square symbols] or without [circles] a distraction task interposed during the delays. Solid lines: Subjects were given a card stating the nonmatching principle. Dashed lines: On a second test, 6 months later, the card was not provided. ALC = alcoholics, $n = 6$; AMN = amnesics, $n = 8$.)

This difference was significant in the case of the first set of scores (solid lines) obtained by the amnesic patients, $F(1, 17) = 16.0$, $p < .01$, and just short of significance in the case of the second set of scores (dashed lines), $F(1, 7) = 4.7$, $p < .07$.

Effect of Practice

Although the amnesic patients performed poorly across all three delay intervals when they were first tested on the nonmatching-to-sample task (Figure 1), they performed better when they were tested for the last time (Figures 2A and 2B, dashed lines). The fairest comparison is between the data for the amnesic patients in Figures 1 and 2B (dashed lines), which were collected under identical conditions but with several hundred intervening test trials on the nonmatching task. Overall, averaging across the five conditions (three unfilled delays and two filled delays), the patients initially averaged 66.0% correct (Figure 1) and later averaged 81.5% correct (Figure 2B, dashed lines), $F(1, 7) = 3.4$, $p < .10$. When they were last tested (Figure 2B, dashed lines), the amnesic patients seemed able to perform reliably on the basis of the nonmatching principle, and performance was now influenced by the length of the delay interval.

Verbalizing the Principle for Delayed Nonmatching to Sample

Subjects were asked to state the nonmatching principle a total of 8 times during the course of testing. The ability to verbalize the principle paralleled performance on the task itself. Thus every alcoholic subject was able to state the principle on every occasion that it was asked for. By contrast, only 1 amnesic patient was able to state the principle consistently. In addition, within the amnesic group, those who were able to state the principle performed better than those who could not state it. When performance was averaged across all the conditions on which the amnesic patients were tested (excluding the tests on which they were provided with the principle on a card), the correlation between nonmatching performance and the ability to state the nonmatching principle was $r = .71$ ($p < .05$).

Experiment 2: Object–Reward Association

Method

Subjects

Amnesic patients. This group consisted of the 8 patients tested in Experiment 1 plus an additional 2 patients with Korsakoff's syndrome (1 male and 1 female). They were 59 and 76 years of age when they completed testing in 1984, they had 16 and 12 years of education, and they had WAIS scores of 103 and 101 (WAIS-R scores of 95 and 94, respectively). Their WMS scores were 64 and 80, and their WAIS – WMS difference scores were 39 and 21. WMS-R scores are not available for these two patients. In addition, these 2 patients averaged 2.7 and 0.5 for immediate and delayed story recall, 2.8 for three trials of paired-associate learning, and 31.0 and 2.5 for copy and delayed reconstruction of the Rey-Osterreith figure.

Alcoholic control group. This group consisted of 11 men who were current or former participants in alcoholic treatment programs in San Diego County. They had an average drinking history of 21.5 years and had abstained from alcohol an average of 5 weeks prior to participating in this study. They averaged 50.5 years of age, had 13.4 years of education, and had WAIS-R subtest scores of 22.4 for Information (18.4 for the 10 amnesic patients) and 56.9 for Vocabulary (48.6 for the amnesic patients). They averaged 7.1 on the test of immediate story recall and 6.2 after a 12-min delay.

Materials

Stimuli were presented in the WGTA (see Experiment 1).

Procedure

Each trial consisted of three parts. First, an object was placed over the center well of the stimulus tray. The screen was then raised, and the subject displaced the object. Then, after 5 s with the screen lowered, a second object was presented. One of the two objects always had a penny hidden beneath it. Finally, after another 5-s delay, the same two objects were presented together, each covering a lateral well of the stimulus tray, and a penny placed beneath the previously rewarded object. On each trial, a new pair of objects was used (intertrial interval = 5 to 10 s), and the position of the rewarded object on the choice phase (left or right) was varied according to a pseudorandom schedule (Gellermann, 1933). The two objects for each trial were drawn randomly from a pool of 100 junk objects. Subjects were trained on this basic task to a learning criterion of 9 out of 10 consecutive trials correct or until 200 trials had been given. When subjects reached criterion performance (or the 200-trial limit), they were asked to state the principle that determined placement of the penny: "When there are two objects, the penny will always be under the object that had the penny under it before."

On the next day, subjects were again trained to criterion on the basic task (or to the 200-trial limit) and were once again asked to state the principle. (One subject who completed 200 trials without reaching criterion on either day was told the principle at this point and then given 10 additional trials to verify that it was understood.)

Immediately afterwards, subjects were given 50 trials: 10 trials with a 5-s delay, 10 unfilled trials with a 15-s delay, 10 filled trials with a 15-s delay, 10 unfilled trials with a 60-s delay, and 10 filled trials with a 60-s delay. The order of the trials and the filler task were the same as in Experiment 1A.

Results

The results are shown in Figure 3. Subjects learned the basic task on the first day (trials to criterion, not including the run of 9 out of 10 correct criterion trials: alcoholic subjects, median = 1 trial; amnesic patients, median = 5.5 trials; $p > .1$). On the next day, subjects relearned the task (alcoholic subjects, median = 0 trials; amnesic patients, median = 2.5 trials; $p > .1$). When the delay between presentation of the second object and the choice trial was then varied between 5, 15, and 60 s, the amnesic patients were markedly impaired. A 2 × 3 ANOVA (Groups × Unfilled Delays) revealed a significant effect of group, $F(1, 19) = 19.5$, $p < .01$, an effect of delay, $F(2, 38) = 5.1$, $p < .02$, and a Group × Delay interaction, $F(2, 38) = 5.0$, $p < .02$. These findings show that the amnesic patients were impaired overall and that the impairment increased as the delay increased.

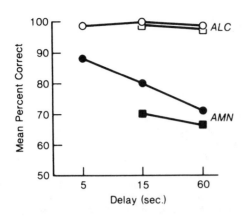

Figure 3. Object–reward association. (After reaching criterion with a 5-s delay, subjects were tested at three different delays, with [square symbols] or without [circles] a distraction task interposed during the delay intervals. ALC = alcoholics, $n = 11$; AMN = amnesics, $n = 10$.)

Effect of the Distraction Task

A second ANOVA evaluated the effects of the distraction task, which was included in half of the 15-s and 60-s delay intervals (2 groups × 2 delays × 2 conditions [distraction task vs. no distraction task]). The effects of group and distraction condition were significant, $Fs(1, 19) > 8.2$, $ps < .01$. The only significant interaction was the effect of Group × Distraction Condition, $F(1, 19) = 7.3$, $p < .05$. These results show that performance was adversely affected by the distraction task and that the amnesic patients were affected by distraction more than the control subjects.

Verbalizing the Principle for Object–Reward Association

The ability to state the principle that determined the location of the penny was related to performance on the object-reward-association task itself. Thus every alcoholic subject was able to state the principle correctly both times that it was asked for. By contrast, only 5 of the 10 amnesic patients were able to state it consistently, and 4 patients were never able to state it. (One patient could state the principle on one occasion but not on the other.) To evaluate these observations statistically, each subject was assigned a score representing the number of times that the principle could be stated (0, 1, or 2). The difference between the amnesic and alcoholic groups was significant, $t(19) = 3.0$, $p < .01$.

In addition, within the amnesic group, those who were able to state the principle tended to perform better than those who could not. Thus, averaging performance across all five test conditions (3 unfilled intervals and 2 filled intervals), the 5 patients who were always able to verbalize the principle scored 81% correct, and the other 5 patients scored 69% correct. One patient in the latter group had been told the principle after failing twice to reach criterion on the basic task within the 200-trial limit and twice failing to state the principle. When this individual was excluded, the 4 remaining patients who were unable to state the principle averaged 62% correct, and the difference (81% vs. 62%) fell just short of significance, $t(7) = 2.2$, $p = .06$.

Experiment 3: 8-Pair Concurrent Discrimination Learning

Method

Subjects

Amnesic patients. See Experiment 2.

Alcoholic control group. This group (7 males and 1 female) consisted of current or former participants in alcoholic treatment programs in San Diego County. They had an average drinking history of 24 years but had abstained from alcohol for 30.5 weeks prior to participating in the study. They averaged 48.6 years of age, had 11.6 years of education, and had WAIS-R subtest scores of 17.9 for Information and 46.3 for Vocabulary (18.4 and 48.6, respectively, for the 10 amnesic patients). For story recall, they scored 8.0 and 7.0 on immediate recall and 12-min delayed recall, respectively.

Materials

Stimuli were presented in the WGTA (see Experiment 1).

Procedure

Days 1–3. Eight pairs of junk objects were used. On each trial, subjects saw a pair of objects covering the lateral wells of the stimulus tray. One of the two objects had been arbitrarily designated as the positive object, and that object could be displaced to reveal a penny reward. The same object remained correct throughout testing, and the position of the correct object (left or right) varied according to a pseudorandom schedule (Gellermann, 1933). The eight pairs were presented in a mixed fashion, so that the eight different discriminations had to be learned simultaneously. Each pair was presented 5 times each day. Testing consisted of 3 consecutive daily sessions of 40 trials each (intertrial interval = 5 to 10 s). At the end of testing on the third day, subjects were asked to state the principle that determined where the penny was hidden; for example, "The penny will always be under the same, previously rewarded object of each pair." Subjects were also asked how many different pairs of objects they had seen, and they were asked to describe any of the objects they could remember.

Day 30. Thirty days after Day 3, subjects were given 40 additional trials with the same eight pairs, following the same procedure described above.

Day 37. Seven days later, subjects were given a recognition test for the eight pairs of objects. On each of 16 trials, one object that had been used in the task (one of the eight correct or eight incorrect objects) was placed together with two other objects that had not been seen before. The position of the familiar object on the tray was random, as was the order of presentation of the objects. Subjects were asked in each case to identify the familiar object.

Results

The results are shown in Figure 4. The amnesic patients were impaired at learning eight object discriminations con-

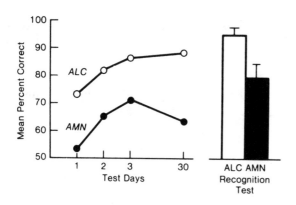

Figure 4. Eight-pair concurrent discrimination. (Learning of eight object discriminations on Days 1, 2, 3, and 30 [40 trials/day]. On Day 37, subjects were given a recognition test for the 16 objects they had seen. ALC = alcoholics, $n = 6$; AMN = amnesics, $n = 10$.)

currently during three consecutive days of testing, $F(1, 14) = 5.2$, $p < .05$, and performance was similarly impaired on the 30-day retention test, $t(14) = 3.2$, $p < .01$. Nevertheless, the amnesic patients were capable of some learning, as indicated by the fact that on Day 30, they performed significantly above chance, $t(9) = 2.7$, $p < .05$. Finally, on Day 37, the amnesic patients were impaired on a test that asked them to recognize each of the 16 objects they had previously seen, $t(14) = 2.2$, $p < .05$.

At the end of the third day of testing, when subjects were asked to verbalize the principle that determined where the penny was hidden, 5 of the 6 control subjects, but only 3 of the 10 amnesic patients, correctly stated that the penny was always under the same object. In general, the ability of subjects at this time to answer questions about the task (specifically, how many pairs of objects had been presented and what objects had been presented) paralleled performance on the discrimination task itself. Thus, at the end of the third day of testing, the control subjects reported that 6.0 different object pairs had been presented (range = 5 to 12), and the amnesic patients reported that 13.7 different object pairs had been presented (range = 5 to 50). Similarly, the control subjects successfully described an average of 8.0 of the 16 objects (range = 5 to 12), whereas the amnesic patients could describe only 3.7 of them (range = 0 to 7), $t(14) = 3.7$, $p < .01$.

Finally, within the amnesic group, the ability at the end of the third test day to answer the two questions about the task (How many pairs of objects? What were the objects?) was correlated with the score obtained up to that point on the discrimination task, that is, during the first 3 days of testing. The Spearman rank-order correlations were $r = .43$ (percent correct score on the discrimination test vs. accuracy on the question of how many object pairs had been presented) and $r = .50$ (percent correct score on the discrimination test vs. the number of objects described correctly). When the data for these two questions were combined by averaging the two ranks assigned to each patient, the ability of the patients to answer the questions correlated significantly with their scores on the discrimination test ($r = .71$, $p < .05$).

Experiment 4: Object Discrimination Learning

Method

Subjects

Amnesic patients. This group consisted of the same 7 patients with Korsakoff's syndrome tested in Experiments 2 and 3 and 2 of the 3 amnesic patients (AB and GD) described in Experiment 1.

Alcoholic control group. This group consisted of 13 men and 1 woman who were current or former participants in alcoholic treatment programs in San Diego County. They had an average drinking history of 18.1 years but had abstained from alcohol an average of 7.1 weeks prior to participating in this study. They averaged 51.2 years of age, had 12.8 years of education, and had WAIS-R subtest scores of 19.4 for Information (17.9 for the amnesic patients) and 44.7 for Vocabulary (47.2 for the amnesic patients). They scored 6.9 and 6.5, respectively, on tests of immediate and delayed story recall.

Materials

The stimuli were presented in the WGTA (see Experiment 1).

Procedure

Subjects learned in sequence three different object discriminations, involving three different pairs of easily discriminable junk objects. For each pair, one object was positive and was consistently rewarded throughout testing. Subjects were told that each time the screen was raised, a penny would always be under one of the two objects, and that they should try to obtain the penny every time. The screen was then raised, and the subject could obtain a penny by displacing the positive object. The placement of the positive object over either the left or right lateral well of the stimulus tray was pseudorandomized (Gellermann, 1933). Testing continued with an intertrial interval of 5 to 10 s until a run of 9 correct out of 10 consecutive trials was achieved. The next day, subjects were again retrained to the criterion of 9 out of 10 correct trials. Training on the second pair of objects began one week later, following the same procedure. Subjects were trained to criterion during one day of testing, and on the next day they were retrained to the same criterion. Finally, one week later, the third pair of objects was presented. The procedure for this third pair of objects was the same as for the first two pairs, except that 9 to 11 days intervened between training and retraining.

Results

Amnesic patients took longer to acquire the first object discrimination than did control subjects (trials to criterion, not including the run of 9 out of 10 correct criterion trials: amnesic patients, median = 17 trials; alcoholic subjects, median = 4 trials; $p = .05$). Relearning the same object pair one day later also took longer for the amnesic patients than the control subjects (amnesic patients, median = 2 trials; alcoholic subjects, median = 0 trials; $p < .01$). Apparently, subjects readily acquired the principle that one of the two objects was consistently rewarded. As a result, the second and third object pairs were learned (and relearned) quickly by both groups in a median of 0 trials to criterion. (Because one error is permitted in a run of 9 out of 10 criterion trials, a subject who makes one error would be scored as reaching criterion in 0 trials.) If a subject understands the principle governing discrimination learning, no more than one error should ever be made in learning any discrimination pair. The first trial serves as an instruction trial to indicate which member of the pair is the rewarded one, and performance should be correct after the first trial.

Given these characteristics of object discrimination tasks, the performance of amnesic patients is best evaluated by considering the first trial of the retention test. Figure 5 shows results for all three object discrimination tasks. The alcoholic subjects averaged 77% correct on the first trial of the three retention tests (71.4%, 78.6%, and 78.6% for each of the three object pairs), whereas the amnesic patients averaged only 51.9% correct (55.6%, 55.6%, and 44.4%). This difference between the two groups was significant, $F(1, 21) = 4.1$, $p < .05$. Thus the alcoholic control subjects exhibited differential choice behavior on the first trial of the retention tests, but the amnesic patients responded at chance levels.

Experiment 5: 24-Hour-Concurrent-Discrimination Learning

Method

Subjects

Amnesic patients. Two groups of amnesic patients were tested. The first group consisted of 3 patients with Korsakoff's syndrome described in Experiment 1 (Patients K3, K4, and K5) and the additional male Korsakoff patient described in Experiment 2. The second group consisted of 3 patients with Korsakoff's syndrome described in Experiment 1 (Patients K1, K4, and K6) and the additional female Korsakoff patient described in Experiment 2.

Alcoholic control group. This group consisted of 4 men who were current or former participants in alcoholic treatment programs in San Diego County. They had an average drinking history of 23.4 years but had abstained from alcohol an average of 2.8 weeks prior to participating in this study. They averaged 54.8 years of age, had 13.3 years of education, and had WAIS-R subtest scores of 18.3 for Information and 51.3 for Vocabulary (18.3 and 45.8, respectively, for the first group of amnesic patients). On immediate and delayed story recall, they scored 7.5 and 5.5, respectively.

Materials

The stimuli were presented in the WGTA (see Experiment 1).

Procedure

The first group of 4 amnesic patients and the alcoholic control group were trained on 20 object pairs simultaneously, according to the procedure of Malamut, Saunders, and Mishkin (1984). During daily sessions, each of the 20 object pairs was presented only once. One object of each pair was designated the positive object and was rewarded throughout testing. The 20 pairs were always presented in the same order. On each day, 10 of the correct objects were on the left and 10 were on the right. The position of the rewarded object (left or right) during the 20 trials of each day was varied according to a pseudorandom schedule (Gellermann, 1933). Subjects were instructed only that each time the screen was raised, a penny would be under one of the objects, and that they should try to obtain the penny every time. Alcoholic control subjects were tested 5 days a week, Monday through Friday, until they had received 8 days of training. Amnesic patients were similarly tested 5 days each week until they had received 20 days of testing. Because the amnesic patients had exhibited little improvement by the end of the first 10 days of testing, on Test Day 11 they were given a card that stated the principle governing the placement of the penny: "The penny will always be under the object that has been rewarded previously." This card was placed in front of the amnesic patients during Test Days 11 to 20.

On another occasion, separated by 24 months, the second group of 4 amnesic patients received training for 15 days (5 days each week for 3 consecutive weeks) on 20 simultaneous pattern discriminations. The procedure was identical to the one just described, except that 20 pairs of geometric figures or nonsense shapes were used instead of 20 object pairs. Subjects were told only that they should try to obtain the penny on every trial. The figures were taken from a test of nonverbal memory described by Kimura (1963) and were copied onto 7.6 × 12.7-cm white cards.

Results

The results are shown in Figure 6. The control subjects gradually learned the 20 object pairs, achieving almost perfect performance by the end of 8 days of testing. By contrast, the amnesic patients were unable to learn the 20 object pairs and performed close to the chance level of 50% throughout 20 days of testing. Performance was measurably better on Days 11 to 20 than on Days 1 to 10, $F(1, 3) = 11.3$, $p < .05$, perhaps as a result of the principle having been provided beginning on the 11th day. Performance averaged 51.8% correct on Days 1–10 and 60.5% correct on Days 11–20. Amnesic patients similarly failed during 15 testing days to exhibit any learning of the 20 pairs of patterns. Performance averaged 52.1% correct during the 15 test days.

General Discussion

The results can be summarized by the statement that the amnesic patients were impaired on all five tasks. Four of the

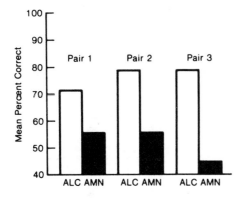

Figure 5. Object discrimination. (Performance on the first trial of the retention test for three sequentially learned object discriminations. The interval between learning and retention was 1 day for Pairs 1 and 2, and 10 days for Pair 3. ALC = alcoholics, $n = 14$; AMN = amnesics, $n = 9$.)

tasks (delayed nonmatching to sample, object-reward association, 8-pair concurrent discrimination learning, and object discrimination learning) are also performed poorly by monkeys with surgical lesions that reproduce the pattern of brain damage found in amnesic patients. The fifth task (24-hour concurrent discrimination learning) can be learned well by operated monkeys. The findings for each task will be discussed separately.

Delayed Nonmatching to Sample

Amnesic patients were initially tested in a condition where delay trials filled with a distraction task were mixed with unfilled delay trials. On this test, patients were markedly impaired on both distraction and no-distraction trials, and the impairment was equivalent across all three delays (Figure 1). The finding that performance was poor even at the 5-s delay, and that performance did not vary as a function of the delay interval, suggested that patients tended to forget the nonmatching principle. That is, performance may have been determined by whether patients could remember the nonmatching principle (choose the novel object), irrespective of how well they could remember the sample object across a delay.

Approximately 4 years later, after several hundred trials of intervening practice on the nonmatching problem (and on the day following a test day without any distraction trials), the amnesic patients were tested again in the identical way: delay trials filled with a distraction task were mixed with unfilled delay trials. At this time (Figure 2B, dashed lines), patients were able to perform reliably on the basis of the nonmatching principle. They performed well at short delays, and they were impaired at long delays.

Thus the performance of the amnesic patients improved with practice. On the first occasion that they were tested (Figure 1), they performed poorly regardless of the delay interval (5 s to 60 s). After intervening practice, however, they performed well at the 5-s delay, and performance at longer delays was related to the delay interval (Figure 2B). The poor performance exhibited initially by the patients contrasts with the performance exhibited by operated monkeys the first time that they were tested (see Zola-Morgan & Squire, 1985, Figure 6). Indeed, the monkeys performed rather like the amnesic patients who had been given additional practice on the task (Figures 2A and 2B): they scored well at the shorter delays and poorer at the longer delays.

One explanation for this observation may lie in the fact that amnesic patients initially reached criterion on the basic task, with a 5-s delay, in relatively few trials (median = 29). When delays were then extended beyond 5 s, performance on the nonmatching task may have deteriorated, because the nonmatching principle was not sufficiently well learned. Operated monkeys typically require hundreds of trials on the same basic task in order to reach learning criterion. The greater amount of training given the monkeys may benefit them when the delays are subsequently extended beyond 5 s. In addition, when operated monkeys are tested on delayed-nonmatching-to-sample task, the delays are typically extended gradually and distraction tasks are not used. By contrast, the amnesic patients in the present study were not presented

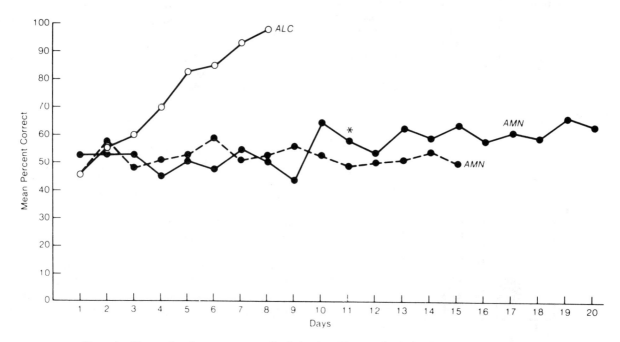

Figure 6. Twenty-four-hour concurrent discrimination. (Twenty discriminations were learned simultaneously, with each of the 20 pairs presented once each day. ALC = alcoholics, $n = 4$; AMN = amnesic, $n = 4$. Solid lines = learning of 20 pairs of objects; dashed lines = learning of 20 pairs of geometric designs. * = the point at which subjects were provided a card during daily sessions, which stated the principle that one of the objects was always the correct one.)

delays in a gradually increasing fashion. They were given a mixture of different delays, some of which included an interpolated distraction task. This difference between the procedures used for testing patients and monkeys may have contributed to the relatively poor performance by the patients when they first encountered the delays.

Another finding was that the distraction task disrupted performance of amnesic patients on the unfilled as well as on the filled trials. Thus performance on unfilled trials was poorer on a test day that included distraction trials (Figure 2B) than on a preceding test day without any distraction trials (Figure 2A). Stated differently, the disruptive effects of the distraction task spilled over onto the unfilled trials of the same test session. The question arises whether this finding is a reflection of the amnesic deficit itself, or whether it could reflect other cognitive deficits that sometimes occur in memory-impaired patients. Patients with Korsakoff's syndrome exhibit cognitive deficits not found in other etiologies of amnesia (Squire, 1982). Is the disruptive effect of the distraction task on unfilled trials limited to patients with Korsakoff's syndrome?

This possibility can be ruled out on two grounds. First, the phenomenon was observed not only in 6 patients with Korsakoff's syndrome but also in 3 other patients who had amnesia resulting from an anoxic or ischemic episode. These patients appear to have a more circumscribed memory impairment than the patients with Korsakoff's syndrome (Squire & Shimamura, 1986). Second, we had the opportunity to observe the same phenomenon during separate testing of Patient R. B., who had memory impairment associated with a bilateral lesion limited to the CA1 field of hippocampus (Zola-Morgan, Squire, & Amaral, 1986). This individual had no detectable cognitive deficit except amnesia. R. B. initially required eight trials to acquire the basic nonmatching-to-sample task. He then performed poorly when distraction and no-distraction trials were presented in a mixed design (50%, 40%, and 40% correct, for unfilled delays of 5 s, 15 s, and 60 s, respectively; 40% and 60% for filled delays of 15 s and 60 s). One month later, after having received a total of 158 trials on the nonmatching problem, R. B. was tested without any distraction task. His scores improved (100%, 100%, and 90% at 5-s, 15-s, and 60-s unfilled delays, respectively); and he performed similarly to the amnesic patients in the present study, when they were tested in the same condition (Figure 2A, dashed lines). However, on the following day, when a distraction task was included for half of the 15-s and 60-s delays (as in Figure 2B, dashed lines), R. B. again performed poorly: 70%, 80%, and 40% at the 5-s, 15-s, and 60-s unfilled delays; 60% and 60% at the 15-s and 60-s filled delays. Thus the disruptive effect of the distraction task on nonmatching-to-sample performance is observed in patients other than those with Korsakoff's syndrome. Moreover, the effect appears to be related to amnesia and not to other cognitive deficits.

In summary, the performance of amnesic patients corresponded to the performance of monkeys with medial temporal (Mahut, Zola-Morgan, & Moss, 1982; Mishkin, 1978; Zola-Morgan & Squire, 1985a) or diencephalic (Aggleton & Mishkin, 1983a, 1983b; Zola-Morgan & Squire, 1985b) lesions. Performance of amnesic patients resembled that of operated monkeys both when no distraction task was used (Figure 2A; this is the standard method for administering the delayed nonmatching-to-sample task to monkeys) and also when a distraction task was used (Figure 2B; compare to Zola-Morgan & Squire, 1985a, Figure 5). Performance was normal or close to normal at the short (5 s) delay, and it was increasingly impaired as the delay interval was lengthened.

Object–Reward Association

After learning the basic object association task with a 5-s delay, the amnesic patients exhibited an impairment that increased with the length of the delay (Figure 3). In addition, distraction had a small but significant effect on performance. Unlike the results for the delayed-nonmatching-to-sample task, once the basic object association task was acquired, with a 5-s delay, the distraction task did not disrupt performance equally across all three delays and across both filled and unfilled delays. Instead, performance was poorer at the long delays than at the short delays, and performance was poorer across filled intervals than across unfilled intervals. In correspondence with these findings in patients, monkeys with diencephalic lesions also exhibited impairment on this task across delays (Aggleton & Mishkin, 1983b). Moreover, monkeys with medial temporal lesions were impaired at acquiring this task (Phillips & Mishkin, 1984); data are not yet available for delays.

Eight-Pair Concurrent Discrimination Learning

Amnesic patients were markedly impaired at learning the eight object pairs during 3 days of testing, and they were impaired in a relearning test 1 month later. In addition, the amnesic patients performed poorly on a test that required them to recognize the 16 objects that had been used during testing. These findings extend to other kinds of amnesic patients an earlier observation that patients with Korsakoff's syndrome were impaired at concurrent visual discrimination learning (Oscar-Berman & Zola-Morgan, 1980b). The present findings also parallel the finding that monkeys with medial temporal lesions (Moss, Mahut, & Zola-Morgan, 1981; Zola-Morgan & Squire, 1985b) are impaired on this same 8-pair concurrent discrimination task.

Object Discrimination

Amnesic patients were impaired at learning the first of three object discrimination problems. This finding confirms and extends the report that patients with Korsakoff's syndrome were impaired at learning the first two in a series of object discrimination problems (Oscar-Berman & Zola-Morgan, 1980b). The present findings also show that amnesic patients performed at chance levels on the first retention trial for all three problems. Thus they were not able to remember which object had previously been rewarded.

This finding for the object discrimination task contrasts with an informal observation made by Gaffan (1972), which had suggested that a two-object discrimination problem could be remembered normally. However, his observation was based on one amnesic patient and one discrimination problem. In our study of nine patients and three discrimination problems, the results showed clearly that amnesic patients tended to forget from one day to the next which of two objects had been associated with reward. Moreover, these results correspond to the finding that monkeys with medial temporal lesions (Mahut, Moss, & Zola-Morgan, 1981; Zola-Morgan & Squire, 1985a) were impaired at learning and retaining object discrimination problems.

Twenty-Four-Hour Concurrent-Discrimination Learning

Amnesic patients evidenced no ability to learn 20 object pairs when the 20 pairs were all presented once daily. Thus the findings for patients differ from the findings for monkeys with medial temporal lesions, who successfully acquired the 20-pair concurrent problem (Malamut et al., 1984). It has been proposed that the ability to learn this task depends on a kind of memory (i.e., a habit system) different from the kind of memory subserved by the medial temporal region (Malamut et al., 1984). Although monkeys with medial temporal lesions did not learn this task at a normal rate when they were first tested (mean test days to criterion for four medial temporal monkeys = 16 days; mean test days to criterion for four normal monkeys = 10 days; $p = .014$), the monkeys did subsequently achieve normal learning scores on another set of 20 object pairs (mean = 10 days for the medial temporal group; mean = 8 days for the normal group). By contrast, amnesic patients exhibited little or no improvement on this task during 20 test days.

Monkeys and humans may approach this task differently. Although monkeys may be able to acquire this task incrementally in a skill-like way, humans appear to approach the task in the same way that they learn a list of paired associates; that is, for humans, the task appears to be a task of declarative memory, in which an explicit attempt is made to memorize the correct stimuli. The finding with amnesic patients does not of course exclude the possibility that the concurrent task could in principle be acquired by amnesic patients; for example, successful learning might be achieved with a much greater than 24-hour interval between test sessions, so that a declarative learning strategy would not so readily be engaged.

Do the Tasks Failed by Amnesic Patients Require Declarative Memory?

Three of the tasks in the present study (delayed nonmatching to sample, object-reward association, and 8-pair concurrent learning) included measures to assess the extent to which amnesic patients acquired explicit knowledge about the task that they were performing. In addition, one task (8-pair concurrent learning) included a task of recognition memory. If amnesic patients can perform above chance levels because they have acquired a skill (i.e., procedural knowledge), then they should not be able to display much explicit knowledge about what they have learned. They should not be able to state the principle that determined which responses were rewarded, or describe the materials they saw, or recognize them as familiar. Alternatively, if amnesic patients perform above chance because they have acquired some declarative knowledge, albeit less than control subjects, then whatever learning is possible should be accompanied by explicit, declarative knowledge about the task. We found that the performance of amnesic patients was consistently related to their ability to verbalize the principles of the tasks. Moreover, in 8-pair concurrent learning, performance was related as well to the ability to describe and recognize the test materials. These findings are consistent with the idea that the learning of these tasks depends to a significant degree on the ability to acquire declarative knowledge.

In conclusion, several memory tasks, which were designed originally for the monkey, were found to be sensitive to human amnesia. The findings provide direct evidence that these tasks are valid measures of memory impairment in the monkey. In addition, it seems likely that monkeys with medial temporal or diencephalic lesions and amnesic patients fail these tasks for the same reason, namely, because of damage to neural structures important for the formation and storage of declarative memory. Finally, one task did not appear to demonstrate good correspondence between monkeys and humans, perhaps because monkeys and humans approach this task in a different way.

References

Aggleton, J. P., & Mishkin, M. (1983a). Visual recognition impairment following medial thalamic lesions in monkeys. *Neuropsychologia, 21,* 189-197.

Aggleton, J. P., & Mishkin, M. (1983b). Memory impairment following restricted medial thalamic lesions in monkeys. *Experimental Brain Research, 52,* 199-209.

Aggleton, J. P., Nicol, R. M., Huston, A. E., & Fairbairn, A. F. (in press). The performance of amnesic subjects on tests of experimental amnesia in animals: Delayed matching-to-sample and concurrent learning. *Neuropsychologia.*

Brooks, D. N., & Baddeley, A. (1976). What can amnesic patients learn? *Neuropsychologia, 14,* 111-122.

Coblentz, J. M., Mattis, S., Zingesser, L. H., Kasoff, S. S., Wiesniewski, H. M., & Katzman, R. (1973). Presenile dementia: Clinical aspects and evaluation of cerebrospinal fluid dynamics. *Archives of Neurology, 29,* 299-308.

Cohen, N. J. (1981). *Neuropsychological evidence for a distinction between procedural and declarative knowledge in human memory and amnesia.* Unpublished doctoral dissertation, University of California, San Diego.

Cohen, N. J. (1984). Preserved learning capacity in amnesia: Evidence for multiple memory systems. In L. R. Squire & N. Butters (Eds.), *Neuropsychology of memory* (pp. 83-103). New York: Guilford Press.

Corkin, S. (1968). Acquisition of motor skill after bilateral medial temporal lobe excision. *Neuropsychologia, 6,* 225-268.

Corkin, S. (1984). Lasting consequences of bilateral medial temporal lobectomy: Clinical course and experimental findings in H. M. *Seminars in Neurology, 4,* 249-259.

Gaffan, D. (1972). Loss of recognition memory in rats with lesions of the fornix. *Neuropsychologia, 10,* 327-341.

Gellermann, L. W. (1933). Chance orders of alternating stimuli in visual discrimination experiments. *Journal of General Psychology, 42*, 207-208.

Jones, J. (1974). Imagery as a mnemonic aid after left temporal lobectomy: Contrast between material-specific and generalized memory disorders. *Neuropsychologia, 12*, 21-30.

Kimura, D. (1963). Right temporal lobe damage. *Archives of Neurology, 8*, 264-271.

Mahut, H., Moss, M., & Zola-Morgan, S. (1981). Retention deficits after combined amygdalo-hippocampal and selective hippocampal resections in the monkey. *Neuropsychologia, 19*, 201-225.

Mahut, H., Zola-Morgan, S., & Moss, M. (1982). Hippocampal resections impair associative learning and recognition memory in the monkey. *Journal of Neuroscience, 2*, 1214-1229.

Malamut, B. L., Saunders, R. C., & Mishkin, M. (1984). Monkeys with combined amygdalo-hippocampal lesions succeed in object discrimination learning despite 24-hour intertrial intervals. *Behavioral Neuroscience, 98*, 759-769.

Mishkin, M. (1978). Memory in monkeys severely impaired by combined but not by separate removal of amygdala and hippocampus. *Nature, 273*, 297-298.

Mishkin, M. (1982). A memory system in the monkey. *Philosophical Transactions of the Royal Society of London, 298*, 85-95.

Moss, M., Mahut, H., & Zola-Morgan, S. (1981). Concurrent discrimination learning of monkeys after hippocampal, entorhinal, or fornix lesions. *Journal of Neuroscience, 1*, 227-240.

Oscar-Berman, M., & Zola-Morgan, S. (1980a). Comparative neuropsychology and Korsakoff's syndrome: I. Spatial and visual reversal learning. *Neuropsychologia, 18*, 499-512.

Oscar-Berman, M., & Zola-Morgan, S. (1980b). Comparative neuropsychology and Korsakoff's syndrome: II. Two-choice visual discrimination learning. *Neuropsychologia, 18*, 513-525.

Oscar-Berman, M., & Zola-Morgan, S. (1982). Comparative neuropsychology and Korsakoff's syndrome: III. Delayed response, delayed alternation, and DRL performance. *Neuropsychologia, 20*, 189-202.

Phillips, R. R., & Mishkin, M. (1984). Further evidence of a severe impairment in associative memory following combined amygdalo-hippocampal lesions in monkeys. *Society of Neuroscience Abstracts, 10*, 136.

Salmon, D. P., Zola-Morgan, S., & Squire, L. R. (1987). Retrograde amnesia following combined hippocampal-amygdala lesions in monkeys. *Psychobiology, 15*, 37-47.

Scoville, W. B., & Milner, B. (1957). Loss of recent memory after bilateral hippocampal lesions. *Journal of Neurological Psychiatry, 20*, 11-21.

Squire, L. R. (1982). Comparisons between forms of amnesia: Some deficits are unique to Korsakoff's syndrome. *Journal of Experimental Psychology: Learning, Memory, and Cognition, 8*, 560-571.

Squire, L. R. (1986). Mechanisms of memory. *Science, 232*, 1612-1619.

Squire, L. R., & Cohen, N. J. (1984). Human memory and amnesia. In G. Lynch, J. L. McGaugh, & N. M. Weinberger (Eds.), *Neurobiology of learning and memory* (pp. 3-64). New York: Guilford Press.

Squire, L. R., & Shimamura, A. P. (1986). Characterizing amnesic patients for neurobehavioral study. *Behavioral Neuroscience, 100*, 866-877.

Squire, L. R., & Zola-Morgan, S. (1983). The neurology of memory: The case for correspondence between findings for man and nonhuman primate. In J. A. Deutsch (Ed.), *The physiological basis of memory* (pp. 199-268). New York: Academic Press.

Victor, M., Adams, R. D., & Collins, G. H. (1971). *The Wernicke-Korsakoff syndrome*. Philadelphia: Davis.

Zola-Morgan, S., & Squire, L. R. (1984). Preserved learning in monkeys with medial temporal lesions: Sparing of motor and cognitive skills. *Journal of Neuroscience, 4*, 1072-1085.

Zola-Morgan, S., & Squire, L. R. (1985a). Amnesia in monkeys after lesions of the mediodorsal nucleus of the thalamus. *Annals of Neurology, 17*, 558-564.

Zola-Morgan, S., & Squire, L. R. (1985b). Medial temporal lesions in monkeys impair memory in a variety of tasks sensitive to human amnesia. *Behavioral Neuroscience, 99*, 22-34.

Zola-Morgan, S., Squire, L. R., & Amaral, D. G. (1986). Human amnesia and the medial temporal region: Enduring memory impairment following a bilateral lesion limited to field CA1 of the hippocampus. *Journal of Neuroscience, 6*, 2950-2967.

Received October 17, 1986
Revision received February 3, 1987
Accepted February 5, 1987 ∎

Glossary

anoxia - an interruption of the normal supply of oxygen to tissue

anterograde amnesia - amnesia for things learned after the amnesia-inducing event (e.g., after the head injury)

bilateral - both sides

declarative memory - this term has been widely used to refer to the kinds of memory that are commonly disrupted in cases of human amnesia; in general, declarative memory refers to any memory that can be verbally expressed (or declared)

Dementia Rating Scale - a test battery designed to assess the degree of general intellectual deterioration in neuropsychological patients

diencephalon - the brain structure that is at the top of the brainstem; it comprises the thalamus and hypothalamus

H.M. - a famous amnesic patient; he experienced severe anterograde amnesia and mild retrograde amnesia, but no intellectual impairment, after the bilateral removal of his medial temporal lobes

ischemia - an interruption of the normal supply of blood to tissue

Korsakoff's syndrome - a disorder commonly observed in chronic alcoholics; advanced cases are characterized by severe intellectual deterioration and both anterograde and retrograde amnesia; Korsakoff's amnesia seems to be associated with damage to the medial diencephalon, in particular to the mediodorsal nuclei of the thalamus

medial - near the midline

mediodorsal nuclei - a pair of medial thalamic nuclei; lesions of the mediodorsal nuclei have been shown to disrupt the performance of memory tasks by monkeys; the mediodorsal nuclei are commonly damaged in advanced cases of Korsakoff's syndrome

procedural memory - this term has been widely used to refer to the kinds of memory that commonly survive in cases of human amnesia; human amnesics can often learn and retain new sensorimotor skills or new tasks that involve tiny increments of improvement over many trials, even though they have no conscious awareness of ever having performed them

pseudorandomized - an order of items that is constructed so that it has no systematic sequences and so that each item (e.g., left or right) occurs an equal number of times; random orders, by their very randomness, do not always meet these criteria

retrograde amnesia - amnesia for things learned before the amnesia-inducing event (e.g., before the head injury)

Rey-Osterreith figure - a complex, straight-line drawing that is used to assess the nonverbal memory of neuropsychological patients; patients first copy the figure and then later they are asked to draw it from memory

temporal lobes - the most lateral lobes of the cerebral hemispheres; the major subcortical structures of the temporal lobes are the hippocampus and the amygdala

Wechsler Adult Intelligence Scale (WAIS) - a test battery that is commonly used to assess the general intelligence of human subjects

Wechsler Memory Scale (WMS) - a test battery that is commonly used to assess the memory of human subjects

Wisconsin General Testing Apparatus - a testing apparatus for monkeys; it consists of a stimulus tray, which can be slid forward within the reach of the monkey in its cage, and a sliding screen, which can block the monkey's view of the stimulus tray; the stimulus tray typically contains three food wells, one of which is usually baited with food; stimulus objects are typically placed over some or all of the food wells, and the monkey's task is to learn to select the correct object to obtain the food

Essay Study Questions

1. Describe the subjects in this study. Why did Squire, Zola-Morgan, and Chen use alcoholics as control subjects? Why were they so thorough in describing the ability of their subjects to perform various objective tests of neuropsychological function?

2. What is the difference between the concept of declarative memory and the concept of procedural memory?

3. Describe the five tasks used in these experiments.

4. What was the effect of the distraction task on the performance of the delayed nonmatching-to-sample test and the object-reward-association test?

5. On which tests was performance substantially improved by providing the subjects with the principle of the task?

6. Summarize the results of this series of experiments.

Multiple-Choice Study Questions

1. The mediodorsal nuclei are
    a. part of the thalamus.
    b. part of the diencephalon.
    c. damaged in advanced cases of Korsakoff's syndrome.
    d. all of the above
    e. none of the above

2. In experimental studies of brain-damage-produced amnesia in monkeys, bilateral lesions are often created in the medial
    a. regions of the temporal lobes.
    b. diencephalon, including the mediodorsal nuclei.
    c. hypothalamus.
    d. all of the above
    e. a and b

3. The only difference between the delayed nonmatching-to-sample task and the object-reward-association task is that in the latter
    a. there is reinforcement.
    b. there is no reinforcement.
    c. the subject is reinforced on each test trial for selecting the sample.
    d. there are rarely any amnesic deficits.
    e. there is no simple way of stating the principle.

4. In the 8-pair concurrent-discrimination experiment,
    a. 16 objects were used.
    b. the same 8 objects always indicated the position of the penny for a particular subject.
    c. each pair of objects was presented five times each day.
    d. all of the above
    e. none of the above

5. In the delayed nonmatching-to-sample task, the disruptive effects of the distraction task
    a. were negligible.
    b. were specific to the patients with Korsakoff's syndrome.
    c. spilled over onto nondistraction trials.
    d. were specific to the alcoholic controls.
    e. were restricted to the 5-second interval.

6. Squire, Zola-Morgan, and Chen speculated that amnesic monkeys and humans perform differently on the 24-hour concurrent-discrimination task because they use different memory systems to solve it. They suggest that monkeys and humans use

    a. procedural and declarative memory systems, respectively.
    b. declarative and procedural memory systems, respectively.
    c. short-term and long-term memory systems, respectively.
    d. long-term and short-term memory systems, respectively.
    e. anterograde and retrograde memory systems, respectively.

The answers to the preceding questions are on page 290.

Food-For-Thought Questions

1. Some people believe that research on laboratory animals has nothing to tell us about human cognitive abilities; others indiscriminately apply the results of any ostensibly relevant animal experiment to humans. The experiments of Squire, Zola-Morgan, and Chen show that neither of these extreme positions is defensible. Explain and discuss.

2. Hypotheses are of little value unless they can tested. How would you test Squire, Zola-Morgan, and Chen's hypothesis that monkeys and humans solve the 24-hour concurrent-discrimination task by using different memory systems?

ARTICLE 16

Hippocampus and Memory for Food Caches in Black-Capped Chickadees

D.F. Sherry and A.L. Vaccarino
Reprinted from <u>Behavioral Neuroscience</u>, 1989, volume 103, 303-318.

Some species of birds, such as black-capped chickadees, regularly perform prodigious feats of memory. They may hide as many as several hundred food items (e.g., seeds) in a day, each in a separate unmarked location, and later they recover most of them. Accordingly, food-caching birds have proven to be particularly interesting subjects for memory research.

Research on humans and other mammalian species has established that the hippocampus plays an important role in memory, but there has been considerable debate about exactly what its role is. One theory of hippocampal function suggests that its primary function is the storage of information about spatial location, that is, the storage of cognitive maps. Another theory suggests that the primary function of the hippocampus is the mediation of working memory. Working memory is memory for information that pertains to only the current performance of a particular task; in contrast reference memory is memory for information that pertains to all performances of a particular task.

In Experiment 1, bilateral hippocampal lesions reduced the accuracy of seed cache recovery by black-capped chickadees to chance levels. The fact that the hippocampal lesions did not reduce the number of seeds cached or the number of attempts to recover them indicated that the deficit in seed cache recovery reflected a memory deficit, rather than a sensory, motor, or motivational deficit.

The purpose of Experiment 2 was to test the cognitive-map and reference-memory theories of hippocampal function. Black-capped chickadees were trained to find seeds hidden by the experimenter under one of two different conditions. In one condition, seeds were hidden by the experimenter in the same six holes each day (place condition); in the other condition, the six holes in which the seeds were hidden by the experimenter varied from day to day but the locations of the seeds were always indicated by cue cards next to them (cue condition). After learning their respective version of the task, the black-capped chickadees in each of the two groups received bilateral hippocampal lesions. The fact that the hippocampal lesions reduced the number of seeds discovered in the first six searches by the chickadees tested in the place condition, but not by those tested in the cue condition, supported the cognitive-map theory of hippocampal function. However, the fact that the lesioned chickadees in both conditions paid many return visits to sites from which seeds had already been taken during that test supported the working-memory theory.

The cognitive-map and working-memory theories are often discussed as if they are mutually exclusive. These experiments suggest that the hippocampus may have multiple memory-related functions.

Hippocampus and Memory for Food Caches in Black-Capped Chickadees

David F. Sherry and Anthony L. Vaccarino
University of Toronto, Toronto, Ontario, Canada

Black-capped chickadees and other food-storing birds recover their scattered caches by remembering the spatial locations of cache sites. Bilateral hippocampal aspiration reduced the accuracy of cache recovery by chickadees to the chance rate, but it did not reduce the amount of caching or the number of attempts to recover caches. In a second experiment, hippocampal aspiration dissociated performance of a task requiring memory for places from performance of a task requiring memory for cues associated with food, disrupting the former but not the latter. On both tasks, however, hippocmpal aspiration increased the frequency of revisiting errors to sites previously searched. These experiments show that the structure in the avian brain that is neuroanatomically and embryologically homologous to the mammalian hippocampus shares some functions with the mammalian hippocampus. The results indicate that memory for places and working memory are both disrupted by hippocampal damage in birds. Finally, it was possible to show that these memory capacities are essential for cache recovery by black-capped chickadees.

Black-capped chickadees (*Parus atricapillus*) store food in concealed locations scattered throughout their home range. They store seeds and invertebrate prey, and they put each food item in a separate cache site. Hollow stems, crevices in bark, moss, leaf buds, dry leaves, and natural cavities are used as storage sites. Chickadees, and the congeneric marsh tit (*Parus palustris*), may store as many as several hundred food items per day and place neighboring caches at least several meters apart. They recover their stored food a few days after caching it, and in the wild they do not reuse cache sites (Cowie, Krebs, & Sherry, 1981; Sherry, Avery, & Stevens, 1982; Stevens & Krebs, 1986). Stored food is recovered by remembering the spatial locations of caches (Cowie et al., 1981; Sherry, 1984a; Sherry, Krebs, & Cowie, 1981; Shettleworth & Krebs, 1982).

The evidence that memory for spatial locations is the principal means of cache recovery comes from a number of experimental field and laboratory studies, reviewed by Sherry (1985). In the field, seeds are taken from cache sites at a higher rate than they are taken from identical control sites 10 cm away (Cowie et al., 1981). Although this is not unequivocal evidence of memory, it shows that cache recovery is spatially precise and is not based on exhaustive search of all sites resembling the cache site. An experiment in which magnetic leg bands worn by birds triggered detectors placed at cache sites confirmed that the individual that caches a food item is the same one that returns later to collect it (Stevens & Krebs, 1986).

In the laboratory, chickadees and marsh tits consistently relocate their caches more accurately than expected by chance, and methods of cache recovery that do not require memory can be shown by control procedures to be inadequate for accurate relocation of stored food. The birds do not detect stored food directly by some sensory means, because they can successfully relocate cache sites that have been previously emptied of food by the experimenter (Sherry, 1984a; Sherry et al., 1981). Furthermore, they cannot detect food placed in other sites by the experimenter or by other birds (Baker et al., 1988; Shettleworth & Krebs, 1982). There is no behavioral indication that marsh tits or black-capped chickadees mark their caches in any way (although some other food-storing birds may; Haftorn, 1974). The birds do not retrace the route followed during caching in order to relocate cache sites. Correlations between caching and recovery sequences are sometimes positive, sometimes negative, and generally not statistically significant (Sherry, 1982, 1984b). After food has been stored at a particular spatial location, the bird is more likely to return to that place than would be expected from previous preferences or biases to search that place (Sherry, 1984a; Sherry et al., 1981). Even for highly preferred sites, birds are more likely to search a site if food has been recently stored there than if it has not (Shettleworth & Krebs, 1982). Selection of cache sites with common features could be used as a mnemonic device, but there is no indication of such cache site selection (Sherry, 1984a). An experiment exploiting the properties of interocular transfer in birds showed that visual information about the cache site and its surroundings must be available during cache recovery for accurate performance to occur (Sherry et al., 1981). Comparable results have been obtained with a variety of storable food types,

---

This research was supported by the Natural Sciences and Engineering Research Council of Canada.

We thank Alison Fleming, Michael Leon, Larry Squire, and Franco Vaccarino for their advice and encouragement, and Elisabeth Murray and Derek van der Kooy for their comments on an earlier draft of the manuscript. We are grateful to Karen Buckenham and Siegfried Schulte for their technical assistance and to Jim Hare for his help catching chickadees.

Correspondence concerning this article should be addressed to David F. Sherry, Department of Psychology, University of Toronto, Toronto, Ontario, Canada M5S 1A1.

cache sites, and delays of up to 72 hr between caching and recovery. Not only can the birds accurately remember the locations of cache sites, but they are also able to avoid revisiting caches from which they themselves have removed the stored food (Sherry, 1982, 1984a).

The results of field and laboratory studies of food storing in another family of birds, the corvids, support the general conclusion that memory for the spatial locations of caches is the principal means by which stored food is recovered (Balda, Bunch, Kamil, Sherry, & Tomback, 1987; Kamil & Balda, 1985; Tomback, 1980; Vander Wall, 1982).

The processing of spatial information is one proposed function of the hippocampus, at least in mammals. This cognitive mapping view of hippocampal function (O'Keefe & Nadel, 1978) is supported by the relation between the firing patterns of single units in the hippocampus and an animal's position in space (Best & Ranck, 1982; O'Keefe, 1979; Olton, Branch, & Best, 1978) and by disruptions in spatial orientation produced by disruption of hippocampal functioning (Jarrard, Okaichi, Steward, & Goldschmidt, 1984; Morris, Anderson, Lynch, & Baudry, 1986; Morris, Garrud, Rawlins, & O'Keefe, 1982; O'Keefe & Nadel, 1978; Sutherland, Whishaw, & Kolb, 1983).

The working memory view of Olton and his colleagues (Olton, Becker, & Handelmann, 1979) presents quite a different picture of hippocampal function and is supported by results of a different kind. *Working memory* is defined as memory for information that is relevant to only the current performance of a task (Honig, 1978), such as the most recent choices in a radial-arm maze. It is distinguished from *reference memory* for information that is relevant to all performances of the same task, such as the number of arms in the maze. The working memory theory is supported by the observation that memory for recent performance of a task (spatial or otherwise) is disrupted by hippocampal damage whereas memory for the nature of the task is not (Becker & Olton, 1981; Olton et al., 1979; Walker & Olton, 1984).

Because memory for spatial locations can be shown to be the mechanism of cache recovery in black-capped chickadees, hippocampal damage should disrupt cache recovery in these birds if the avian hippocampus functions similarly to the mammalian hippocampus. Neuroanatomical and embryological evidence indicate that the mammalian and avian hippocampus are structurally homologous (Benowitz, 1980; Casini, Bingman, & Bagnoli, 1986; Källén, 1962; Krayniak & Siegel, 1978). Cache recovery would be disrupted by hippocampal damage according to the cognitive mapping view because of the resulting deficit in the birds' ability to process spatial locations. Cache recovery would be disrupted according to the working memory view because of failure to recall information relevant to the most recent episode of food caching, whether or not this information was spatial in nature. Indeed, disruption of cache recovery as a result of hippocampal damage seemed particularly likely in light of an earlier report that lesions of the hyperstriatum produced cache recovery errors in another food-storing bird, the Eurasian nutcracker (*Nucifraga caryocatactes*; Krushinskaya, 1966).

In the first experiment the effect of bilateral hippocampal aspiration on cache recovery was examined. In the second experiment the nature of the disruption produced by hippocampal aspiration was determined by observing performance on a spatial and a nonspatial task.

Experiment 1

Chickadees cached and recovered seeds in an indoor aviary, and their subsequent accuracy in locating cache sites was measured. The hippocampus was aspirated bilaterally, the birds were allowed to recover from surgery, and observations on caching and cache recovery were repeated.

Method

Subjects and environment. Black-capped chickadees ($n = 9$) were caught in the wild as adults on the Erindale campus of the University of Toronto and held in captivity under a Canadian Wildlife Service Scientific Capture permit. The birds were housed individually in wire mesh home cages (80 × 80 × 100 cm) on a 14:10 hr light/dark cycle (onset 0800). They were maintained on an ad-lib mixed diet, with a vitamin supplement added to drinking water. Sunflower seeds were not available except when provided during caching trials, as described below.

The home cage communicated through a trap door to a 6 × 3 × 2.5 m indoor aviary that contained six artificial trees. Trees were branches approximately 0.6 m in length held upright in metal bases. Each tree consisted of a main trunk with 2–4 principal side branches and many smaller branches. Cache sites were 72 small holes, 12 on each tree, 1 cm deep and 0.5 cm in diameter, with a dowel perch 0.5 cm in diameter inserted in the branch near each hole. There was considerable variation in the nature of the branch at each cache site, and no two cache sites were alike in all respects. Some sites were at a fork in the branch; others at the end of a branch. Branch diameter, color, and texture varied among cache sites, and cache sites were in various orientations with respect to the base of the tree (Figure 1). The trees stood in six predetermined positions in the aviary, and trees were randomly assigned to these six positions at the beginning of each trial in order to vary the spatial arrangement of holes.

Procedure. On each trial, birds that had been deprived of food for 3 hr were allowed into the aviary individually and were observed for 10 min before being given any food to store. The location and duration of searches at all holes were entered on a microcomputer event recorder. A search was defined as close visual inspection or probing with the bill at a hole. Behavior during this prestorage control period provided an estimate of each bird's preference to search particular holes, trees, or parts of the aviary before any food had been stored. Each bird was then given 12 husked sunflower seeds and allowed to eat and cache seeds for 10 min. The location of each cache was recorded. The bird was returned to its home cage for 3 hr after which it was allowed to search the aviary once again for 10 min and attempt to recover its caches. All stored food had been removed in the interval, to prevent direct visual or olfactory detection of caches. During the cache recovery stage of the trial, all searches at holes and their durations were again recorded. The number of visits to cache sites during the recovery phase of a trial, expressed as a percentage of all visits, was compared with the percentage of visits to the same holes occurring in the prestorage phase. In this way it could be determined whether the bird was more likely to return to its cache sites during recovery than would be expected from initial biases or preferences shown before storing food.

After five trials of this procedure, subjects were randomly assigned to three experimental groups (n = 3 per group). The hippocampal group (Hp) received bilateral aspiration of the hippocampus. In the

Figure 1. Cache sites, showing typical variation in branch size and shape and in bark texture.

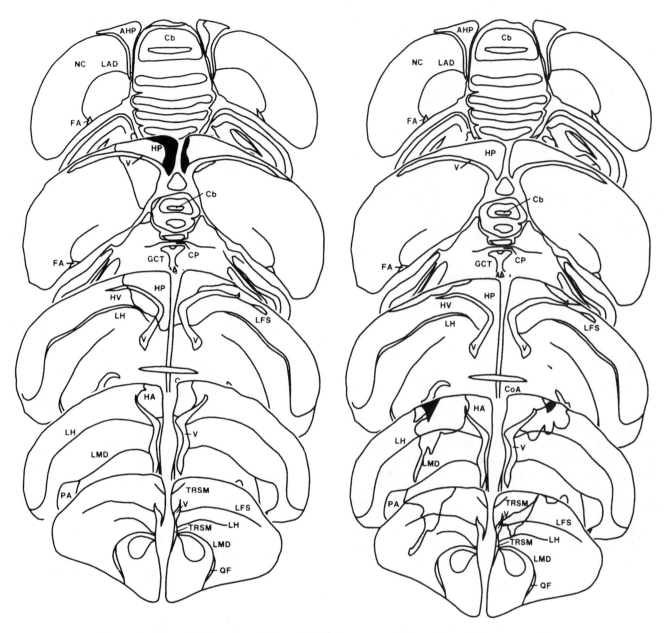

Figure 2. Reconstruction of aspirations of the hippocampus (left) and hyperstriatum accessorium (right) for subjects in Experiment 1. (The five coronal sections, shown rostro-caudally, are at intervals of 1,500 μm. Black: Aspiration common to all subjects; gray: maximum extent of aspiration in any subject. AHP = area parahippocampus; Cb = cerebellum; CoA = anterior commissure; CP = posterior commissure; FA = tractus fronto-archistriatalis; GCT = substantia grisea centralis; HA = hyperstriatum accessorium; HP = hippocampus; HV = hyperstriatum ventrale; LAD = lamina archistriatalis dorsalis; LFS = lamina frontalis superior; LH = lamina hyperstriatica; LMD = lamina medullaris dorsalis; NC = neostriatum caudale; PA = paleostriatum augmentatum; QF = tractus quintofrontalis; TRSM = tractus septomesencephalicus; V = ventricle.)

surgical control group (Ha), bilateral aspirations of a comparable size were made in the hyperstriatum accessorium, rostral and lateral to the site of the hippocampal aspiration. The 3 remaining subjects served as an unoperated control group. After a 3-day postoperative recovery period, five further caching and recovery trials were run with each bird.

Surgery and histology. Birds were anesthetized with sodium pentobarbital (Nembutal, 6 μg/g) and placed in a modified stereotaxic instrument (Vaccarino, 1986), under a binocular dissecting microscope. A small region of skull overlying the hippocampus (or hyperstriatum accessorium for the Ha group) was removed, and bilateral aspirations were made with a 20-ga. syringe connected to a vacuum pump. A small piece of dissolvable sponge was placed over the wound and covered with a drop of dental cement. Prior to the experiment, aspirations had been performed on four pilot subjects, by using the canary brain atlas of Stokes, Nottebohm, and Leonard (1974) as a guide, to ensure accurate placement of aspirations.

Following behavioral observations, the size and placement of lesions were verified in all operated subjects. Birds were sacrificed by Nembutal overdose and perfused with physiological saline followed by 10% formalin. The brain was removed and placed in sucrose formalin for 3 days. Frozen 40-μm sections were stained with cresyl violet. The location and volume of lesions were determined by enlarging and tracing these sections. The avian hippocampus is bounded medially by the midline, dorsally by the brain surface, and ventrally by the septum and ventricle. A cytoarchitectonic criterion was used for the lateral and rostro-caudal limits of the hippocampus. In black-capped chickadees these boundaries can be recognized by an abrupt change from the low cell density of the hippocampus to the higher cell density of the hyperstriatum. Use of this criterion causes some of the region labeled area parahippocampus in the atlases of Karten and Hodos (1967) and Stokes et al. (1974) to be included in the hippocampus.

The size of aspirations and the total volume of the hippocampus were determined by tracing enlarged coronal sections on a Talos Series 600 digitizer (Talos Systems Inc, Scottsdale, Arizona) and computing the volume between sections with the formula for a truncated cone.

Results

Reconstructions from brain sections showed that a mean volume of 3.75 mm³ ± 1.97 (SE), or 37.6% ± 19.8% of the total volume of the hippocampus, was aspirated (Figure 2). Aspiration extended rostral to the hippocampus in 1 Hp subject, and included part of the right striatum, ventral to the ventricle in another Hp subject. Performance did not differ markedly among Hp subjects, however, as shown by the standard errors of the mean in Figure 3. Hyperstriatal aspirations had a mean volume of 3.50 mm³ ± 0.41 and did not differ in size from hippocampal lesions, $t(6) = 0.125$ (Figure 2). For 1 Ha subject aspiration damaged a narrow band of right striatum. Performance did not differ markedly among Ha subjects, however, as shown by standard errors of the mean in Figure 3.

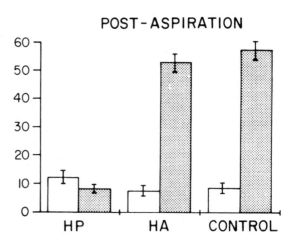

Figure 3. Hippocampal aspiration and cache recovery. (Searches at cache sites, as a percentage of searches at all sites, during prestorage and cache recovery. Differences between the prestorage and recovery phases of a trial indicate changes in the likelihood of searching a site after food has been stored in it. Hp: Hippocampal aspiration; Ha: Hyperstriatum accessorium aspiration; Control: Unoperated control. The $n = 3$ per group. Error bars equal ± 1 SE.)

The effects on cache recovery accuracy of aspiration of hippocampus or hyperstriatum accessorium are shown in Figure 3. Analysis of variance of recovery accuracy showed a three-way interaction between aspiration, site of aspiration, and phase of the caching trial, $F(2, 6) = 78.47$, $p < .01$ (arcsine transformed percentages). This indicates that the difference between prestorage and cache recovery behavior is affected by the surgical procedure in some groups but not others. Post hoc tests showed significant differences by Tukey's honestly significant difference test, $HSD(6) > 16.93$, $p < .01$, in all prestorage versus recovery comparisons except the hippocampal group postaspiration. There were no sig-

Table 1
Effect of Hp and Ha Aspiration on Seeds Stored, Seeds Eaten, and Searches for Caches

Group	n	Stored[a]		Eaten[b]		Searches[c]	
		Pre	Post	Pre	Post	Pre	Post
Hp	3	5.3	6.3	2.8	2.2	11.0	11.0
Ha	3	4.8	4.6	3.1	2.9	15.7	9.3
Control	3	4.6	4.3	3.4	3.2	8.3	10.7

Note. Hp = hippocampal aspiration; Ha = hyperstriatum accessorium aspiration; Pre = preaspiration; Post = postaspiration.
[a] Mean numbers of seeds stored per trial. There is no significant effect of aspiration, $F(1, 6) = 0.07$, site of aspiration, $F(2, 6) = 1.62$, or any interaction, $F(2, 6) = 0.42$.
[b] Mean numbers of seeds eaten per trial. There is no significant effect of aspiration, $F(1, 6) = 3.24$, site of aspiration, $F(2, 6) = 1.52$, or any interaction, $F(2, 6) = 0.66$.
[c] Mean numbers of searches during the recovery phase of each trial. There is no significant effect of aspiration, $F(1, 6) = 0.67$, site of aspiration, $F(2, 6) = 0.85$, or any interaction, $F(2, 6) = 2.53$.

nificant differences among prestorage levels before or after aspiration. The rate of recovering caches by chance is between 6% and 9% for all groups, which can be seen in Figure 3 to be close to the level of prestorage accuracy. These results indicate that although the birds can normally relocate caches more accurately than prior preferences or chance could produce, hippocampal aspiration reduced the cache recovery accuracy of this group of birds to the prestorage or chance levels. The performance of Ha subjects was unaffected by aspiration, and control birds performed consistently accurately in all trials.

Hippocampal aspiration did not reduce the number of seeds stored, the number eaten, or the amount of searching during the prestorage and cache recovery phases of the trial (Table 1). Thus neither gross impairment in the initial caching of seeds nor a reduction in the number of attempts to recover caches can account for the results. It can also be seen from Table 1 that surgery and anesthesia, when followed by a 3-day recovery period, did not by themselves disrupt eating, caching, or searching for caches.

Discussion

Birds in the Hp group searched for caches as intensely as birds in other groups, but failed to find them. It seems likely that Hp aspiration disrupted memory for the locations of caches, because caching performance, feeding, and other behavior were not affected. Any nonspecific motivational, sensory, or motor effects would be likely to produce observable effects on behavior, in addition to the effects on cache recovery accuracy. Failure to locate caches, however, provides little information on the kind of memory impairment produced by hippocampal aspiration. In order to answer this question, two additional groups of chickadees were trained to perform tasks that required searching for concealed food but differed in the kind of memory required for each task.

Experiment 2

In order to distinguish between cognitive mapping and working memory explanations of the results of the previous experiment, chickadees were trained to perform two tasks that required searching for food concealed in the holes used as cache sites in the previous study. The place task required birds to learn that a subset of six holes consistently contained food. This was regarded as a spatial task, because the spatial locations of baited holes did not change from trial to trial. The cue task required birds to learn that a cue was consistently paired with six sites containing food. The sites where cues appeared changed from trial to trial so that learning of spatial locations would not serve as a solution to this task. According to the cognitive mapping view, the former task but not the latter should be disrupted by hippocampal aspiration.

For both tasks it was also possible to assess the working memory effects of hippocampal damage. Working memory errors were defined as repeat visits to sites, whether baited or unbaited. Chickadees are normally able to avoid revisiting sites where they have previously collected their stored food (Sherry, 1982, 1984a), and during storing they can avoid revisiting places where they have already concealed a food item (Shettleworth & Krebs, 1982). According to the working memory view, revisiting errors should occur in both place and cue tasks with a higher frequency following hippocampal aspiration. In order to distinguish cognitive mapping from working memory accounts however, it is necessary to show that revisiting errors are not simply a consequence of disruptions in spatial orientation. If spatial locations cannot be distinguished, then revisiting errors on the cue task could be due to an inability to discriminate sites sharing a common cue. To deal with this problem, we followed the logic behind the Jarrard version of the radial-arm maze (Jarrard, 1983). In the Jarrard procedure, the arms of the maze differ not only in spatial location but also in color, pattern, and texture. Thus a revisit to a particular arm indicates a failure to recall

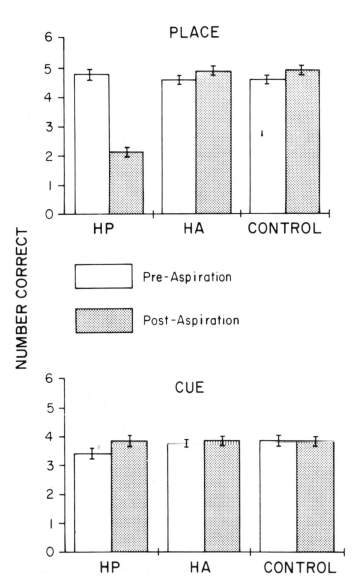

Figure 4. Number correct in the first six searches pre- and postaspiration for the place and the cue tasks. (The $n = 4$ per group. Error bars equal ± 1 *SE*.)

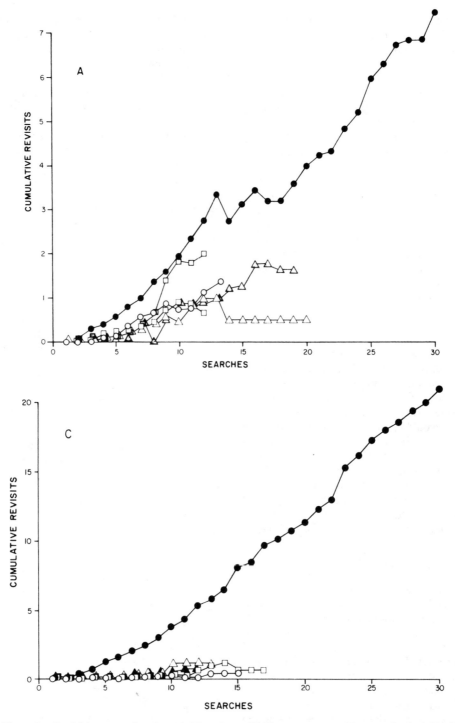

Figure 5. Revisiting errors for place (A, B) and cue (C, D) groups, pre- and postaspiration. (Panels A and C [this page] show revisiting errors to correct holes; panels B and D [opposite page] to incorrect holes. Note change of scale on the ordinate in panel C. As the number of searches increases, *n* in some groups decreases because birds made different numbers of searches, which produced decreases in the mean cumulative number of revisits in some groups.)

that a particular distinctive maze arm was previously visited on the current trial. As mentioned earlier, cache sites differed in branch diameter and configuration, bark color, texture, and other features. To make a revisiting error at a cued site, for example, a bird would have to fail to recall that it had previously searched a cued site of a particular color, texture, and branch diameter regardless of its spatial location. Thus naturally occurring variation in cache sites, of the kind shown

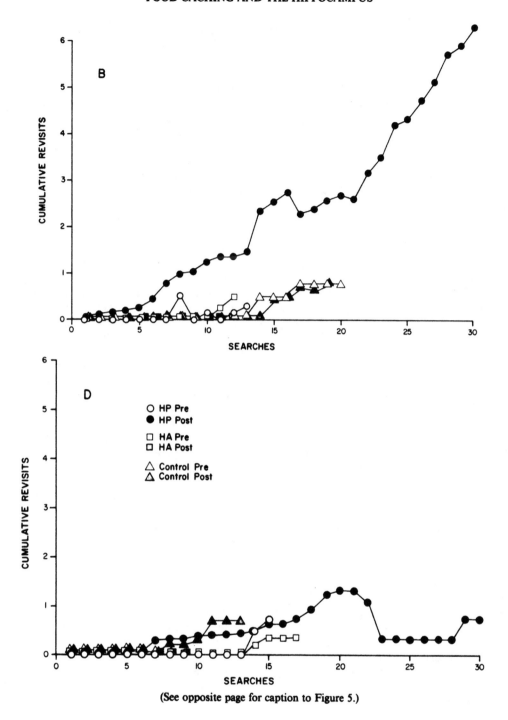

(See opposite page for caption to Figure 5.)

in Figure 1, took the place in our task of the variously colored and textured inserts of the Jarrard radial-arm maze.

Method

Subjects and tasks. Subjects were 24 adult black-capped chickadees maintained as described in the first experiment. None had served previously as subjects, and all were naive with respect to the apparatus and procedures. The arrangement of trees and holes was as in the first experiment except that the holes did not serve as cache sites but instead as sites where food was concealed by the experimenter. In the place task, 12 chickadees were trained to find six small pieces of sunflower seed, each placed in a different hole. The same holes were used for all birds on all trials and were selected at random from among the holes used for caching by chickadees in the first experiment. Seed pieces were of a size that they could not be seen except when viewed from directly in front of the hole. In the cue task, a second group of 12 chickadees was trained to find six sunflower seeds that were placed in six different holes on every trial. Holes used in the cue task were selected at random from the same population of holes that had been used for caching in the previous experiment. Holes containing a seed were indicated by cues, which were 5 × 5 cm cards placed on the perch at all holes. For half of the subjects, white cards indicated holes baited with seeds, and all other holes had black cards. For the remaining birds, the colors were reversed.

Procedure. For both tasks, each bird had one 5-min trial per day. Birds were trained until they were able to locate three or more seeds in the first six attempts for 5 consecutive days. Once this criterion had been met, birds in both groups were assigned at random to Hp, Ha, and control groups. There were thus six treatment groups, with $n = 4$ in each group. On the third day following surgery, testing recommenced and continued for 5 daily trials. Performance accuracy was measured as the number of baited sites located in the first six search attempts. A search was defined as in the previous experiment—close visual inspection of a site or probing with the bill—and surgical, histological, and lesion reconstruction procedures were as in the previous experiment.

Results

The place group achieved a mean of 4.7 ± 0.4 (SE) correct choices in their five criterion trials, compared with 3.6 ± 0.1 for the cue group, $F(1, 22) = 50.51$, $p < .01$. The probability of locating three or more seeds by chance in the first six choices, when 6 out of 72 holes are baited, is .006. The expected number correct is 0.5 ± 0.23 (variance). Both place and cue groups therefore performed well above chance on their criterion trials. The place task was acquired more quickly, taking 7.2 ± 0.5 (SE) trials, compared with 14.0 ± 1.8 for the cue group, $F(1, 22) = 13.71$, $p < .01$.

Reconstructions from brain sections showed a placement of lesions similar to that found in the first experiment. In the place group, 4.58 mm³ ± 1.13, or 45.9% ± 11.3% of the total volume of the hippocampus, was aspirated; in the cue group, 3.98 mm³ ± 0.83, or 39.9% ± 8.3% of the hippocampus, was aspirated. Lesion size did not differ significantly between groups: volume, $t(6) = 0.42$, percentage of hippocampus, $t(6) = 0.39$ (arcsine transformed percentages). The location and volume of hyperstriatal aspirations in the place and cue groups (2.79 mm³ ± 0.75 and 2.57 mm³ ± 0.60, respectively) were similar to those in the caching study, including a narrow penetration of the striatum in some but not all animals. The mean volume of hippocampal and hyperstriatal lesions did not differ significantly in either the place, $t(6) = 1.32$, or the cue, $t(6) = 1.39$, group.

The effects of hippocampal and hyperstriatal aspirations are shown in Figure 4, which gives the number of correct choices of baited holes in the first six search attempts. Repeat visits to a hole are not counted as searches in this measure. There is a significant interaction between aspiration and site of aspiration on the place task, $F(2, 9) = 77.04$, $p < .01$. Groups do not differ preaspiration, HSD(9) < 1.06, and the only significant effect within place groups is in performance by the Hp group before and after aspiration, HSD(9) > 1.06, $p < .01$. Following aspiration, the place Hp group differs significantly from all other place groups. There are no significant effects within or between groups on the cue task.

Hippocampal aspiration reduced the number of baited holes successfully located for the place task but not for the cue task. Ha aspiration was without effect on either task. Thus hippocampal aspiration disrupts the task that requires relocating well-known places but has no effect on the task that requires orienting to well-learned cues.

Tabulation of revisiting errors permits a comparison of working memory among groups. Figure 5 shows cumulative revisiting errors, over the entire trial, both at correct holes that contained seeds and at incorrect holes that did not. A score consisting of revisits divided by total searches was calculated on each trial for each bird and all statistical tests were performed on arcsine transformations of these data. There are significant interactions between aspiration and site of aspiration in the place task for revisits to correct and incorrect sites, $F(2, 9) = 4.63$, $p < .05$ and $F(2, 9) = 7.16$, $p < .01$, respectively, and in the cue task for correct sites only, $F(2, 9) = 15.80$, $p < .01$. These results indicate that surgery affected revisiting in some groups but not others. There is no significant interaction in revisits to incorrect sites on the cue task, $F(2, 9) = 3.83$, because there were few visits to incorrect sites by any group on the cue task (see Figure 4). Post hoc tests of the significant interactions showed that the hippocampal groups made significantly more revisiting errors after aspiration than before at correct and incorrect sites in the place task, HSD(9) > 11.99, $p < .05$ and HSD(9) > 20.34, $p < .01$, respectively, and at correct sites on the cue task, HSD(9) > 21.51, $p < .01$. Birds in the Hp group make more revisiting errors postaspiration than any other group at correct and incorrect sites in the place task, HSD(9) > 11.99, $p < .05$ and HSD(9) > 20.34, $p < .01$, respectively, and at correct sites in the cue task, HSD(9) > 21.51, $p < .01$.

There were also significant interactions between aspiration and site of aspiration in the number of searches for both the place, $F(2, 9) = 10.70$, $p < .01$, and the cue, $F(2, 9) = 8.3$, $p < .01$, task. In all cases, hippocampal subjects make significantly more searches postaspiration than any other group, place: HSD(9) > 7.82, $p < .05$; cue: HSD(9) > 6.24, $p < .05$, and this effect can be seen clearly in all panels of Figure 5.

Discussion

The dissociation between performance of the place and cue tasks produced by hippocampal aspiration is as predicted by the cognitive mapping theory of hippocampal function. This dissociation is unlikely to be a result of task difficulty, because the place task, which was disrupted by hippocampal aspiration, was acquired more quickly and showed a higher level of performance on criterion trials than did the cue task. The observed increase in revisiting errors regardless of the nature of the task is a prediction of the working memory theory. Revisiting errors are not likely to be due to the disruption of spatial abilities alone, especially in the cue task, because cued sites differed not only in their spatial location but also in the nature of the local features around the site.

Hippocampal aspirations varied somewhat in both size and location. There was little variation, however, in the behavioral effects of hippocampal damage on performance of the place task, on the occurrence of revisiting errors, or on cache recovery in the previous experiment. It is not possible to determine from our results whether the common effects of hippocampal aspiration depend on a common region of hippocampal damage or on the aspiration of a minimum volume of hippocampus regardless of the exact site of damage.

General Discussion

The dissociation between the place and cue tasks produced by hippocampal aspiration supports the cognitive mapping theory of hippocampal function, and the occurrence of revisiting errors on both tasks supports the working memory theory. Jarrard et al. (1984) observed a similar outcome on the radial-arm maze after disrupting hippocampal connections in rats. The cognitive mapping and working memory theories are often discussed as mutually exclusive accounts of hippocampal function. The present result suggests that in birds the hippocampus may have multiple functions, and results with mammals can also be interpreted in this way.

The effect of damage to the hippocampus on cache recovery by chickadees is similar to the effects of hippocampal damage on the performance of comparable tasks by mammals (Becker & Olton, 1981; Jarrard et al., 1984; Morris et al., 1982, 1986; Olton et al., 1979; Sutherland et al., 1983; Walker & Olton, 1984). These results, and others with homing pigeons and nutcrackers (Bingman, Bagnoli, Ioalè, & Casini, 1984; Bingman, Ioalè, Casini, & Bagnoli, 1985; Krushinskaya, 1966), indicate that common functions accompany the structural homology between the avian and mammalian hippocampi (Benowitz, 1980; Casini et al., 1986; Källén, 1962; Krayniak & Siegel, 1978).

Finally, the results show that the ability to perform the cue task, which is spared by hippocampal damage, is not sufficient to support accurate cache recovery by black-capped chickadees. This provides neurophysiological confirmation of previous behavioral findings that cache recovery in food-storing birds is not accomplished by memory for cues common to cache sites. Instead, memory for places, working memory, and probably both, are essential to the process of relocating stored food and avoiding revisits to previously harvested food caches.

References

Baker, M. C., Stone, E., Baker, A. E. M., Shelden, R. J., Skillicorn, P., & Mantych, M. D. (1988). Evidence against observational learning in storage and recovery of seeds by black-capped chickadees. *Auk, 105,* 492–497.

Balda, R. P., Bunch, K. G., Kamil, A. C., Sherry, D. F., & Tomback, D. F. (1987). Cache site memory in birds. In A. C. Kamil, J. R. Krebs, & H. R. Pulliam (Eds.), *Foraging behavior* (pp. 645–666). New York: Plenum Press.

Becker, J. T., & Olton, D. S. (1981). Cognitive mapping and hippocampal system function. *Neuropyschologia, 19,* 733–744.

Benowitz, L. (1980). Functional organization of the avian telencephalon. In S. O. E. Ebbesson (Ed.), *Comparative neurology of the telencephalon* (pp. 389–421). New York: Plenum Press.

Best, P. J., & Ranck, J. B., Jr. (1982). Reliability of the relationship between hippocampal unit activity and sensory-behavioral events in the rat. *Experimental Neurology, 75,* 652–664.

Bingman, V. P., Bagnoli, P., Ioalè, P., & Casini, G. (1984). Homing behavior of pigeons after telencephalic ablations. *Brain, Behavior and Evolution, 24,* 94–108.

Bingman, V. P., Ioalè, P., Casini, G., & Bagnoli, P. (1985). Dorsomedial forebrain ablations and home loft association behavior in homing pigeons. *Brain, Behavior and Evolution, 26,* 1–9.

Casini, G., Bingman, V. P., & Bagnoli, P. (1986). Connections of the pigeon dorsomedial forebrain studied with WGA-HRP and ³H-proline. *Journal of Comparative Neurology, 245,* 454–470.

Cowie, R. J., Krebs, J. R., & Sherry, D. F. (1981). Food storing by marsh tits. *Animal Behaviour, 29,* 1252–1259.

Haftorn, S. (1974). Storage of surplus food by the boreal chickadee *Parus hudsonicus* in Alaska, with some records on the mountain chickadee *Parus gambeli* in Colorado. *Ornis Scandinavica, 5,* 145–161.

Honig, W. K. (1978). Studies of working memory in the pigeon. In S. H. Hulse, H. Fowler, & W. K. Honig (Eds.), *Cognitive processes in animal behavior* (pp. 211–248). Hillsdale, NJ: Erlbaum.

Jarrard, L. E. (1983). Selective hippocampal lesions and behavior: Effects of kainic acid lesions on performance of place and cue tasks. *Behavioral Neuroscience, 97,* 873–889.

Jarrard, L. E., Okaichi, H., Steward, O., & Goldschmidt, R. B. (1984). On the role of hippocampal connections in the performance of place and cue tasks: Comparisons with damage to hippocampus. *Behavioral Neuroscience, 98,* 946–954.

Källén, B. (1962). Embryogenesis of brain nuclei in the chick telencephalon. *Ergebnisse der Anatomie und Entwicklungs Geschichte, 36,* 62–82.

Kamil, A. C., & Balda, R. P. (1985). Cache recovery and spatial memory in Clark's nutcrackers (*Nucifraga columbiana*). *Journal of Experimental Psychology: Animal Behavior Processes, 11,* 95–111.

Karten, H. J., & Hodos, W. (1967). *A stereotaxic atlas of the brain of the pigeon (Columba livia).* Baltimore, MD: Johns Hopkins Press.

Krayniak, P. F., & Siegel, A. (1978). Efferent connections of the hippocampus and adjacent regions in the pigeon. *Brain, Behavior and Evolution, 15,* 372–388.

Krushinskaya, N. L. (1966). Some complex forms of feeding behaviour of nutcracker *Nucifraga caryocatactes*, after removal of old cortex. *Zhurnal Evoluzionni Biochimii y Fisiologgia, 2,* 563–568.

Morris, R. G. M., Anderson, E., Lynch, G. S., & Baudry, M. (1986). Selective impairment of learning and blockade of long-term potentiation by an N-methyl-D-aspartate receptor antagonist, AP5. *Nature, 319,* 774–776.

Morris, R. G. M., Garrud, P., Rawlins, J. N. P., & O'Keefe, J. (1982). Place navigation impaired in rats with hippocampal lesions. *Nature, 297,* 681–683.

O'Keefe, J. (1979). A review of the hippocampal place cells. *Progress in Neurobiology, 13,* 419–439.

O'Keefe, J., & Nadel, L. (1978). *The hippocampus as a cognitive map.* Oxford: Clarendon Press.

Olton, D. S., Becker, J. T., & Handelmann, G. E. (1979). Hippocampus, space, and memory. *Behavioral and Brain Sciences, 2,* 313–365. (Includes commentary)

Olton, D. S., Branch, M., & Best, P. J. (1978). Spatial correlates of hippocampal unit activity. *Experimental Neurology, 58,* 387–409.

Sherry, D. F. (1982). Food storage, memory and marsh tits. *Animal Behaviour, 30,* 631–633.

Sherry, D. F. (1984a). Food storage by black-capped chickadees: Memory for the location and contents of caches. *Animal Behaviour, 32,* 451–464.

Sherry, D. F. (1984b). What food-storing birds remember. *Canadian Journal of Psychology, 38,* 304–321.

Sherry, D. F. (1985). Food storage by birds and mammals. *Advances in the Study of Behavior, 15,* 153–188.

Sherry, D. F., Avery, M., & Stevens, A. (1982). The spacing of stored food by marsh tits. *Zeitschrift für Tierpsychologie, 58,* 153–162.

Sherry, D. F., Krebs, J. R., & Cowie, R. J. (1981). Memory for the location of stored food in marsh tits. *Animal Behaviour, 29,* 1260–1266.

Shettleworth, S. J., & Krebs, J. R. (1982). How marsh tits find their hoards: The roles of site preference and spatial memory. *Journal of Experimental Psychology: Animal Behavior Processes, 8,* 354–375.

Stevens, T. A., & Krebs, J. R. (1986). Retrieval of stored seeds by marsh tits *Parus palustris* in the field. *Ibis, 128,* 513–525.

Stokes, T. M., Nottebohm, F. & Leonard, C. M., (1974). The telencephalon, diencephalon, and mesencephalon of the canary, *Serinus canaria,* in stereotaxic coordinates. *Journal of Comparative Neurology, 156,* 337–374.

Sutherland, R. J., Whishaw, I. Q., & Kolb, B. (1983). A behavioural analysis of spatial localization following electrolytic, kainate- or colchicine-induced damage to the hippocampal formation in the rat. *Behavioural Brain Research, 7,* 133–153.

Tomback, D. F. (1980). How nutcrackers find their seed stores. *Condor, 82,* 10–19.

Vaccarino, A. L. (1986). An inexpensive and reliable rat stereotaxic adaptor for small bird neurosurgery. *Physiology and Behavior, 38,* 735–737.

Vander Wall, S. B. (1982). An experimental analysis of cache recovery in Clark's nutcracker. *Animal Behaviour, 30,* 84–94.

Walker, J. A., & Olton, D. S. (1984). Fimbria–fornix lesions impair spatial working memory but not cognitive mapping. *Behavioral Neuroscience, 98,* 226–242.

Received November 19, 1987
Revision received April 5, 1988
Accepted April 11, 1988 ■

Glossary

aspiration - removing a structure by suction; for example, hippocampal lesions are often made by drawing off the tissue through a fine glass pipette connected by tubing to a small vacuum pump
avian - pertaining to birds
aviary - a place for housing birds
brain atlas - a book that contains a series of two-dimensional maps of various slices of the brain of a particular species; it is used by surgeons to determine the coordinates of target brain structures prior to surgery
cache site - a location where an animal has hidden an object such as a seed
cognitive-map theory - the theory that the hippocampus stores memories about spatial location
congeneric - of the same genus
corvids - a family of birds that includes crows, jays, and magpies
cytoarchitectonic - pertaining to the arrangement of cells in tissue
embryologically - pertaining to the embryo
hippocampus - a neural structure that plays an important role in memory
homologous - of similar structure
hyperstriatum accessorium - a neural structure adjacent to the hippocampus in birds
interocular - between eyes
mark - to place an identifying stimulus on a particular object or site; many species mark home territories and routes with specialized odoriferous secretions
mnemonic - pertaining to memory
post hoc - after the fact
radial-arm maze - a maze in which several arms radiate out from a central starting area; in one version of the radial-arm-maze test, all arms are baited and rats learn to obtain all of the food without ever re-entering an arm that has already been visited during that particular test
reference memory - memory that pertains to all performances of a particular task; for example, memory for the fact that food can always be found at the end of each arm of a radial-arm maze on each trial
working memory - memory for information that pertains to only the current performance of a task; for example, memory for what arms of a radial arm maze have already been emptied of food during a particular test

Essay Study Questions

1. Why are black-capped chickadees interesting subjects for memory experiments?

2. What evidence is there that the food-caching behavior of birds depends on their mnemonic abilities?

3. Describe two prominent theories of hippocampal function.

4. Compare working and reference memory. How can working memory and reference memory be measured in a radial-arm maze?

5. Describe Experiment 1. What were its results, and what was its major conclusion?

6. Experiment 2 provided support for both the cognitive-map and working-memory theories of hippocampal function. Explain.

7. In Experiment 2, the experimenter, rather than the chickadees, hid the seeds. Why?

Multiple-Choice Study Questions

1. Which of the following characterizes the recovery of food caches in chickadees?
   a. In the wild, individual chickadees do not repeatedly cache seeds in the same cache sites.
   b. In the wild, seeds are recovered from cache sites at a higher rate than they are taken from similar control sites.
   c. In the wild, the individual chickadee that caches a food item is the same one that later returns to collect it.
   d. In the laboratory, individual birds relocate their cache sites even when the food has been removed from them by the experimenter.
   e. all of the above

2. Magnetic leg bands were used to confirm that
   a. the bird that caches a piece of food in the wild is the same one that retrieves it.
   b. birds detect cached food by smelling the food.
   c. birds do not detect cached food by smelling the food.
   d. there is no interocular transfer of memory in birds.
   e. during retrieval, birds do not retrace the route that they took while caching.

3. Rats that repeatedly enter the same arm of a radial-arm maze during a single test are said to have deficits in
   a. reference memory.
   b. cognitive mapping.
   c. working memory.
   d. interocular transfer.
   e. free recall.

4. After bilateral hippocampal aspiration, the chickadees in Experiment 1 visited just as many potential cache sites as they had before surgery, but only about
   a. 1% of their visits were to sites where they had cached food.
   b. 8% of their visits were to sites where they had cached food.
   c. 27% of their visits were to sites where they had cached food.
   d. 55% of their visits were to sites where they had cached food.
   e. 72% of their visits were to sites where they had cached food.

5. Aspiration of the hyperstriatum accessorium significantly reduced the number of
   a. visits to cache sites.
   b. seeds eaten.
   c. visits to noncache sites.
   d. visits to sites during the prestorage phase.
   e. none of the above

6. Experiment 2 provided support for the
   a. cognitive-mapping theory of hippocampal function.
   b. reference-memory theory of hippocampal function.
   c. working-memory theory of hippocampal function.
   d. both a and b
   e. both a and c

The answers to the preceding questions are on page 290.

Food-For-Thought Questions

1. The "spatial task" in Experiment 2 was not totally a spatial task; it could be solved on the basis of spatial cues or on the basis of the distinctive surroundings of the baited holes. Explain.

2. Despite having no neocortex (see Figure 2), food-caching birds routinely perform feats of memory that would be difficult for most primates. Discuss this lack of neocortex in relation to the widely held assumption that the neocortex is the basis of human intellectual supremacy.

ARTICLE 17

Spatial Selectivity of Rat Hippocampal Neurons: Dependence on Preparedness for Movement

T.C. Foster, C.A. Castro, and B.L. McNaughton

Reprinted from Science, 1989, volume 244, 1580-1582.

Although evidence from many sources indicates that the hippocampus plays an important role in memory, two seemingly incompatible views of this role have emerged, both of which have received support from the study of hippocampal neurons—in particular, the neurons of the pyramidal cell layer. Such studies in freely moving animals have found the activity of pyramidal cells to be place-specific; particular cells become active only when the subject is in a particular location of the test environment. These results have led to the view that the hippocampus plays an important role in memory for places. In contrast, in studies of immobilized animals, the activity of pyramidal cells has been shown to be related to learned associations between stimuli and responses. These results have led to the alternative view that the hippocampus plays a role in memory for learned associations.

Foster, Castro, and McNaughton have a view of hippocampal function that is an integration of these seemingly contradictory theories. They proposed that the hippocampus develops a spatial representation of the environment in terms of learned relations between various environmental stimuli and the movements that are required to get the animal from one place to another. Accordingly, they hypothesized that hippocampal neurons will display place-specific activity only when the animal is freely moving.

Foster, Castro, and McNaughton tested their motor-programming theory of hippocampal function by comparing the activity of hippocampal pyramidal neurons in freely moving and restrained rats. In freely moving rats, pyramidal neurons displayed the place specificity reported in previous studies: individual neurons increased their activity when the rat was in one location of the test environment but not when it was in others. In contrast, when the same rats were restrained and placed in different locations of the test environment by the experimenter, the same pyramidal neurons displayed little or no place specificity.

This finding provides strong support for the view that preparedness for movement plays an important role in the encoding of spatial location by the hippocampus, and it explains why experiments in freely moving and restrained animals have led to two different views of hippocampal function.

SCIENCE

Reprint Series
30 June 1989, Volume 244, pp. 1580–1582

Copyright © 1989 by the American Association for the Advancement of Science

Spatial Selectivity of Rat Hippocampal Neurons: Dependence on Preparedness for Movement

TOM C. FOSTER,* CARL A. CASTRO, BRUCE L. MCNAUGHTON

The mammalian hippocampal formation appears to play a major role in the generation of internal representations of spatial relationships. In rats, this role is reflected in the spatially selective discharge of hippocampal pyramidal cells. The principal metric for coding spatial relationships might be the organism's own movements in space, that is, the spatial relationship between two locations is coded in terms of the movements executed in getting from one to the other. Thus, information from the motor programming systems (or "motor set") may contribute to coding of spatial location by hippocampal neurons. Spatially selective discharge of hippocampal neurons was abolished under conditions of restraint in which the animal had learned that locomotion was impossible. Therefore, hippocampal neuronal activity may reflect the association of movements with their spatial consequences.

IN RATS, THE HIPPOCAMPAL FORMATION plays a major role in the encoding of spatial memory (1). Specifically, hippocampal pyramidal cells recorded in freely moving rats display both selectivity and memory for spatial location (2). In contrast, studies with restrained rabbits and primates emphasize hippocampal cellular involvement in associative learning (3), suggesting that neuron activity is related to learned stimulus-response contingencies (4). There is little sign of place-specific neuronal activation in such studies. However, the same cells that develop representations of conditioned responses also engage in spatial coding in extended environments (5), suggesting that the two types of activity may reflect processing of fundamentally similar kinds of information.

Two hippocampal cell types, complex spike (CS) and theta cells, can be identified electrophysiologically as pyramidal cells and interneurons, respectively (6). In the freely moving rat, both CS and theta cells discharge in phase with the rhythmic (theta) electroencephalogram (EEG) that accompanies orienting or translational movements,

Department of Psychology, University of Colorado, Boulder, CO 80303.

*To whom correspondence should be addressed.

that is, movements that carry the animal from one place to another (7). Furthermore, the responsiveness of CS cells to spatial location is modulated by the velocity and direction of movement (8), indicating a possible influence of "motor set" (9).

We investigated the possible role of motor set for location-specific discharge activity in freely moving and restrained rats. Four animals were trained to tolerate restraint, which was implemented by snugly wrapping the body and limbs in a towel fastened with clips. This procedure allowed the animal to observe the environment by head movement and exploratory myostatial sniffing, while inhibiting attempts at displacement movements. Single units were isolated and recorded with "stereotrodes" (10) mounted in miniature manipulators. The manipulators were permanently implanted over the CA1 region of the hippocampus in rats under pentobarbital anesthesia. Several weeks after surgery, well-differentiated units were identified as CS or theta cells according to established criteria. In unrestrained, or free, rats the discharge specificity of the cells was repeatedly tested by manually transporting the animal alternately to an identified place field for 5 s, then to a neutral location for 5 s (11) (Fig. 1). The short interval between each transportation ensured that animals maintained an alert behavioral state. In some cases animals were left in one location long enough for us to examine unit activity during the large amplitude, irregular EEG activity (LIA) (voluntary prolonged immobility), which replaces the theta state after cessation of orienting or translational behaviors (12). Animals were then restrained and cells were again tested for place specificity and relation to EEG (Fig. 1). A recovery session was included to ensure the recording integrity. Behavior during the free, restraint, and recovery conditions was very similar in that animals engaged in head movements and myostatial sniffing. During free and recovery conditions animals also extended their limbs in anticipation of contact with the tabletop and made small shifts of posture. However, they were not actually locomoting during the 5-s sampling epochs.

Of 66 units recorded, 12 were classified as theta cells. The remaining 54 were identified as CS cells, of which 38 (70%) exhibited place specificity in the experimental environment. Thirty-one of these cells could be monitored during the conditions of free, restraint, and recovery. There was an almost

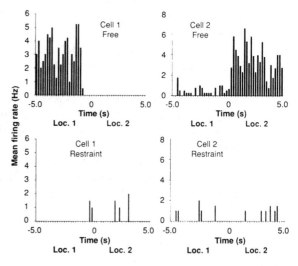

Fig. 1. Time histograms of two simultaneously recorded CS cells (cell 1 and cell 2) from one animal as it was manually transported from the place field of one cell (Loc. 1) into the place field of the other cell (Loc. 2) during the free (top panels) and restraint (lower panels) conditions; each place field thus served as a neutral location for the other place field (11). Spatially selective firing was abolished under the restraint condition.

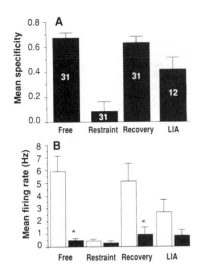

Fig. 2. (**A**) Mean place specificity scores (11) for CS cells in the four behavioral conditions: free, restraint, recovery, and LIA. Numbers in bars represent units per treatment condition. There is a loss in specificity as a result of restraint and a slight decrease during LIA. (**B**) Mean discharge rate for 12 cells as animals were manually moved into the previously determined place field (open bars) or neutral location (solid bars) under the four behavioral conditions. The loss of specificity during restraint is entirely due to decreased discharge in the place field. Bars, mean + SEM; asterisks, significant difference ($P < 0.05$) between place field and neutral location.

complete suppression of CS cell place specificity during restraint, compared to free and recovery conditions [$F(2,30) = 55.35$, $P < 0.0001$] (Fig. 2A). There was also an overall decrease in the discharge rate of CS [$F(2,30) = 22.85$, $P < 0.001$] and theta cells during restraint [$F(2,11) = 14.18$, $P < 0.005$]. The decrease in mean firing rate for CS cells was due almost entirely to decreased discharge in the previously determined place field [$F(2,30) = 31.64$, $P < 0.0001$]. No cell was observed to increase its firing rate during restraint. Although discharge activity in the place field was not significantly different from the neutral location in the restrained state (Fig. 2B), there was a small, but significant correlation between the specificity scores in the two states, indicating a very slight residual specificity. Place specificity scores for 12 CS units examined during LIA were slightly reduced, compared to free and recovery conditions, confirming an earlier report (13) (Fig. 2A).

The hippocampal EEG under the different conditions exhibited a restraint-induced decrease in spectral power for type I (movement) theta (at about 7 to 10 Hz) and increased power at lower frequencies (1 to 4 Hz) (Fig. 3C). Furthermore, type II (sensory) theta (about 6 Hz) (12) was not eliminated during the restraint condition, suggesting that the reduction in specificity was not due to inattention to environmental stimuli. Elimination of this sensory theta by atropine (25 to 50 mg/kg) had no discernable effect on place specificity in either the free or restraint conditions. Thus, there was a dissociation between reduction of theta power and loss of place specificity (Fig. 3, A and B).

Our results indicate that motor set makes a major contribution to spatially selective activity in CA1 cells. This contribution may be simply a gating mechanism. Alternatively, information about actual movements or possible movements may play a more fundamental role in the representation of spatial location. The persistence of head movement and exploratory sniffing in addition to type II (sensory) theta EEG activity recorded during restraint indicates that the animals were attending to environmental cues and also that such activity is not responsible for place-specific discharge. Although the loss of spatial firing was accompanied by a reduction of type I (movement-related) theta activity, the latter effect must be coincidental rather than causal, because inactivation of the medial septal projection to the hippocampus by local anesthesia completely abolishes both types of theta activity, with no effect on CA1 place selectivity in freely moving animals (14). When free, animals could have moved, but they rarely did so apart from limb extension and some head and sniffing movements that were common to all conditions.

We thus favor the hypothesis that information on the preparedness for movement must be an intrinsic component of the information projected to the hippocampus by way of its cortical afferents, and this information on preparedness must do more than simply gate hippocampal output. The data are consistent with two related hypotheses concerning the role of hippocampal activity in spatial representation: a proposal that the spatial role of the hippocampus may be primarily that of learning about spatially directed movements (15) and a proposal that spatial representation involves the formation of conditional associations between representations of movements and representations of locations (16). In addition, other investigators (17) found that stimulus-evoked unit activity in human hippocampus is decreased if subjects are instructed not to respond to the stimuli. Finally, the dependence of location-specific discharge on the animal's perceived ability to engage in movements through space may partly account for the relatively small number of spatially selective neurons recorded from hippocampus in primates and rabbits under conditions of restricted translational movement (4).

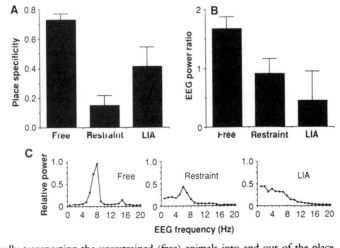

Fig. 3. (**A**) Mean place specificity scores (11) and (**B**) mean EEG power ratio scores for 12 CS cells recorded from the CA1 region of the hippocampus during the three behavioral conditions of free, restraint, and LIA. EEG scores were computed as the ratio of total spectral power in the 6- to 9-Hz (theta) band divided by that for the 1- to 4-Hz (LIA) band. (**C**) Typical power spectra recorded in the CA1 region of the hippocampus for one animal during the three behavioral conditions. Manually transporting the unrestrained (free) animals into and out of the place field was associated with high unit place specificity, intense EEG activity in the theta band, and lower power in the LIA band. Under restraint, the same manipulation was associated with an abolition of place specificity and only a moderate reduction in the EEG ratio (due to both a reduction of the higher frequency theta components and an increase in the LIA band). Under LIA the EEG power ratio was lowest, but there was only a partial reduction in place specificity. Thus, the loss of place specificity is not accounted for by the change in EEG state. Bars, mean + SEM.

REFERENCES AND NOTES

1. J. O'Keefe and L. Nadel, *The Hippocampus as a Cognitive Map* (Clarendon Press, Oxford, 1978); J. O'Keefe and A. Speakman, *Exp. Brain. Res.* **68**, 1 (1987).
2. J. O'Keefe and J. Dostrovsky, *Brain Res.* **34**, 171 (1971); J. O'Keefe, *Prog. Neurobiol.* (Oxford) **13**, 419 (1979).
3. R. F. Thompson, T. W. Berger, S. D. Berry, F. K. Hoehler, *Neural Mechanisms of Behavior* (Springer-Verlag, New York, 1980); M. Mishkin, *Philos. Trans. R. Soc. London Ser. B* **298**, 85 (1982).
4. R. F. Thompson, T. W. Berger, J. Madden, *Annu. Rev. Neurosci.* **6**, 447 (1983); T. Wantanabe and H. Niki, *Brain Res.* **325**, 241 (1985); E. T. Rolls, Y. Miyashita, P. Cahusac, R. P. Kesner, *Soc. Neurosci. Abstr.* **13**, 525 (1987).

5. P. J. Best and R. F. Thompson, *Soc. Neurosci. Abstr.* **10**, 125 (1984).
6. S. E. Fox and J. B. Ranck, Jr., *Exp. Neurol.* **49**, 299 (1975); T. C. Foster *et al.*, *Brain Res.* **408**, 86 (1987).
7. B. H. Bland, P. Andersen, T. Ganes, O. Sveen, *Exp. Brain Res.* **38**, 205 (1980); G. Buzsaki, L. S. Leung, C. H. Vanderwolf, *Brain Res. Rev.* **6**, 139 (1983).
8. B. L. McNaughton, C. A. Barnes, J. O'Keefe, *Exp. Brain Res.* **52**, 41 (1983).
9. Motor set is used in the general sense of preparedness for a particular movement [E. V. Evarts, Y. Shinoda, S. P. Wise, *Neurophysiological Approaches to Higher Brain Functions* (Wiley, New York, 1984)].
10. Stereotrodes [B. L. McNaughton, J. O'Keefe, C. A. Barnes, *J. Neurosci. Methods* **8**, 391 (1983)] consisted of two closely spaced (20 μm) wires. Up to five cells at different distances from the two tips were discriminated at one time by their signal amplitude ratios on the two channels.
11. Unit data were collected as animals traversed a raised platform. On-line analysis identified the location of maximum discharge, that is, the "place field." Event flags inserted into the data stream marked corresponding spatial location for computation of spatial specificity scores. The EEG was filtered, digitized, and stored for off-line Fourier analysis. Mean firing rates for each location, for eight repetitions of each treatment condition, were used to calculate specificity scores. The mean rate in the neutral location was subtracted from the mean rate in the place field, and this difference was divided by the sum of the two means. Treatment effects were assessed by one-way analysis of variance. Location of units in the CA1 region was verified by histology.
12. C. H. Vanderwolf, *Electroencephalogr. Clin. Neurophysiol.* **26**, 407 (1969); R. Kramis, C. H. Vanderwolf, B. H. Bland, *Exp. Neurol.* **49**, 58 (1975).
13. R. U. Muller, J. L. Kubie, J. B. Ranck, Jr., *J. Neurosci.* 7, 1935 (1987).
14. S. J. Y. Mizumori, B. L. McNaughton, C. A. Barnes, *J. Neurosci.*, in press.
15. D. Gaffan, *Philos. Trans. R. Soc. London Ser. B.* **308**, 87 (1985); N. M. Rupniak and D. Gaffan, *J. Neurosci.* **7**, 2331 (1987).
16. B. L. McNaughton, *Neurosci. Lett.* **29**, S143 (1987); in *Neural Connections and Mental Computations*, L. Nadel, L. Cooper, P. Culicover, R. Harnish, Eds. (Academic Press, New York, 1989), pp. 285–350.
17. E. Halgren, T. L. Babb, P. H. Crandall, *Electroencephalogr. Clin. Neurophysiol.* **45**, 585 (1987).
18. We thank C. A. Barnes for assistance and comments. Supported by PHS grants NS20331 and T32-HD07288.

26 January 1989; accepted 18 April 1989

Glossary

afferents - neurons that carry signals to the structure in question; for example, hippocampal afferents are neurons that carry signals to the hippocampus

associative learning - the learning of associations between stimuli, between responses, or between stimuli and responses; Pavlovian and operant conditioning are examples of associative learning

atropine - a drug that blocks the activity at a subset of acetylcholine receptors; one effect of atropine is to block type II hippocampal theta activity

CA1 - a particular region of the hippocampus

complex spike cells - hippocampal neurons that tend to fire in well-defined bursts; complex spike neurons have been shown to be pyramidal neurons

conditioned responses - responses elicited by conditioned (conditional) stimuli; in Pavlov's experiments, the salivation elicited by the bell was a conditioned (conditional) response

hippocampus - one of the major structures of the medial temporal lobes; the hippocampus is also called the hippocampal formation

large-amplitude irregular EEG activity (LIA) - the characteristic hippocampal EEG activity of a rat that is neither moving nor attending to stimuli

motor set - the preparedness to make a particular motor response or class of motor responses

myostatial sniffing - investigatory sniffing, which is accompanied by vibrissae (whisker) movement

orienting - the characteristic response of an organism to any sudden unexpected stimulus; the subject reflexively turns to face the stimulus

place-specific neurons - neurons whose activity changes when the subject is in one location but not others; many of the pyramidal neurons of the rat hippocampus are place specific

power spectrum - any raw EEG signal is composed of many different frequencies; a power spectrum is a graph of the relative amount of power associated with the various frequencies in a raw EEG signal

stereotrodes - a two-wire electrode designed for recording the activity of single neurons; by analyzing the relation between the signals picked up at the tip of the two fine insulated wires, it is possible to extract the individual signals of several different neurons in the vicinity

theta activity - strictly speaking theta activity refers to any EEG activity whose predominant frequency is 4 to 7 Herz (Hz; cycles per second); however, in the hippocampus there is a rhythmic, high-amplitude EEG wave that is commonly referred to as hippocampal theta activity, despite the fact that it occurs at frequencies between 5 and 10 Hz; hippocampal theta activity has been widely studied because it occurs in readily recognizable, well-defined bursts that are correlated with certain aspects of the subjects' behavior

theta cells - hippocampal neurons that fire in relation to hippocampal theta EEG activity; unlike complex spike cells, theta cells do not tend to fire in well-defined bursts; theta cells have been shown to be interneurons

translational movements - movements that carry an animal from one location to another

type I theta - hippocampal theta activity that tends to occur in association with translational movements; type I theta is not blocked by atropine, and its frequency tends to be slightly greater than that of type II theta

type II theta - hippocampal theta activity that occurs when a subject is not moving but is paying attention to something (i.e., is orienting); type II theta is blocked by atropine, and its frequency tends to be slightly less than that of type I theta

Essay Study Questions

1. Describe two different theories that have emerged concerning the role of the hippocampus in memory. What kinds of single-neuron recording studies have supported these respective theories?

2. Compare complex spike cells and theta cells.

3. Compare type I and type II hippocampal theta activity.

4. Summarize Foster, Castro, and McNaughton's theory of hippocampal function.

5. Describe the results of Foster, Castro, and McNaughton's experiment. What do these results suggest?

Multiple-Choice Study Questions

1. The rats in Foster, Castro, and McNaughton's experiment were restrained by
    a. locking them in a very small chamber.
    b. reinforcing them to stay still.
    c. gluing their feet to a board.
    d. shocking them when they moved.
    e. none of the above

2. When animals are not orienting or moving, their hippocampal EEG is usually dominated by
    a. type I theta activity.
    b. type II theta activity.
    c. large-amplitude, irregular EEG activity.
    d. low-amplitude, regular EEG activity
    e. rhythmic theta activity.

3. Most of the neurons studied by Foster, Castro, and McNaughton were
    a. in the hippocampal formation.
    b. in the pyramidal cell layer.
    c. CA1 cells.
    d. complex spike cells.
    e. all of the above

4. What proportion of the complex spike cells studied by Foster, Castro, and McNaughton displayed place specificity in the experimental environment?
    a. 70%
    b. 0%
    c. 100%
    d. 50%
    e. 94%

5. Atropine
    a. blocked type I theta.
    b. had no effect on place specificity.
    c. eliminated place specificity.
    d. elicited type II theta.
    e. both a and c

6. In one study, stimulus-evoked neuronal activity was reduced when the human subjects were instructed to
    a. engage in myostatial sniffing.
    b. not respond to the stimulus.
    c. shut their eyes.
    d. memorize the layout of the room.
    e. do the lambada.

The answers to the preceding questions are on page 290.

Food-For-Thought Questions

1. It is important to remember that correlation does not imply causation. Discuss with respect to the relation between type I hippocampal theta activity and the place specificity of hippocampal neurons.

2. Most neurophysiological studies are conducted in chemically or physically restrained animals. The experiment of Foster, Castro, and McNaughton illustrates that it is often important to study neural activity in freely moving subjects in order to appreciate its function. Discuss.

ARTICLE 18

Organizational Changes in Cholinergic Activity and Enhanced Visuospatial Memory as a Function of Choline Administered Prenatally or Postnatally or Both

W.H. Meck, R.A. Smith, and C.L. Williams
Reprinted from Behavioral Neuroscience, 1989, volume 103, 1234-1241.

As we learn more about the neural bases of memory, it may become possible to develop methods of improving memory In the following article, Meck, Smith, and Williams reported that the exposure of rat pups to large doses of choline in the perinatal period improved their performance of a visuospatial memory task in adulthood.

There were four groups of subjects: (1) one group received choline supplements prenatally through their mothers' drinking water and postnatally until weaning by injection into their stomachs through a fine esophageal tube; (2) one group received only the prenatal supplement; (3) one group received only the postnatal supplement; and (4) a control group received no supplement. The assessment of the subject's performance on a 12-arm radial-arm-maze task began when they reached 60 days of age. On each test day, the same eight arms were baited with a food pellet, and the number of trials required for each subject to recover all eight pellets was recorded. Following the testing phase of the experiment, the frontal neocortex and the hippocampus were dissected from the brains of the control subjects and the prenatal-plus-postnatal choline-exposure subjects. The ability of [^{3}H] quinuclidinyl benzilate (QNB) to bind to these two samples of tissue and the level of choline acetyltransferase (ChAT) activity in them were then determined—QNB binding is indicative of the number of muscarinic cholinergic receptors, and ChAT is the enzyme that stimulates the conversion of choline to acetylcholine.

The results were consistent with the hypothesized memory-enhancing effect of choline: the subjects that had received choline supplements both before and after birth made significantly fewer total errors than did the control subjects, and the performance of the prenatal-choline and the postnatal-choline groups was intermediate. An analysis of the performance of the prenatal-plus-postnatal choline-exposure subjects indicated that both their working memories (fewer repeat visits to arms already visited on a given test) and reference memories (fewer visits to arms that were always unbaited) were significantly superior to those of the control subjects. Furthermore, QNB binding was greater in the frontal cortex and the hippocampus of the prenatal-plus-postnatal choline-exposure subjects than in the control subjects, and their hippocampi displayed lower levels of ChAT activity. These results suggest that early exposure to choline is capable of effecting long-term changes in their cholinergic function and improvements both in their working memory and their reference memory.

Organizational Changes in Cholinergic Activity and Enhanced Visuospatial Memory as a Function of Choline Administered Prenatally or Postnatally or Both

Warren H. Meck and Rebecca A. Smith
Columbia College, Columbia University

Christina L. Williams
Barnard College, Columbia University

This experiment was an examination of the effects of supplemental dietary choline chloride given prenatally (to the diet of pregnant rats) and postnatally (intubed directly into the stomachs of rat pups) on memory function and neurochemical measures of brain cholinergic activity of male albino rats when they became adults. The data demonstrate that perinatal choline supplementation causes (a) long-term facilitative effects on working and reference memory components of a 12-arm radial maze task, and (b) alternations of muscarinic receptor density as indexed by [^{3}H]quinuclidinyl benzilate (QNB) binding and choline acetyltransferase (ChAT) levels in the hippocampus and frontal cortex of adult rats. An analysis of the relationship between these organizational changes in brain and memory function indicated that the ChAT-to-QNB ratio in the hippocampus is highly correlated with working memory errors, and this ratio in the frontal cortex is highly correlated with reference memory errors.

Neurochemical evidence has demonstrated that under certain conditions, increasing the availability of choline or lecithin (precursors for acetylcholine) may enhance cholinergic (Ch) transmission (Cohen & Wurtman, 1975; Haubrich, Wang, Clody, & Wedeking, 1975). Moreover, when Ch neurons are physiologically active, the rate at which they synthesize and release acetylcholine varies directly with the amount of free choline available to them (Bartus, Dean, Sherman, Friedman, & Beer, 1981; Blusztajn & Wurtman, 1983). This information has stimulated research on the behavioral effects of manipulating the choline supply to the brain during various periods of development. An increasing amount of evidence indicates that Ch neurons are important for short- and long-term memory processes in various species (Bartus, Dean, Pontecorvo, & Flicker, 1985). There are numerous reports of improved memory function in adult animals following dietary supplementation with various forms of choline (e.g., Bartus, Dean, Goas, & Lippa, 1980; Golczewski, Hiramoto, & Ghanta, 1982; Leathwood, Heck, & Mauron, 1982; Meck & Church, 1987; Mizumori et al., 1985).

Although increasing the amounts of choline or lecithin can improve performance on certain behavioral tasks in mature subjects, attempts to improve memory performance in aged humans or animals generally have not succeeded (for review, see Bartus et al., 1985). Reasons for this failure in geriatric subjects may be that the blood-brain barrier transport of choline is reduced in aged subjects (Mooradian, 1988) or that the aged brain is unable to incorporate extra amounts of choline into acetylcholine, as reportedly occurs in younger brain tissue (Bartus, Dean, & Beer, 1980). Also, other factors in aged brains may need to be improved before substantial increases in presynaptic Ch effects are obtained with precursor loading. For example, although normal Ch activity is dependent on intact oxidative metabolism, several variables that reflect energy production are decreased in the aged central nervous system (Meier-Ruge, Hunziker, Wangoff, Reichmeier, & Sandoz, 1978; Sims et al., 1980; Sylvia & Rosenthal, 1979). Furthermore, although the conversion of choline into acetylcholine occurs more readily under conditions of increased neuronal activity, recent evidence suggests that the activity of certain Ch pathways progressively decreases with age (Decker, Pelleymounter, & Gallagher, 1988; Meck, Church, & Wenk, 1986, in press; Sherman, Kuster, Dean, Bartus, & Friedman, 1981). Thus, either of these (or similar) factors could contribute to a situation in the aged brain that would prohibit extra choline from being effectively used for the synthesis of additional acetylcholine, which, in turn, would explain the negative results obtained with precursor studies in aged animals and humans.

One way to compensate for these possible age-related deficits would be to administer abundant amounts of choline not to mature or aged animals, but to young animals during prenatal and postnatal stages of development. Dietary intake of choline contributes to high serum choline concentrations in the neonatal rat (Zeisel & Wurtman, 1981), and the neonatal blood-brain barrier readily transports choline into the brain (Pardridge, Cornford, Braun, & Oldendorf, 1979). Therefore, high serum choline concentrations should be reflected in high brain choline levels and subsequently in high brain acetylcholine concentrations. At these high concentrations, choline may also function directly as a Ch agonist and modify neuronal organization through developing feedback

---

This investigation was supported in part by research grants from the National Institute of Neurological and Communicative Disorders and Stroke to Warren H. Meck (NS 24794) and Christina L. Williams (NS 20671). Additional support was provided by an Alfred P. Sloan Foundation Research Fellowship awarded to Warren H. Meck.

We acknowledge Hyesung Ha and Matthew Kleiderman for their excellent assistance in testing the animals and analyzing the data.

Correspondence concerning this article should be addressed to Warren H. Meck, Department of Psychology, Columbia University, New York, New York 10027.

mechanisms (Krnjevic & Reinhardt, 1979). However, no attempt has been made to examine whether supplemental dietary choline during early development can cause long-term changes in Ch function and memory that persist in adulthood. Organizational effects produced by the short-term administration of Ch agents (e.g., choline) during neonatal development may produce long-term modification of memory processes that remain functional during later stages of development and aid in the lessening or prevention of age-related memory impairments.

We recently reported that choline supplementation during prenatal and postnatal development has long-term facilitative effects on visuospatial memory that extend well into adulthood (Meck, Smith, & Williams, 1988; Meck & Williams, in press; Smith, Meck, & Williams, 1986). These effects on visuospatial memory were examined with a 12-arm radial maze food-searching task, with 8 baited and 4 unbaited arms. The task includes working memory and reference memory components (Olton, Becker, & Handelmann, 1979; Olton & Papas, 1979). During each test, on the first approach to each baited arm, the rat is expected to choose that arm and get the food from the end of it. In all subsequent approaches during that test, the rat should not select a previously chosen arm because the food has been removed. Consequently, the rat must remember each choice for the duration of a test, but at the end of the test, the rat ought to forget these choices so that they will not interfere with its performance in subsequent tests. Thus, errors made to previously visited arms constitute a measure of the reliability of working memory. There are also reference memory components of the radial arm maze task. These include remembering that food is found on the ends of the arms, the motor coordination that is required to stay on the arms, and particularly, the set of arms in the maze that are consistently baited or not baited.

Choline supplementation before weaning led to a significant reduction in the number of working memory errors and the number of reference memory errors made during acquisition and at steady-state performance. These facilitative effects persisted up to the conclusion of behavioral testing, when the rats were approximately 8 months of age. This behavioral facilitation observed in the visuospatial memory task was due, in part, to an increase in memory capacity and a reduction in proactive interference. These changes in memory capacity and precision allowed for greater discrimination and less generalization among spatial locations both within and between test trials for animals treated perinatally with choline chloride.

This experiment was designed to extend our previous behavioral results with the radial arm maze task in two ways. First, using a factorial design, we investigated the relative contributions of pre- and postnatal choline exposure to determine the most effective period of treatment. Second, neurochemical measures of choline acetyltransferase (ChAT) levels and muscarinic receptor density as indexed by [^{3}H]quinuclidinyl benzilate (QNB) binding in the hippocampus and frontal cortex were obtained after the testing to complement the behavioral measures. The major question is whether choline administered to animals during prenatal and postnatal periods of development can lead to long-term organizational changes in cholinergic function that are associated with the facilitation of visuospatial memory.

Method

Subjects

The subjects were 40 male albino rats from eight litters born in our breeding colony. Initially, two groups ($n = 6$) of Sprague-Dawley CD-strain female rats were exposed to 0 or 5 ml/L solution of 70% choline chloride (Syntex Agribusiness, Springfield, MO). The choline chloride was delivered in 0.05 M saccharin solution in tap water that was given freely to the rats as their only source of drinking water. Choline supplementation for female breeders began approximately 2 days before conception and was maintained throughout pregnancy. Animals were housed 2 per cage, and female pairs in the untreated group (0 ml/L choline chloride) consumed 70 ± 6 ml/day ($M ± SE$) of solution, and female pairs in the treated groups (5 ml/L choline chloride) consumed 58 ± 4 ml/day of solution, $t(4) = 1.18$, $p > .10$, ns.

At birth, litters were culled to 10 pups per litter and one litter per cage, with pups from each of the experimental conditions evenly divided among four untreated foster mothers. During the next 24 days before weaning, pups received 70% choline chloride solution in concentrations of 0 or 25.0 ml/L saline once per day at a volume of approximately 0.05 ml for the first 5 days of life and 0.1 ml thereafter. Pups were hand held while the solution was intubed into the stomach through an esophageal cannula consisting of a 0.64-mm-diameter, 7-cm length of Silastic tubing (Dow Corning). The intubation procedure took approximately 30 s to complete and appeared to be relatively stressless for the young pup. When animals were weaned at approximately 24 days of age, choline supplementation was terminated, and the animals were placed in individual cages and given free access to food and water. Rats from the four experimental conditions were randomly assigned to home cages on two single-sided racks.

The treatments described above were combined in a factorial design such that 20 rats were exposed to choline prenatally and 20 were exposed to choline postnatally. Half of the rats exposed to choline prenatally were also exposed to choline postnatally (PRE + POST), whereas the remaining rats in the prenatally choline-treated group were given physiological saline postnatally (PRE). In a similar fashion, half of the rats given the control treatment prenatally were given physiological saline postnatally (CON), whereas the remaining rats in this group were given choline postnatally (POST). The group designations indicate the developmental period(s) during which choline supplementation was given.

Behavioral testing began when the rats reached 60 days of age. The rats were placed on a 24-hr food deprivation schedule during which they were fed approximately 12 g/day of Purina Rat Chow after daily behavioral testing. This food source contains approximately 2.3 mg of choline chloride per gram. A light–dark cycle of 12:12 hr was maintained in the vivarium, with fluorescent lights on from 6:00 a.m. to 6:00 p.m. eastern standard time.

Apparatus

The apparatus was a radial arm maze with 12 arms extending away from a central platform at equal angles. A stand placed under the central platform raised the maze 80 cm above the floor of the room and allowed rotation of the maze. Each arm of the maze was 83 cm long and 7.6 cm wide, with an edge 1.2 cm high along each side. At the far end of each arm, a hole 2.5 cm in diameter and 0.06 cm deep served as a food cup. The central platform was 36.5 cm in diameter, and the entire maze was painted flat gray. Rats could travel to different arms of the maze only by returning to the central platform. The maze was placed in the vivarium (4.6 × 4.8 m) in which the rats lived. The room contained two racks of rat cages arranged along two of the walls, a window, and other extramaze stimuli that remained in a relatively constant position throughout the experiment.

Procedure

Shaping. Beginning at 60 days of age, rats were shaped to run down the arms of the maze to obtain food from the food cups, which were baited with 10–15 45-mg Noyes food pellets. Five to 7 rats were placed on the maze together after a period of 24 hr of food deprivation and allowed to explore the maze for 20–30 min. This pretraining was conducted for 2 days, after which all rats failing to obtain food were forced to the ends of the arms and allowed to eat. When each rat ran readily to the ends of the arms and ate the food pellets, training was begun.

Training Phase 1 (Sessions 1–12). To investigate the nature of choline's effects on a visuospatial memory task, a mixed-pattern paradigm of baited (S⁺) and unbaited (S⁻) arms was used to distinguish between working and reference memory components of the radial arm maze task (e.g., Olton & Papas, 1979). One session was given each day, 7 days a week. At the beginning of each session, two Noyes food pellets were placed in the food well at the end of 8 of the 12 arms. Ten patterns of S⁺ and S⁻ arms were randomly selected, and the 10 rats in each of the four experimental groups were each randomly assigned to 1 of the 10 patterns. Once assigned, each rat's pattern was maintained throughout testing.

At the beginning of each training session, the rat was placed on the central platform and allowed to choose arms in the maze until each baited arm was visited at least once. All rats typically completed the maze by finding all eight food locations within 5–10 min. The order of arms chosen was recorded, and a *response* was defined as a rat advancing more than half way down an arm. Rats from each of the four experimental groups were randomly assigned to a particular chronological test order (1–40) that was kept constant across training.

Training Phase 2 (Sessions 13–24). A period of approximately 2 weeks intervened between Training Phase 1 and Training Phase 2. During this interval, the rats were confined to their home cages and were maintained on their 12-g/day feeding schedule. Training was resumed so that rats would be behaviorally active before being killed for neurochemical analysis. The procedure was identical to that described for the first training phase except that only rats in the CON and PRE + POST groups received training.

Neurochemistry. Approximately 24 hr after behavioral training, when rats were approximately 3.5 months old, each rat in the CON and the PRE + POST groups was weighed and killed by decapitation. The frontal neocortex and entire hippocampus were dissected on ice and stored at −40 °C. The tissues were homogenized in 0.5% Triton X-100 and 10 mM EDTA (pH 7.4) and assayed for ChAT activity (Fonnum, 1969). Protein was measured according to Lowry, Rosebrough, Fair, and Randall (1951). The frontal neocortical sample (4 mm²; approximately 50–75 mg) was taken from a region immediately lateral to cingulate cortex in motor cortex (Areas 2, 5, and 10). [³H]QNB receptor binding assays were performed according to the method of Yamamura and Snyder (1974). An aliquot of sucrose homogenate was rehomogenized in sodium–potassium phosphate buffer, centrifuged at 50,000 × g for 10 min, and resuspended in the buffer. The final concentration of [³H]QNB (30.2 Ci/mmol, New England Nuclear) was 1.6 nM. Nonspecific binding was determined in the presence of 10 μM atropine. The membranes were collected

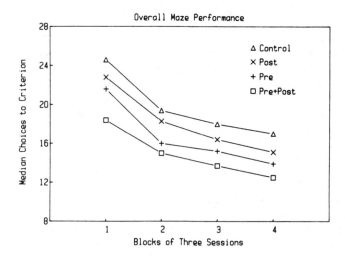

Figure 1. Overall maze performance (combined working and reference memory errors) represented by the median number of choices to criterion as a function of blocks of three sessions for rats in the PRE + POST, PRE, POST, and CON treatment conditions. (Group designations indicate development periods during which choline was given. CON = control.)

on a Brandel cell harvester (Gaithersburg, Maryland) with Whatman GF/B glass filters dipped in 3% polyethylenamine (pH 7.0).

Results

Training Phase 1 (Sessions 1–12)

The median number of choices required to obtain the food pellets in each of the eight S⁺ arms for each of the four experimental groups was plotted as a function of blocks of three training sessions (Figure 1). The general pattern seen in the number of choices to criterion was a separation by choline treatment as a function of prenatal or postnatal exposure or both and an improvement over blocks of sessions at approximately equal rates for all groups. The PRE + POST group showed the most accurate performance, followed by the PRE group, the POST group, and the CON group showing the least acurate performance. An analysis of the sequences of choices made during these sessions was conducted. Each choice made on a training trial was scored on a scale from 0 to 6; 0 corresponded to reentrance into the arm just exited, and 1–6 corresponded to entrances into arms that were one to six units away from the arm just exited. The left (−) or right (+) direction of each turn was recorded. The mean frequencies for entrance into each of the relative arm positions (or proportions for the relative turn magnitudes) showed that rats in each of the four groups generally turned more frequently into arms one or two units from the arm just departed (Figure 2). These findings are in general agreement with other data on turning tendencies in rats and pigeons (Dale, 1982; Roberts & Dale, 1981; Roberts & Van Veldhuizen, 1985).

The fact that the frequencies of arm entrances were not evenly distributed over turn magnitudes suggests that simple calculations of expected chance performance are not entirely appropriate for comparison with the observed empirical per-

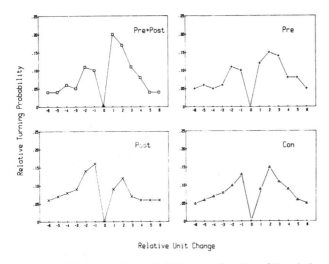

Figure 2. Median turning probabilities as a function of the relative unit change in the 12-arm radial maze task for rats in the PRE + POST, PRE, POST, and CON treatment conditions. (Negative values indicate left turns, and positive values indicate right turns. See Figure 1 for group identifications.)

formance of the rats. That is, a rat could score much higher than chance by entering adjacent arms in succession. However, such behavior would require only a response algorithm and no memory for previously entered arms. To deal with this potential problem, we performed Monte Carlo runs in which sampling from the pool of responses was biased according to the rat's turning preferences but independent of any working or reference memory information (see Eckerman, 1980). Thus, if a rat entered the +1 arm 50% of the time, then half of the responses in the response pool were +1. For each of our rats, we performed 500 similar computerized Monte Carlo runs. The mean observed and expected levels of accuracy generated by this procedure showed no reliable differences in the expected levels of accuracy among any of the treatment groups, $F(3, 36) < 1$, ns (Table 1). Because no group differences were found in the expected number of choices to criterion based on the levels of accuracy predicted by the Monte Carlo simulations, all succeeding analyses were conducted with the observed number of choices to criterion.

A Treatment × (Blocks × Subjects) analysis of variance with treatment condition as one variable and subjects nested with blocks of three sessions as another indicated a significant effect of treatment and session blocks, $F(3, 36) = 5.32, p < .01$, and $F(3, 108) = 70.61, p < .001$, respectively. Post hoc Tukey contrasts revealed that all comparisons for the ordering of treatment conditions were significant at $p < .05$. For additional analysis, the combined performances of rats were broken down into working memory and reference memory components (Figures 3 an 4, respectively). A two-way analysis of variance again indicated a significant effect of treatment and session blocks for both types of memory. For working memory, $F(3, 36) = 4.96, p < .01$, and $F(3, 108) = 59.55, p < .001$, as a function of treatment condition and session blocks, respectively; and for reference memory, $F(3, 36) = 4.74, p < .01$, and $F(3, 108) = 65.14, p < .001$, as a function of treatment condition and session blocks, respectively.

Training Phase 2 (Sessions 13–24)

Rats were considered to have reached steady-state levels of performance when they showed no further improvement in choice behavior from the previous training phase (Training Phase 1) for at least 6 sessions. On the basis of this criterion, all subjects in both the CON and the PRE + POST groups reached a steady-state level of maze performance before Training Session 20. The mean level of performance for the subjects in the CON group during the last 3 sessions (Sessions 22–24) was 15.7 ± 0.8 ($M \pm SE$) choices, and for subjects in the PRE + POST group, it was 12.4 ± 0.5 choices, $t(18) = 3.34, p < .01$.

Neurochemistry

The long-term neurochemical effects of perinatal choline supplementation for rats in the PRE + POST group were compared with measures taken from control rats in the CON group. QNB receptor binding and ChAT levels in the frontal cortex and the hippocampus are presented as a function of treatment condition in Figures 5 and 6, respectively. The results indicated that there was a significant increase in QNB receptor binding in both areas of the brain in choline-treated animals, $F(1, 18) = 10.19, p < .01$. In addition, there was a reliable decrease in ChAT levels in the hippocampus, but not in the frontal cortex, of choline-treated animals as indicated by the significant factor interaction, $F(1, 18) = 4.66, p < .05$, and post hoc comparisons.

An analysis was made to determine relations between maze performance and regional measures of Ch function. A correlational analysis was conducted for all of the animals in the CON and PRE + POST groups. The number of choices repeated each session averaged over the 12 training sessions was used as a measure of working memory, and the number of S⁻ arms entered each session averaged over the 12 training sessions was used as a measure of reference memory. These measures were correlated for each individual with respect to ChAT, QNB, and ChAT-to-QNB ratios in the hippocampus and frontal cortex. The correlations between performance and regional measures of Ch function indicated that only the

Table 1
Observed and Expected Choices to Criterion by Monte Carlo Simulations

Group	Observed[a]	Expected[b]
PRE + POST	15.6 ± 0.7	27.5 ± 0.7
PRE	17.7 ± 0.6	28.2 ± 0.4
POST	19.5 ± 0.7	27.9 ± 0.2
CON	21.8 ± 0.8	28.6 ± 0.2

Note. Values are $M \pm SE$. Groups are as described in Method.
[a]Mean number of obtained choices to criterion averaged over individual subjects in each group for the first 6 sessions of maze performance. [b]Mean number of predicted choices to criterion averaged over the 500 Monte Carlo simulations per subject, which were based on the turning biases obtained for individuals in each group during the first 6 sessions.

Table 2
Correlation Between Mean Number of Working Memory and Reference Memory Errors and Regional Measures of Cholinergic Function

Region	ChAT		QNB		ChAT:QNB	
	WM	RM	WM	RM	WM	RM
Hippocampus	.28	.25	.21	.26	.62**	.43
Frontal cortex	.32	.13	.24	.23	.37	.49*

Note. Choline acetyltransferase (ChAT) levels and muscarinic receptor density as indexed by [^{3}H]quinuclidinyl benzilate (QNB) binding in the hippocampus and frontal cortex, expressed as nanomoles per milligram of protein per hour and femtomoles per milligram of protein, respectively. Working memory (WM) errors are indexed by the mean number of arms repeated per session, and reference memory (RM) errors are indexed by the mean number of unbaited arms entered per session.
*$p < .05$. **$p < .01$.

ChAT-to-QNB ratio reached statistical significance (Table 2). In general terms, the ratio of ChAT to QNB was inversely related to choice accuracy. When overall performance was separated into different memory components, the ChAT-to-QNB ratio in the hippocampus was positively correlated with the number of working memory errors, and the ratio in the frontal cortex was positively correlated with the number of reference memory errors. The ratio in the hippocampus for CON rats was 0.040 ± 0.002 ($M \pm SE$), and for PRE + POST rats it was 0.027 ± 0.002. The ChAT-to-QNB ratio in the frontal cortex for CON rats was 0.028 ± 0.003, and for PRE + POST rats it was 0.019 ± 0.003; the effects of treatment and brain region were significant, $F(1, 18) = 10.63$ $p < .01$, and $F(1, 18) = 12.24$, $p < .01$, respectively.

Body Weight

When the rats were killed, there was no difference between body weight in the CON group (344.0 ± 8.5 g) and that in

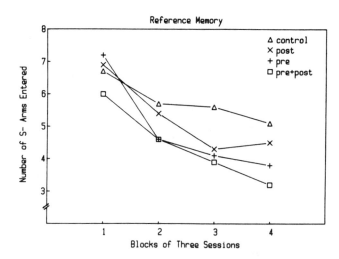

Figure 4. Reference memory performance represented by the mean number of unbaited (S$^-$) arms entered during a session as a function of blocks of three sessions for rats in the PRE + POST, PRE, POST, and CON treatment conditions. (See Figure 1 for group identifications.)

the PRE + POST group (349.6 ± 5.3 g), $t(18) = 0.06$, $p > .10$, *ns.*

Discussion

These data provide evidence that long-term neurochemical changes in Ch activity are produced by prenatal or postnatal choline supplementation or both. Furthermore, these organizational changes in brain function continue to modify working memory and reference memory processes of adult rats long after the dietary supplementation is discontinued. Behavioral support for these conclusions comes from the observation that rats given choline early in development made fewer errors in working memory and reference memory components of a radial arm maze task and that the extent to which these different types of errors are made is correlated with Ch activity. Changes in choice behavior were apparent in the first few training sessions, remained throughout the 2 weeks of training, and were not due to alterations in rats' turning preferences. Although there are numerous reports of improved memory function in adult animals after increases in dietary choline (see Bartus et al., 1985, for review), this study is the first demonstration of long-term organizational effects of perinatal choline supplementation on brain and behavior processes.

Our results did not enable us to point to a specific developmental time frame for the occurrence of organizational effects of choline supplementation on memory. Within the dose range we used, prenatal and postnatal treatment produced the largest behavioral facilitation, followed by prenatal treatment alone, postnatal treatment alone, and finally the no-choline (control) group. Clearly, the sensitive period (assuming that one exists) spans both prenatal and postnatal development. The fact that prenatal treatment was more effective than postnatal treatment may be a dose effect rather than a timing effect.

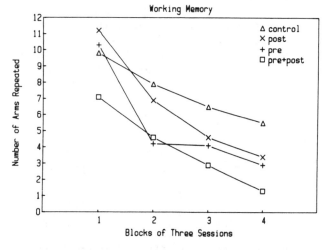

Figure 3. Working memory performance represented by the mean number of arms repeated during a session as a function of blocks of three sessions for rats in the PRE + POST, PRE, POST, and CON treatment conditions. (See Figure 1 for group identifications.)

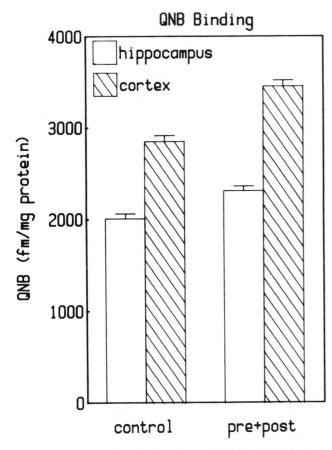

Figure 5. [³H]Quinuclidinyl benzilate (QNB) muscarinic receptor binding in the hippocampus and cortex of rats in the PRE + POST and CON treatment conditions. (Values are expressed as $M \pm SE$ femtomoles per milligram. See Figure 1 for group identifications.)

As noted earlier, not only did choline supplementation during early development cause long-term changes in memory, but the treatment also led to alterations in biochemical measures of Ch activity in the frontal cortex and the hippocampus. Specifically, QNB-indexed receptor binding was higher in both brain areas in the rats that received choline prenatally and postnatally, and there was a reliable decrease in ChAT levels in the hippocampus but not in the frontal cortex of choline-treated animals. Note that although we interpret the increase in QNB binding to reflect an increase in receptor density, a change in receptor affinity might also be responsible for this effect. We also could not determine whether the changes in ChAT activity reflect a simple regulatory effect within Ch neurons or a decrease in actual terminal numbers. Furthermore, it is impossible to determine with the methods we used whether there are significant regional differences within the hippocampus and frontal cortex in the magnitude of the observed neurochemical changes. All of these possibilities are further complicated by our uncertainty about whether these changes in Ch function caused the behavioral facilitation described earlier or whether the behavioral changes altered Ch function. Neurochemical measures from neonatally choline-treated rats that did not receive any behavioral training on the radial arm maze are necessary for an investigation of the causal relationships between Ch activity and behavior.

Regardless of the direction of this relation, we examined whether correlations existed between the function of Ch neurons in the hippocampus and the frontal cortex and the performance of rats in the radial arm maze in which both working and reference memory can be evaluated. This type of analysis has been used previously to elucidate possible neurobiological mechanisms of attention and memory when treatment-related differences in the neurobiological parameters have been relatively small or insignificant (Altschuler, Morton, Gold, Lawton, & Miller, 1971; Ingram, London, & Goodrick, 1981; Ingram, London, Waller, & Reynolds, 1983; Lowy et al., 1985; Meck, 1987). Some of this previous work has shown individual maze performance and spatial memory processes to be correlated significantly with hippocampal ChAT activity in aged Wistar rats and with cortical glutamic acid decarboxylase activity in Wistar rats regardless of age (Ingram et al., 1981). These types of correlational analyses are useful to suggest which brain regions and neurotransmitter systems might be involved in the memory components required by a particular task.

This study revealed that choline-induced improvement in

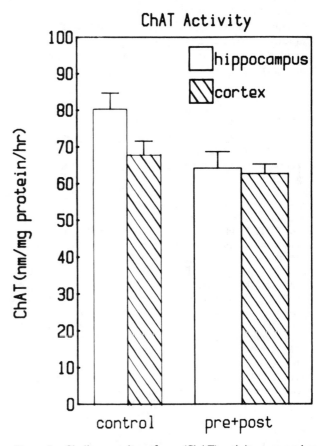

Figure 6. Choline acetyltransferase (ChAT) activity expressed as nanomoles per milligram of protein per hour ($M \pm SE$) in the hippocampus and frontal cortex of rats in the PRE + POST and CON treatment conditions. (See Figure 1 for group identifications.)

visuospatial memory is significantly correlated with the ChAT-to-QNB ratio in the hippocampus for working memory and with the ChAT-to-QNB ratio in the frontal cortex for reference memory. No other correlations between performance and regional measures of Ch function reached statistical significance. A similar relation between the ratios and behavior has been observed for age-related memory impairments in temporal memory (Meck et al., 1986).

Taken together, these findings suggest that prenatal and postnatal choline supplementation increases muscarinic receptor density (as measured by QNB) along with concomitantly decreasing ChAT levels. In contrast, the aging process leads to a decrease in muscarinic receptor density along with a concomitant increase in ChAT levels (Meck et al., 1986). In both cases, there is a similar relation between memory function and ChAT-to-QNB ratios. That is, the ratio of ChAT to QNB in both situations is inversely related to memory ability. Apparently, this ratio is a more sensitive indicator of synaptic efficacy with regard to memory function than either ChAT or QNB measures. The significant correlations between behavioral performance and neurochemical measures reported here are a first step toward demonstrating that a functional relation, rather than a parallel organizational effect of choline supplementation or a parallel age effect, exists. Further research will determine whether perinatal choline supplementation, which clearly improves adult memory and alters Ch function, influences the time of onset or the rate of memory deterioration that is frequently associated with aging.

References

Altschuler, H., Morton, H. K., Gold, M., Lawton, M. P., & Miller, M. (1971). Neurochemical changes in the brains of aged albino rats resulting from avoidance learning. *Journal of Gerontology, 26*, 63–69.

Bartus, R. T., Dean, R. L., & Beer, B. (1980). Memory deficits in aged Cebus monkeys and facilitation with central cholinomimetics. *Neurobiology of Aging, 1*, 145–152.

Bartus, R. T., Dean, R. L., Goas, J. A., & Lippa, A. S. (1980). Age-related changes in passive avoidance retention: Modulation with dietary choline. *Science, 209*, 301–303.

Bartus, R. T., Dean, R. L., Pontecorvo, M. J., & Flicker, C. (1985). The cholinergic hypothesis: A historical overview, current perspective, and future directions. In D. S. Olton, E. Gamzu, & S. Corkin (Eds.), *Annals of The New York Academy of Sciences: Memory dysfunctions: An integration of animal and human research from preclinical and clinical perspectives* (Vol. 444, pp. 332–358). New York: New York Academy of Sciences.

Bartus, R. T., Dean, R. L. III, Sherman, K. A., Friedman, E., & Beer, B. (1981). Profound effects of combining choline and piracetam on memory enhancement and cholinergic function in aged rats. *Neurobiology of Aging, 2*, 105–111.

Blusztajn, J. K., & Wurtman, R. J. (1983). Choline and cholinergic neurons. *Science, 221*, 614–620.

Cohen, E. L., & Wurtman, R. J. (1975). Brain acetylcholine: Increases after systemic choline administration. *Life Sciences, 16*, 1095–1102.

Dale, R. H. I. (1982). Parallel-arm maze performance of sighted and blind rats: Spatial memory and maze structure. *Behaviour Analysis Letters, 2*, 127–139.

Decker, M. W., Pelleymounter, M. A., & Gallagher, M. (1988). Effects of training on a spatial memory task on high affinity choline uptake in hippocampus and cortex in young adult and aged rats. *Journal of Neuroscience, 8*, 90–99.

Eckerman, D. A. (1980). Monte Carlo estimation of chance performance for the radial arm maze. *Bulletin of the Psychonomic Society, 15*, 93–95.

Fonnum, F. (1969). Radiochemical microassays for the determination of choline acetyltransferase and acetylcholinesterase activities. *Biochemical Journal, 115*, 465–472.

Golczewski, J. A., Hiramoto, R. N., & Ghanta, V. K. (1982). Enhancement of maze learning in old C57BL/6 mice by dietary lecithin. *Neurobiology of Aging, 3*, 223–226.

Haubrich, D. R., Wang, P. F. L., Clody, D. E., & Wedeking, P. W. (1975). Increases in rat brain acetylcholine induced by choline and deanol. *Life Sciences, 17*, 975–980.

Ingram, D. K., London, E. D., & Goodrick, C. L. (1981). Age and neurochemical correlates of radial maze performance in rats. *Neurobiology of Aging, 2*, 41–47.

Ingram, D. K., London, E. D., Waller, S. B., & Reynolds, M. A. (1983). Age-dependent correlation between motor performance and neurotransmitter synthetic enzyme activities. *Behavioral and Neural Biology, 39*, 284–298.

Krnjevic, K., & Reinhardt, W. (1979). Choline excites cortical neurons. *Science, 206*, 1321–1323.

Leathwood, P. D., Heck, E., & Mauron, J. (1982). Phosphatidyl choline and avoidance performance in 17 month-old SEC/1ReJ mice. *Life Sciences, 30*, 1065–1071.

Lowry, O. H., Rosebrough, N., Fair, A., & Randall, R. (1951). Protein measurement with the Folin phenol reagent. *Journal of Biological Chemistry, 193*, 265–276.

Lowy, A. M., Ingram, D. K., Olton, D. S., Waller, S. B., Reynolds, M. A., & London, E. D. (1985). Discrimination learning requiring different memory components in rats: Age and neurochemical comparisons. *Behavioral Neuroscience, 99*, 638–651.

Meck, W. H. (1987). Vasopressin metabolite neuropeptide facilitates simultaneous temporal processing. *Behavioral Brain Research, 23*, 147–157.

Meck, W. H., & Church, R. M. (1987). Nutrients that modify the speed of internal clock and memory storage processes. *Behavioral Neuroscience, 101*, 465–475.

Meck, W. H., Church, R. M., & Wenk, G. L. (1986). Arginine vasopressin inoculates against age-related discrepancies in the content of temporal memory and increases in sodium-dependent high-affinity choline uptake. *European Journal of Pharmacology, 130*, 327–331.

Meck, W. H., Church, R. M., & Wenk, G. L. (in press). Temporal memory is sensitive to choline acetyltransferase inhibition in mature and aged rats. *Neurobiology of Aging*.

Meck, W. H., Smith, R. A., & Williams, C. L. (1988). Pre- and postnatal choline supplementation produces long-term facilitation of spatial memory. *Developmental Psychobiology, 21*, 339–353.

Meck, W. H., & Williams, C. L. (in press). Choline supplementation during pre- and postnatal development eliminates proactive interference in spatial memory. *Developmental Brain Research*.

Meier-Ruge, W., Hunziker, O., Wangoff, P. I., Reichmeier, K., & Sandoz, P. (1978). Alternations of morphological and neurochemical parameters of the brain due to normal aging. In K. Nandy (Ed.), *Developments in Neuroscience: Vol. 3. Senile dementia: A Biochemical Approach* (pp. 98–137). New York: Elsevier/North-Holland.

Mizumori, S. J. Y., Patterson, T. A., Sternberg, H., Rosenzweig, M. R., Bennett, E. L., & Timiras, P. S. (1985). Effects of dietary choline on memory and brain chemistry in aged mice. *Neurobiology of Aging, 6*, 51–56.

Mooradian, A. D. (1988). Blood-brain barrier transport of choline is reduced in the aged rat. *Brain Research, 440,* 328–332.

Olton, D. S., Becker, J. T., & Handelmann, G. E. (1979). Hippocampus, space, and memory. *Behavioral and Brain Sciences, 2,* 313–365.

Olton, D. S., & Papas, B. C. (1979). Spatial memory and hippocampal function. *Neuropsychologia, 17,* 669–682.

Pardridge, W. M., Cornford, E. M., Braun, L. D., & Oldendorf, W. M. (1979). In A. Barbeau, J. Growdon, & R. Wurtman (Eds.), *Nutrition and the brain* (pp. 25–34). New York: Raven Press.

Roberts, W. A., & Dale, R. H. I. (1981). Remembrance of places lasts: Proactive inhibition and patterns of choice in rat spatial memory. *Learning and Motivation, 10,* 261–281.

Roberts, W. A., & Van Veldhuizen, N. (1985). Spatial memory in pigeons on the radial maze. *Journal of Experimental Psychology: Animal Behavior Processes, 11,* 241–260.

Sherman, K. A., Kuster, J. E., Dean, R. L., Bartus, R. T., & Friedman, E. (1981). Presynaptic cholinergic mechanisms in brain of aged rats with memory impairments. *Neurobiology of Aging, 2,* 99–104.

Sims, N. R., Smith, C. C. T., Davison, A. N., Bowen, D. M., Flack, R. H. A., Snowden, J. S., & Neary, D. (1980). Glucose metabolism and acetylcholine synthesis in relation to neuronal activity in Alzheimer's disease. *Lancet, 1,* 333–336.

Smith, R. A., Meck, W. H., & Williams, C. L. (1986). Pre- and postnatal choline supplementation modifies cholinergic function and produces long-term facilitation of spatial memory. *International Society for Developmental Psychobiology Abstracts, 19,* 60.

Sylvia, A. L., & Rosenthal, M. (1979). Effects of age on brain oxidative metabolism in vivo. *Brain Research, 165,* 235–248.

Yamamura, H. I., & Snyder, S. H. (1974). Muscarinic cholinergic binding in rat brain. *Proceedings of the National Academy of Sciences (USA), 71,* 1725–1729.

Zeisel, S. H., & Wurtman, R. J. (1981). Developmental changes in rat blood choline concentration. *Biochemical Journal, 198,* 565–570.

Received February 2, 1987
Revision received April 4, 1989
Accepted April 5, 1989 ■

Glossary

[³H] quinuclidinyl benzilate (QNB) - a chemical that binds selectively to muscarinic cholinergic receptors

acetylcholine (Ch) - a neurotransmitter that is thought to play a role in memory

agonist - a chemical that augments or duplicates the effects of a neurotransmitter is said to be an agonist of that neurotransmitter; at high doses, choline is thought to function as an agonist of acetylcholine

cannula - a tube inserted in the body for the purpose of administering drugs or other substances

choline - a precursor of acetylcholine; we normally obtain choline from the food that we eat

choline acetyltransferase (ChAT) - the enzyme that stimulates the conversion of choline to acetylcholine

cholinergic - pertaining to acetylcholine

esophageal - pertaining to the esophagus, the passage through which swallowed food reaches the stomach

factorial design - a research design in which there are two or more independent variables and in which all possible combinations of the levels of these independent variables are assessed; for example, the experiment of Meck et al. is an example of a 2 x 2 factorial design: the subjects received prenatal choline or they didn't and they received postnatal choline or they didn't, in all four possible combinations

frontal cortex - the neocortex of the frontal lobe

geriatric - pertaining to old age

hippocampus - a subcortical structure that is known to play an important role in memory

lecithin - a precursor of acetylcholine

Monte Carlo simulation - the generation by computer of a large number of random samples drawn from a population prescribed by the experimenter under conditions prescribed by the experimenter; in this experiment, a Monte Carlo simulation was used to determine how well each subject would be expected to do by chance given its nonrandom arm-selection tendencies

muscarinic receptors - there are two major subclasses of receptors to which acetylcholine binds, and muscarinic receptors are one of these subclasses

perinatal - around birth; before or after birth

postnatally - after birth

precursor - a chemical from which another chemical is synthesized is said to be a precursor of the synthesized chemical; choline and lecithin are precursors of acetylcholine

prenatally - before birth

proactive interference - a difficulty in learning or remembering a task that is caused by the learning of a previous task

radial arm maze - an elevated maze task in which several arms (e.g., 12 arms) radiate out from a central starting platform; in one application of the radial arm maze, the same arms are baited with food pellets each day and rats learn to enter only the arms that are always baited (a measure of reference memory) and to visit a particular arm only once on a particular test (a measure of working memory)

reference memory - memory for those aspects of a test that do not change from trial to trial

visuospatial memory - memory for locations in the environment that is based on visual cues; the radial arm maze is considered to be a test of visuospatial memory

vivarium - a room designed for housing animals

weaning - when a young mammal stops relying on its mother's milk as its primary source of nutrition; when a young mammal stops suckling

working memory - memory for those aspects of a test that change from trial to trial

Essay Study Questions

1. Describe the radial-arm-maze task and how it can be used to measure working and reference memory.

2. Despite suggestive findings from animal experiments, choline has not been shown to improve the memory of geriatric patients. Discuss.

3. In what two ways did this experiment extend previous work by the same authors?

4. The subjects were reared by foster mothers. Why do you think that Meck, Smith, and Williams instituted this important control procedure?

5. Why did the authors conduct a Monte Carlo simulation?

6. Briefly summarize the behavioral and neurochemical results of this experiment. How were the behavioral and neurochemical results related?

Multiple-Choice Study Questions

1. Choline was administered postnatally
   a. in the mothers' drinking water.
   b. via an esophageal cannula.
   c. in the foster mothers' drinking water.
   d. both a and b
   e. both b and c

2. The precursors of acetylcholine are
   a. ChAT and QNB.
   b. lecithin and choline.
   c. choline and QNB.
   d. choline and Ch.
   e. Ch and lecithin.

3. In this experiment, Meck, Smith, and Williams tried to extend their previous experiments by
   a. investigating the relative contributions of mothers and foster mothers.
   b. investigating the relative contributions of prenatal and postnatal choline exposure.
   c. taking acetylcholine-related neurochemical measures.
   d. all of the above
   e. both b and c

4. If the same arms of a radial arm maze are baited on each trial, entering an unbaited arm is considered to reflect a deficit in
   a. short-term memory.
   b. procedural memory.
   c. reference memory.
   d. acetylcholine.
   e. working memory.

5. Meck, Smith, and Williams conducted a Monte Carlo simulation because
   a. they like to gamble.
   b. the probability of a correct response did not equal the probability of an error.
   c. many of the rats tended to select arms in a systematic pattern, thus causing the number of trials that it would take for them to reach criterion on the basis of chance alone to deviate somewhat from that value calculated on the basis of random selection.
   d. the probability of a working memory error did not equal the probability of a reference memory error.
   e. the levels of accuracy expected on the basis of chance alone differed among the four treatment groups.

6. The neurochemical analyses indicated that perinatal choline produced a statistically significant
   a. increase in QNB binding in the frontal cortex.
   b. increase in QNB binding in the hippocampus.
   c. decrease in ChAT levels in the hippocampus.
   d. all of the above
   e. none of the above

The answers to the preceding questions are on page 290.

Food-For-Thought Questions

1. On the basis of this research would you administer choline supplements to a baby of yours in order to improve its memory? If not, what further research would have to be done before you would take this step?

2. If there were unequivocal evidence that choline or some other chemical could improve memory, many people would consume it in large quantities because it would give them a great advantage in the game of life. Yet many of these same individuals are highly critical of competitive athletes who take performance-enhancing steroids. Discuss.

ARTICLE 19

Alcohol Inhibits and Disinhibits Sexual Behavior in the Male Rat

J. G. Pfaus and J.P.J. Pinel

Reprinted from Psychobiology, 1989, volume 17, 195-201.

Alcohol has been reported to have two diametrically different effects on sexual behavior. It has been reported to disrupt it, and it has been reported to facilitate it—perhaps you have experienced one or both of these effects. Although widely acknowledged, the bidirectionality of alcohol's effects on sexual behavior has never been subjected to experimental analysis.

Pfaus and Pinel began by testing the hypothesis that the bidirectionality of alcohol's effects on sexual behavior is simply a matter of dose, with facilitation occurring at low doses and disruption at high doses. Accordingly, they injected male rats with a wide range of doses of alcohol before test sessions with sexually receptive females. However, instead of the hypothesized bidirectional dose-response function, they found that alcohol had little or no effect at low doses and that it disrupted copulation at high doses. There was no evidence of facilitation.

It has often been suggested that the facilitatory effect of alcohol on human sexual behavior is a consequence of alcohol's ability to release sexual inhibitions. To test this inhibition hypothesis, Pfaus and Pinel tested the effects of alcohol on male rats that had learned to inhibit their sexual advances. First, the male rats were presented on alternate days with either sexually receptive or nonreceptive females. The males began by attempting to mount and copulate with receptive and nonreceptive females alike; however, they soon learned to inhibit their sexual behavior during the tests with nonreceptive females. Eventually they made no effort to mount nonreceptive females but readily copulated with receptive females. At this point, Pfaus and Pinel injected the males with 0.5 kg of alcohol prior to a test with a nonreceptive females. The results were striking: despite the lack of cooperation from the nonreceptive females, over half the males mounted them and ejaculated on their hindquarters. The same dose of alcohol produced a mild disruption of the sexual behavior of the same males when it was injected prior to tests with receptive females.

These results constitute the first experimental support for the hypothesis that sexual disinhibition is the basis for alcohol's facilitatory effect on sexual behavior. Moreover, by establishing that alcohol-produced sexual disinhibition is not a purely human phenomenon, these results demonstrate the accessibility of this social problem to the analytic power of controlled comparative research.

Alcohol inhibits and disinhibits sexual behavior in the male rat

JAMES G. PFAUS and JOHN P. J. PINEL
University of British Columbia, Vancouver, British Columbia, Canada

Anecdotal evidence from human case studies suggests that alcohol exerts opposite effects on sexual behavior: facilitation at low doses and disruption at higher doses. We tested this dose-dependent dual-effect hypothesis in Experiment 1 by assessing the effects of a wide range of doses of alcohol (0.25 to 2.0 g/kg, i.p.) on the copulatory behavior of sexually active male rats. Rather than the predicted dual effect, all doses of alcohol disrupted copulation, with the highest blocking it completely. Because the facilitatory effect of alcohol on human sexual behavior is believed to reflect a release from inhibition, we hypothesized that the facilitatory effect of low doses of alcohol might be found only when sexual responding was inhibited. In Experiment 2, we tested this hypothesis by assessing the effects of two doses of alcohol (0.5 and 1 g/kg), both of which had proven disruptive in Experiment 1, on the sexual behavior of male rats that had learned to inhibit their sexual behavior during tests with sexually nonreceptive females. Following injection with the lower dose, most of the previously nonresponsive males mounted the nonreceptive females and ejaculated without ever gaining vaginal intromission. These results provide the first experimental evidence that alcohol can inhibit or disinhibit the copulatory behavior of male rats depending upon dose and upon the absence or presence, respectively, of sexual inhibition.

The effects of alcohol on sexual behavior are of special interest for two diametrically different reasons. On one hand, alcohol consumption has been repeatedly implicated in the etiology of various sexual disorders; and on the other, it has been reported to enhance sexual arousal and behavior. The general purpose of the present experiments was to contribute to the resolution of this apparent paradox.

Although alcohol is widely believed to increase sexual arousal by a process of disinhibition, that is, by inhibiting cortical centers allegedly responsible for sexual inhibition (Carver, 1948; Ford & Beach, 1951; Hollister, 1975; Kaplan, 1974; Lemere & Smith, 1973; MacDougald, 1967; Masters, Johnson, & Kolodny, 1986), empirical reports of the facilitatory effect of alcohol on subjective and physiological measures of human sexual arousal have been inconsistent. Some experiments on males have demonstrated a slight facilitation of certain physiological measures of sexual arousal (e.g., percent of maximum penile tumescence) at low blood alcohol levels but a marked inhibition of the same measures at higher blood alcohol levels (Farkas & Rosen, 1976; Rubin & Henson, 1976; Wilson & Niaura, 1984). In other experiments, however, alcohol has reduced both subjective and physiological measures of male sexual arousal at all active doses (Briddell & Wilson, 1976; Farkas & Rosen, 1976; Rubin & Henson, 1976). In females, alcohol intoxication has been found to decrease physiological measures of sexual arousal while increasing subjective measures (Wilson & Lawson, 1976, 1978). Complicating the situation even further is the fact that consuming a nonalcoholic drink that is believed to contain alcohol has been found to increase the subjective sexual arousal of males (Marlatt, Demming, & Reid, 1973; Wilson, 1977; Wilson & Lawson, 1976) but not that of females (Wilson & Lawson, 1978).

Inconsistency also pervades the clinical case-study literature on alcohol and human sexual performance. Numerous reports have linked alcohol use to such sexual dysfunctions as inhibited sexual desire, erectile failure, delayed ejaculation, and inhibited orgasm (Hollister, 1975; Kaplan, 1974; Kinsey, 1966; Lemere & Smith, 1973; Masters & Johnson, 1966; McKendry et al., 1983; Pinhas, 1987; Wilson, 1977); however, others have suggested that alcohol consumption can improve sexual performance in some individuals with preexisting sexual dysfunctions, such as premature ejaculation, inhibited sexual desire, or inhibited orgasm (Pinhas, 1987; Smith, Wesson, & Apter-Marsh, 1984).

Only two published experiments exist in which the effects of alcohol on human sexual behavior have been examined directly, one in males (Malatesta, Pollack, Wilbanks, & Adams, 1979) and one in females (Malatesta, Pollack, Crotty, & Peacock, 1982). In both of these experiments, the effects of a range of doses of alcohol were examined on subjects attempting to masturbate to orgasm while viewing a sexually explicit film. In males, alcohol dose-dependently delayed ejaculation, reduced the intensity of sexual orgasm, and decreased both physiological and subjective measures of sexual arousal. Most of these effects were also observed in females: alcohol dose-

---

The authors wish to thank Bruce Christensen for his help in conducting Experiments 1 and 2. This research was supported by a grant from the British Columbia Health Care Research Foundation awarded to John P. J. Pinel. Requests for reprints should be sent to J. P. J. Pinel, Department of Psychology, 2136 West Mall, University of British Columbia, Vancouver, British Columbia V6T 1Y7, Canada.

Copyright 1989 Psychonomic Society, Inc.

dependently decreased physiological measures of sexual arousal, delayed orgasm, and decreased the subjective intensity of orgasm. However, the females reported that alcohol increased their subjective levels of sexual arousal. Although the results of these experiments support clinical observations that alcohol can disrupt sexual activity, their relevance to the reported effects of alcohol on copulatory behavior is unclear.

Another approach to clarifying the nature of alcohol's effect on sexual performance has been to assess its effect on the sexual behavior of laboratory animals copulating under controlled conditions. Moderate to high doses of alcohol have been shown to delay or block erection, ejaculation, and mounting behavior in both male dogs and rats (Dewsbury, 1967; Gantt, 1940, 1952, 1957; Hart, 1968, 1969; Teitelbaum & Gantt, 1958). Unfortunately, the effect of low doses of alcohol on the copulatory behavior of laboratory animals has not been assessed, nor has there been any attempt to document alcohol's putative disinhibitory effect on sexual behavior. The objective of the present study was to take advantage of the experimental control offered by an animal model to test hypotheses about the effects of alcohol on sexual behavior derived from the human literature.

GENERAL METHODS

Subjects and Surgery

Male Long-Evans and female Sprague-Dawley rats were obtained from Charles River Canada, Inc., St. Constant, Quebec. They were housed by sex in groups of 6 in standard wire-mesh cages in a colony room maintained at approximately 21°C on a reversed 12:12-h light:dark cycle. Food and water were continuously available in the home cages. Twenty-four female rats, which served as stimulus females in both experiments, were bilaterally ovariectomized, via lumbar incisions, under ether anesthesia, approximately 1 month prior to Experiment 1. The stimulus females were divided into two cohorts of 12, and each cohort served on alternate test sessions. To render the females sexually receptive, we injected each with 10 μg of estradiol benzoate 48 h before each test and 500 μg of progesterone 4 h before each test.

Drug Treatments

Ethyl alcohol (95%) was diluted with physiological saline to obtain doses of 0.25, 0.5, 1.0, and 2.0 g/kg, each in a 25% aqueous v/v solution. The control solution of physiological saline was administered in a volume equal to that of the highest dose of ethanol. Both the ethanol solutions and the saline vehicle were injected intraperitoneally (i.p.) 45 min before each test. The dose and time-course parameters were chosen on the basis of evidence showing that blood alcohol levels in rodents reach a peak approximately 60 to 90 min after i.p. injections of a 25% v/v alcohol solution (Goldstein, 1983). Estradiol benzoate and progesterone (Steraloids) were dissolved in 0.1 ml of peanut oil and injected subcutaneously (s.c.).

Testing Procedure

Both Experiments 1 and 2 began with 10 baseline tests of copulatory behavior. All baseline tests were 45 min long and were conducted in a dimly lit room in 29×45 cm Plexiglas testing chambers lined with San-i-Cel bedding. The baseline tests occurred once every 4 days during the middle third of the dark phase of the circadian light:dark cycle. Each male was habituated to the testing chamber for 5 min prior to the introduction of a sexually receptive female, which marked the beginning of the test session.

For all tests, the occurrence of each mount, intromission, and ejaculation was entered by an experienced observer on a computerized event recorder, which subsequently calculated the following seven primary measures of masculine copulatory behavior: (1) mount and (2) intromission latencies (times from the introduction of the female to the first mount and the first intromission); (3) ejaculation latency (time from the first intromission to the first ejaculation); (4) the postejaculatory interval (time from the first ejaculation to the next intromission); (5) the number of mounts and (6) the number of intromissions prior to the first ejaculation; and (7) the total number of ejaculations. Calculated from these seven primary measures were two secondary measures: the interintromission interval (the ejaculation latency/number of intromissions) and the intromission rate (the number of mounts with intromission/total number of mounts with and without intromission).

EXPERIMENT 1

In contrast to the numerous anecdotal reports that low doses of alcohol can facilitate human sexual behavior, experiments in laboratory animals had revealed only disruption. We hypothesized that alcohol might facilitate the sexual behavior of animals at doses lower than those used in previous experiments (i.e., at doses lower than 0.5 g/kg). We tested this hypothesis in Experiment 1 by assessing the effects of a wide range of doses of alcohol on the copulatory behavior of sexually active male rats.

Method

Subjects. The subjects were 60 sexually active 400–600-g Long-Evans male rats.

Procedure. Following the 10 baseline copulatory tests, the rats were assigned randomly to one of five alcohol-dose groups (0, 0.25, 0.5, 1.0, or 2.0 g/kg). The rats in each group ($n = 12$) received an i.p. injection of alcohol 45 min before the 45-min copulation test, which occurred 4 days after the last baseline test and was identical to it.

The significance of the differences in the proportion of rats in each group that displayed mounts, intromissions, or ejaculations was assessed using chi-square analyses. The effects of alcohol on the seven primary and two secondary measures of copulatory behavior were subjected to nonparametric Kruskal-Wallis analyses of variance for independent measures ($p < .05$). For each measure of copulatory behavior found to be significantly affected by the alcohol treatment, multiple pairwise comparisons were made between the effect of each dose of alcohol and the effect of saline using nonparametric Mann-Whitney tests. Using the Bonferroni method (Marascuilo & Levin, 1983), all multiple comparisons were corrected for elevated experiment-wise error.

Results

Alcohol produced a dose-dependent disruption of male copulatory behavior. The two lowest doses, 0.25 and 0.5 g/kg, produced a moderate degree of disruption, which was reflected by changes in only a few of the measures. The disruption produced by the 1-g/kg dose was greater and more extensive than that produced by the two lower doses, and the highest dose (2 g/kg) resulted in a complete lack of sexual activity. There was no evidence whatsoever of the hypothesized facilitatory effect at any dose.

As illustrated in Figure 1, alcohol significantly decreased the proportion of rats that mounted [$\chi^2(4) = 13.41, p < .01$], intromitted [$\chi^2(4) = 13.41, p < .01$],

ALCOHOL AND SEXUAL BEHAVIOR

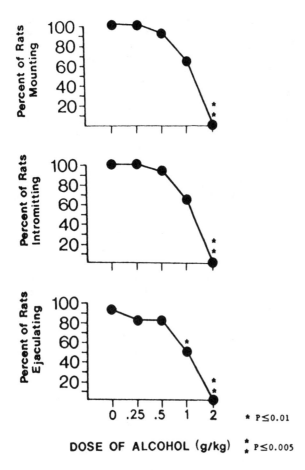

Figure 1. Percentage of rats displaying mounts, intromissions, and ejaculations as a function of alcohol dose in Experiment 1.

reduced the proportion of rats that ejaculated $[\chi^2(1) = 5.56, p < .01]$, and although it substantially reduced the proportion of rats that mounted and intromitted, these latter effects did not reach statistical significance with the alpha adjusted by the Bonferroni method (both chi-square ps $< .05$ but $> .01$). The rats that received the highest dose (2 g/kg) of alcohol were conspicuously ataxic, and none attempted copulation during the 45-min test (all three chi-square ps $< .0001$).

The effects of different doses of alcohol on the primary and secondary measures of copulatory behavior were compared statistically using only the data of those rats in each group that displayed the copulatory behavior under analysis. For example, statistical analyses of the differences in latencies to mount, intromit, and ejaculate were based on the data of only those rats that mounted, intromitted, and ejaculated, respectively; no arbitrary scores were assigned. The effects of alcohol on these measures are shown in Table 1.

In rats that engaged in copulatory behavior, alcohol significantly increased the latencies to first mount ($H = 8.65$, $p < .05$), first intromission ($H = 10.27, p < .025$), and first ejaculation $H = 11.82, p < .01$). In addition, the number of intromissions preceding the first ejaculation was significantly increased by alcohol treatment ($H = 16.73, p < .005$), and alcohol significantly decreased the number of ejaculations achieved during the 45-min test ($H = 25.83, p < .001$). Although alcohol markedly increased the number of mounts without intromission, this effect did not reach statistical significance ($H = 6.23$, $p < .10$). Finally, in rats that copulated, alcohol had no significant effect on the postejaculatory interval, the interintromission interval, or the intromission rate.

Post hoc comparisons revealed that the 1-g/kg dose of alcohol significantly increased the mount latency ($U = 18.5, p < .01$) and the intromission latency ($U = 16$, $p < .01$) of those rats that mounted and intromitted, respectively. Although the 0.5-g/kg dose also produced a substantial increase in the latencies to first mount and intromission, these effects did not reach the statistical sig-

and ejaculated $[\chi^2(4) = 15.74, p < .01]$. Post hoc pairwise comparisons of the proportion of subjects that mounted, intromitted, and ejaculated in each alcohol group and the proportions in the saline control group revealed that the effects of the two lower doses were not statistically significant. In contrast, the 1-g/kg dose significantly

Table 1
Effects of Alcohol on Male Sexual Behavior

	Dose of Alcohol									
	0 g/kg (Control)		.25 g/kg		.50 g/kg		1.0 g/kg		2.0 g/kg‡	
Behavioral Parameters	M	SE	M	SE	M	SE	M	SE	M	SE
Mount latency	47.8	15	57.9	20	154.1	47*	270.0	81†	–	
Intromission latency	75.5	16	103.6	35	186.2	42*	340.4	78†	–	
Ejaculation latency	540.5	143	838.4	155	1039.4	130†	1341.5	144†	–	
Postejaculatory interval	386.4	18	357.9	13	418.1	32	378.2	8	–	
Number of mounts	13.3	3	14.1	4	17.6	2	24.1	4	–	
Number of intromissions	4.8	1	12.3	2†	8.8	1†	9.3	1†	–	
Number of ejaculations	1.8	.2	1.4	.3	1.1	.2*	0.8	.2†	–	
Interintromission interval	152.3	69	104.3	47	328.4	161	146.7	14	–	
Intromission rate	0.34	.06	0.52	.05	0.32	.04	0.30	.02	–	

Note—Table 1 presents the effects of each dose of alcohol on the seven primary and two secondary measures of sexual behavior assessed in Experiment 1. The means and standard errors are from rats that displayed the appropriate copulatory behavior in each group. All latencies and intervals are in seconds. The number of mounts and intromissions are calculated prior to the first ejaculation. *$p < .05$, †$p < .01$, and ‡$p < .005$ from control values.

nificance required by the Bonferroni adjustment (both $ps < .05$ but $> .01$). In contrast, the ejaculation latencies of those rats that ejaculated were increased significantly by both the 0.5-g/kg dose ($U = 23.5, p < .01$) and the 1-g/kg dose ($U = 12, p < .01$). The number of intromissions preceding ejaculation was increased significantly by the 0.25-g/kg dose ($U = 12.5, p < .01$), the 0.5-g/kg dose ($U = 24, p < .01$), and the 1-g/kg dose ($U = 4, p < .01$). Finally, although each dose of alcohol substantially decreased the number of ejaculations in rats that copulated, the decrease was statistically significant only at the 1-g/kg dose ($U = 22.5, p < .01$).

Discussion

The results of Experiment 1 confirmed previous reports that moderate to high doses of alcohol disrupt or block copulatory behavior in sexually active male rats (Dewsbury, 1967; Hart, 1968, 1969), but they did not confirm our hypothesis that low doses of alcohol facilitate copulatory behavior. Moderate to high doses of alcohol (i.e., 1 and 2 g/kg) substantially decreased the proportion of rats that mounted, intromitted, and ejaculated. Moreover, in rats that engaged in copulatory behavior, low to moderate doses (i.e., 0.5 and 1 g/kg) increased the mean latencies to first mount, first intromission, and first ejaculation, and reduced the total number of ejaculations that they achieved during the 45-min test. Although the 1-g/kg dose significantly increased the mean number of intromissions, which might, at first glance, seem to be evidence of facilitation, an increase in the number of intromissions required to trigger an ejaculation is widely considered to be evidence of disruption or desensitization (see Sachs & Barfield, 1976).

Several hypotheses can be advanced to account for the present results. The two most obvious are that alcohol produced either a general disruption of motor activity or a general aversive state incompatible with sexual behavior. The first hypothesis could account for the ataxia observed in rats injected with the highest dose. Although this hypothesis could also account for some of the disruptive effects observed at lower doses (e.g., the decreased percentage of rats that initiated or completed an ejaculatory series), it cannot account for other effects (e.g., the increased number of intromissions that preceded ejaculation). Furthermore, none of the rats that copulated to ejaculation following alcohol treatment appeared sluggish in its general activity or in its attempts to pursue the stimulus females. Although we cannot rule out the second hypothesis, that alcohol might have disrupted sexual behavior by inducing an incompatible aversive state, we note that i.p. injections of alcohol at doses lower than 1.75 g/kg are not sufficient to induce conditioned taste aversions in rats (Jeffreys, Pournaghash, & Riley, 1989).

There are three possible explanations for the disruption of sexual activity that occurred at the lower doses: (1) alcohol may have reduced sexual motivation, (2) it may have interfered with the perception of stimuli essential for eliciting sexual activity, or (3) it may have produced a selective disruption of the motor responses involved in copulation. Although it was not the purpose of this experiment to distinguish among these three interpretations, our results provide some support for the first two over the last. Diminished sexual motivation can be inferred from several of the observed effects, including the decreased proportion of rats that initiated mounts and intromissions, the increased mount and intromission latencies of rats that eventually copulated, and the increased ejaculation latencies of rats that eventually ejaculated. The same effects, together with the increased number of intromissions, can also be taken as evidence that alcohol desensitized the rats to such external cues as estrous odors, female proceptive behaviors, or penile stimulation. In contrast, alcohol did not reduce the intromission ratio or increase the interintromission interval. Thus, there is little support in the present results for the idea that alcohol disrupts copulatory behavior by reducing the ability of rats to perform sexual responses. We tentatively suggest that low to moderate doses of alcohol disrupt male copulatory behavior by reducing sexual motivation, by reducing sensitivity to sexual stimulation, or by some combination of the two.

EXPERIMENT 2

In Experiment 1, all doses of alcohol disrupted the copulatory behavior of sexually active male rats. Despite the numerous anecdotal reports that low doses of alcohol can facilitate human sexual behavior, no evidence of facilitation was observed at any dose. There are two possible explanations for this inconsistency. First, alcohol may affect the sexual behavior of rodents differently than it does the sexual behavior of humans. In this regard, Wilson (1977) has argued against the relevance of animal models to study the effects of alcohol on human sexual behavior. A critical feature of Wilson's argument is his assumption that lower animals lack the "cognitive mediation" commonly associated with the control of sexual behavior in humans. If alcohol facilitates human sexual behavior by an action on cognitive processes specific to human sexual behavior, then attempts to generalize the effects of alcohol on rodent sexual behavior to human sexual behavior—or, as in the present case, from human to rodent—have little merit.

The second explanation for the absence of facilitatory effects in Experiment 1 stems from the traditional belief that alcohol facilitates sexual behavior by releasing it from inhibitory control. Accordingly, it is possible that low doses of alcohol failed to facilitate sexual behavior in Experiment 1, not because the effects of alcohol on sexual behavior are fundamentally different in rodents and humans, but because the sexual behavior of the male rats was not under inhibitory control. The purpose of Experiment 2 was to test this disinhibition hypothesis.

To determine whether alcohol can facilitate the sexual behavior of male rats when it is under inhibitory control, we first induced sexually active male rats to suppress their copulatory behavior by giving them repeated tests with sexually nonreceptive females (Pfaus, Jacobs, & Wong, 1986). We then assessed the effects of a low (0.5 g/kg) and a moderate (1.0 g/kg) dose of alcohol, both of which had disrupted copulatory behavior in Experiment 1, on the sexual behavior of the inhibited males.

Method

Subjects. The subjects were 30 sexually active 400–600-g Long-Evans male rats.

Procedure. Experiment 2 began with 10 baseline tests identical to those of Experiment 1. Following these baseline tests, each rat was trained to inhibit its sexual activity during tests with nonreceptive females. During this training phase, the male rats received seven 45-min tests with sexually nonreceptive females (i.e., inhibitory tests), one every 4 days. Two days prior to each of these seven inhibitory tests, the males received a similar 45-min test with sexually receptive females (i.e., a noninhibitory test). To habituate them to the injection procedure, all of the male rats received a saline injection (2 ml/kg) 45 min before each of the 14 tests of the training phase.

Following the training phase was a drug-testing phase comprising four additional tests: two noninhibitory tests and two inhibitory tests administered on the same alternating 2-day schedule established during the training phase. Prior to the first test (noninhibitory) of the drug-testing phase, the rats were randomly assigned to one of three drug conditions ($n = 10$) and received either saline, 0.5 g/kg, or 1 g/kg of alcohol 45 min before the test. All rats received saline prior to the two intervening tests. On the fourth and final test (inhibitory) of the drug-testing phase, the rats once again were randomly assigned to one of the three drug conditions ($n = 10$) and injected accordingly 45 min before the test. This method made it possible to compare the effects of different doses of alcohol in the same group of rats under conditions of sexual excitation and inhibition.

The data were analyzed as in Experiment 1. However, because the male rats could not gain vaginal intromission during the test with nonreceptive females (i.e., during the inhibitory test), statistical comparisons involving the inhibitory test focused exclusively on mount and ejaculation measures.

Results

Development of sexual inhibition during the training phase. Consistent with the results of Pfaus et al. (1986), the male rats gradually learned to suppress their copulatory behavior during a series of regularly scheduled tests with nonreceptive females while maintaining their baseline rates of copulatory behavior during intervening tests with receptive females. Figure 2 illustrates the proportion of rats that displayed mounts and ejaculations during the seven inhibitory tests and seven noninhibitory tests that composed the training phase of Experiment 2. The monotonic decline in the proportion of rats that attempted mounts during the inhibitory tests is readily apparent: On the first inhibitory test, 80% of the rats mounted the nonreceptive females, but on the fifth, sixth, and seventh inhibitory tests, none of the rats attempted copulation. In marked contrast, all 30 rats mounted and ejaculated during each of the seven noninhibitory tests.

Effects of alcohol during inhibitory and noninhibitory tests. The major result of Experiment 2 is clearly evident in Figure 3. During the inhibitory test, the 0.5-g/kg dose of alcohol significantly increased the proportion of rats that mounted [$\chi^2(2) = 7.91, p < .005$] and ejaculated [$\chi^2(2) = 5.95, p < .01$], despite the fact that the

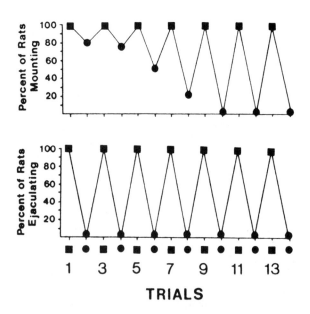

Figure 2. Percentage of rats displaying mounts or ejaculations during the training phase of Experiment 2. Squares represent tests with sexually receptive females; circles represent tests with nonreceptive females.

nonreceptive females were uncooperative. Although 2 of the rats that received the 1-g/kg dose of alcohol attempted to mount the nonreceptive females, this increase above the saline baseline was not statistically significant. Neither of these 2 rats ejaculated.

The effects of alcohol during the noninhibitory test confirmed the results of the comparable drug tests administered in Experiment 1. The 1-g/kg dose of alcohol sig-

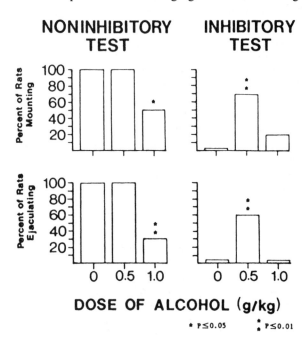

Figure 3. Percentage of rats displaying mounts or ejaculations as a function of alcohol dose during the noninhibitory and inhibitory tests in Experiment 2.

nificantly reduced the proportion of rats that mounted [$\chi^2(2) = 4.27, p < .025$] and ejaculated [$\chi^2(2) = 7.91, p < .005$], whereas the 0.5-g/kg dose did not. Furthermore, as in Experiment 1, the 1-g/kg dose significantly increased the mount, intromission, and ejaculation latencies, and increased the number of mounts and intromissions that preceded ejaculation; and the 0.5-g/kg dose significantly increased the ejaculation latency and the number of mounts and intromissions that preceded ejaculation (all ps $< .01$).

Performance during the two intervening baseline tests of the drug-testing phase indicated that there had been no shift in baseline. None of the rats mounted or ejaculated during the intervening inhibitory test, whereas all rats mounted and ejaculated during the intervening noninhibitory test.

Discussion

In Experiment 2, a low dose of alcohol (0.5 g/kg), which disrupted but did not block the copulatory behavior of male rats during a test with sexually receptive females, released copulatory behavior from the inhibition induced by regular tests with sexually nonreceptive females. A higher dose of alcohol (1 g/kg), sufficient to disrupt or block the copulatory behavior of male rats during a noninhibitory test, was not effective in releasing copulatory behavior from inhibition during an inhibitory test.

These results are consistent with anecdotal reports that low doses of alcohol can facilitate human copulatory behavior through a process of disinhibition (Athanasiou, Shaver, & Tavris, 1970; Hollister, 1975; Lemere & Smith, 1973; Masters et al., 1986). They are also consistent with results showing that low doses of alcohol that decrease physiological measures of sexual arousal in human males under conditions of sexual excitation can impair the ability of human males to voluntarily inhibit their sexual arousal (Rubin & Henson, 1976; Wilson & Niaura, 1984). These similarities challenge Wilson's (1977) contention that the effects of alcohol on the sexual behavior of laboratory animals provide few insights into the effects of alcohol on human sexual activity. It would be of interest in future studies to determine whether doses of alcohol lower than 0.5 g/kg can facilitate the copulatory behavior of males given training with sexually nonreceptive females and to assess the effect of alcohol on other conditions of low sexual activity (e.g., in sexually sluggish males or in males trained to inhibit their sexual advances by different means).

GENERAL DISCUSSION

The present experiments were designed to investigate alcohol's putative inhibitory and facilitatory effects on the copulatory behavior of sexually active male rats under conditions of sexual excitation and inhibition. The results confirmed that a low dose of alcohol can facilitate or disrupt copulatory behavior in male rats depending upon the presence or absence, respectively, of sexual inhibition.

The theory proposed by Steele and Southwick (1985) to account for the disinhibitory effects of alcohol on human social behavior adapts well to the present results. This theory suggests that alcohol disinhibits behavior because intoxicated individuals lack the perceptual cognitive functioning necessary to inhibit inappropriate responses in the face of strong eliciting cues. The theory predicts that the disinhibitory effect of alcohol can occur only in those situations in which there is strong inhibitory conflict, that is, in situations in which inhibitory control is in conflict with strong behavior-eliciting cues. Accordingly, in Steele and Southwick's terms, the low dose of alcohol facilitated copulation in Experiment 2 because there were both strong eliciting cues and inhibitory control. However, we can only speculate about the identity of the eliciting stimuli in this condition. Because the female rats were not sexually receptive, none of the potential stimulus qualities specifically associated with proceptivity or receptivity (e.g., pheremonal cues, soliciting behaviors, or lordosis) were present. When the males attempted to mount, the nonreceptive females generally displayed such defensive behaviors as biting, boxing, kicking, running in circles, and lying on their backs. It is conceivable that simply encountering a female, regardless of its state of estrus, in a testing chamber associated with copulation, might have been a sufficient eliciting cue for copulation. It is also possible, however, that the intoxicated rats misinterpreted the females' defensive behavior for proceptive behavior. Darting, kicking, and running in circles are characteristics of both types of behavior (Beach, 1976; Madlafousek & Hlinak, 1978).

The disinhibitory effects of alcohol are not restricted to sexual behavior. In mice, equivalent doses of alcohol potentiate suppressed aggression against an intruder in an unfamiliar environment but reduce the display of unsuppressed aggression against an intruder in the home cage (Miczek & O'Donnell, 1980). In rats, equivalent doses of alcohol reverse the suppressive effects of corticotropin-releasing factor (CRF) on punished responding for food but act synergistically with CRF to inhibit unpunished responding for food (Thatcher-Britton & Koob, 1986). And in humans, equivalent doses of alcohol decrease the amount of ice cream eaten by nondieters but increase the amount eaten by dieters (Polivy & Herman, 1976).

In summary, the present study provides the first experimental evidence that alcohol can both disrupt and disinhibit the copulatory behavior of male rats. The fact that these results were predicted from a consideration of anecdotal reports of alcohol's effects on human sexual activity lends support to the disinhibitory theory of alcohol's facilitatory effects on human sexual behavior and underscores the value of the comparative approach in the study of drug-induced sexual disinhibition.

REFERENCES

Athanasiou, R., Shaver, P., Tavris, C. (1970). Sex. *Psycholology Today*, **4**, 39-52.

Beach, F. A. (1976). Sexual attractivity, proceptivity, and receptivity in female mammals. *Hormones & Behavior*, **7**, 105-138.

Briddell, D. W., & Wilson, G. T. (1976). Effects of alcohol and expectancy set on male sexual arousal. *Journal of Abnormal Behavior*, **85**, 225-234.

Carver, A. E. (1948). The interrelationship of sex and alcohol. *International Journal of Sexology*, **2**, 78-81.

Dewsbury, D. A. (1967). Effects of alcohol ingestion on copulatory behavior of male rats. *Psychopharmacologia* (Berlin), **11**, 276-281.

Farkas, G. M., & Rosen, R. C. (1976). Effect of alcohol on elicited male sexual response. *Journal of Studies on Alcohol*, **37**, 265-272.

Ford, C. S., & Beach, F. A. (1951). *Patterns of sexual behavior*. New York: Harper & Row.

Gantt, W. H. (1940). Effect of alcohol on sexual reflexes in dogs. *American Journal of Physiology*, **129**, 360.

Gantt, W. H. (1952). Effect of alcohol on the sexual reflexes of normal and neurotic male dogs. *Psychosomatic Medicine*, **14**, 174-181.

Gantt, W. H. (1957). Acute effect of alcohol on automatic (sexual, secretory, cardiac) and somatic responses. In H. E. Himwich (Ed.), *Alcoholism: Basic aspects and treatment* (pp. 73-89). Washington, DC: American Association for the Advancement of Science.

Goldstein, D. B. (1983). *Pharmacology of alcohol*. New York: Oxford University Press.

Hart, B. L. (1968). Effects of alcohol on sexual reflexes and mating behavior in the male dog. *Quarterly Journal of Studies on Alcohol*, **29**, 839-844.

Hart, B. L. (1969). Effects of alcohol on sexual reflexes and mating behavior in the male rat. *Psychopharmacologia* (Berlin), **14**, 377-382.

Hollister, L. E. (1975). The mystique of social drugs and sex. In M. Sandler & G. L. Gessa (Eds.), *Sexual behavior: Pharmacology and biochemistry* (pp. 85-92). New York: Raven Press.

Jeffreys, R., Pournaghash, S., & Riley, A. L. (1989). *Conditioned taste aversions to alcohol in rats*. Manuscript submitted for publication.

Kaplan, H. S. (1974). *The new sex therapy*. New York: Brunner.

Kinsey, B. A. (1966). *The female alcoholic*. Springfield, IL: Thomas.

Lemere, F., & Smith, J. W. (1973). Alcohol-induced sexual impotence. *American Journal of Psychiatry*, **130**, 212-213.

MacDougald, D. (1967). Aphrodesiacs and anaphrodesiacs. In A. Ellis & A. Abarbanel (Eds.), *Encyclopaedia of sexual behavior* (pp. 145-153). New York: Hawthorn Books.

Madlafousek, J., & Hlinak, Z. (1978). Sexual behavior of the female rat: Inventory, patterning, and measurement. *Behavior*, **63**, 129-173.

Malatesta, V. J., Pollack, R. H., Crotty, T. D., & Peacock, L. J. (1982). Acute alcohol intoxication and female orgasmic response. *Journal of Sex Research*, **18**, 1-17.

Malatesta, V. J., Pollack, R. H., Wilbanks, W. A., & Adams, H. E. (1979). Alcohol effects on the orgasmic-ejaculatory response in human males. *Journal of Sex Research*, **15**, 101-107.

Marascuilo, L., & Levin, J. (1983). *Multivariate statistics in the social sciences: A researcher's guide*. Monterey, CA: Brooks/Cole.

Marlatt, G. A., Demming, B., & Reid, J. B. (1973). Loss of control drinking in alcoholics: An experimental analogue. *Journal of Abnormal Psychology*, **81**, 233-241.

Masters, W. H., & Johnson, V. E. (1966). *Human sexual response*. Boston: Little-Brown.

Masters, W. H., Johnson, V. E., & Kolodny, R. C. (1986). *Sex and human loving*. Boston: Little-Brown.

McKendry, J. B. R., Collins, W. E., Silverman, M., Krul, L. E., Collins, J. P., & Irvine, A. H. (1983). Erectile impotence: A clinical challenge. *Canadian Medical Association Journal*, **128**, 653-663.

Miczek, K. A., & O'Donnell, J. M. (1980). Alcohol and chlordiazepoxide increase suppressed aggression in mice. *Psychopharmacology*, **69**, 39-44.

Pfaus, J. G., Jacobs, W. J., & Wong, R. (1986). Conditional olfactory cues facilitate the acquisition of copulatory behavior and affect mate choice in male rats. *Canadian Psychology*, **27**, 470.

Pinhas, V. (1987). Sexual dysfunctions in women alcoholics. *Medical Aspects of Human Sexuality*, **21**, 97-101.

Polivy, J., & Herman, C. P. (1976). Effects of alcohol on eating behavior: Influence of mood and perceived intoxication. *Journal of Abnormal Psychology*, **85**, 601-606.

Rubin, H. B., & Henson, D. E. (1976). Effects of alcohol on male sexual responding. *Psychopharmacology*, **47**, 123-134.

Sachs, B. D., & Barfield, R. J. (1976). Functional analysis of masculine copulatory behavior in the rat. *Advances in the Study of Behavior*, **7**, 91-154.

Smith, D. E., Wesson, D. R., & Apter-Marsh, M. (1984). Cocaine- and alcohol-induced sexual dysfunctions in patients with addictive disease. *Journal of Psychoactive Drugs*, **16**, 359-361.

Steele, C. M., & Southwick, L. (1985). Alcohol and social behavior: I. The psychology of drunken excess. *Journal of Personality & Social Psychology*, **48**, 18-34.

Teitelbaum, H. A., & Gantt, W. H. (1958). The effect of alcohol on sexual reflexes and sperm count in the dog. *Quarterly Journal of Studies on Alcohol*, **19**, 394-398.

Thatcher-Britton, K., & Koob, G. F. (1986). Alcohol reverses the proconflict effect of corticotropin-releasing factor. *Regulatory Peptides*, **16**, 315-320.

Wilson, G. T. (1977). Alcohol and human sexual behavior. *Behavior Research & Therapy*, **15**, 239-252.

Wilson, G. T., & Lawson, D. M. (1976). Expectancies, alcohol, and sexual arousal in male social drinkers. *Journal of Abnormal Psychology*, **85**, 587-594.

Wilson, G. T., & Lawson, D. M. (1978). Expectancies, alcohol, and sexual arousal in women. *Journal of Abnormal Psychology*, **87**, 358-367.

Wilson, G. T., & Niaura, R. (1984). Alcohol and the disinhibition of sexual responsiveness. *Journal of Studies on Alcohol*, **45**, 219-224.

(Manuscript received June 6, 1988;
revision accepted for publication December 15, 1988.)

Glossary

cognitive - this is a general term that refers to complex mental processes such as thinking and remembering

conditioned taste aversion - if an animal is made ill after it consumes a novel food, it often develops a strong aversion to the taste of that food; this form of learning is called conditioned taste aversion

disinhibition - a facilitation produced by the release from the influence of some inhibitory factor

ejaculation - the ejection of sperm

empirical - based on careful observation

estradiol - a hormone released by the gonads (i.e., by the ovaries and testes), primarily in females

etiology - the factors that contribute to the cause of disease

intromission - insertion of the penis into the vagina

mount - the males of many mammalian species climb onto the hindquarters of receptive females for the purpose of copulation; this is called mounting

ovariectomize - to remove the ovaries; ovariectomized female rats are made sexually receptive by injecting them with estradiol and progesterone

penile - referring to the penis

postejaculatory interval - the interval between an ejaculation and the next intromission

proceptive behaviors - behaviors of sexually receptive females that attract the sexual advances of males

progesterone - a hormone released by the gonads (i.e., by the ovaries and testes), primarily in females

putative - hypothesized

Essay Study Questions

1. The effects of alcohol on sexual behavior have been studied only rarely. Briefly review the research on this topic that existed prior to Pfaus and Pinel's article.

2. Why do you think that the effects of alcohol on sexual behavior not been more widely studied?

3. Describe the technique that was used by Pfaus and Pinel to induce sexual inhibition in male rats.

4. Compare the effects of the same doses of alcohol, 0.5 and 1.0 g/kg, in the two experiments.

5. Why did Pfaus and Pinel not report intromission data in Experiment 2?

6. The disinhibitory effects of alcohol are not restricted to sexual behavior. Explain.

Multiple Choice Questions

1. The only two published experimental studies of the effects of alcohol on overt human sexual behavior were studies of
    a. premature ejaculation.
    b. masturbation.
    c. erectile failure.
    d. penile (no relation to the second author) tumescence.
    e. none of the above

2. Which of the following is the usual chronological order of one copulatory sequence?
    a. mount, intromit, ejaculate
    b. mount, ejaculate, intromit
    c. ejaculate, intromit, mount
    d. intromit, mount, ejaculate
    e. intromit, ejaculate, mount

3. In Experiment 2, 0.5g/kg of alcohol produced a striking disinhibition of copulatory behavior, whereas in Experiment 1, it
    a. had no effect.
    b. produced a slight but statistically significant disinhibition.
    c. produced a slight statistically insignificant disinhibition.
    d. produced a slight disruption, which was reflected by statistically significant changes in only a few of the measures.
    e. abolished copulation.

4. The receptive females that participated in Experiment 1 and in the noninhibitory tests of Experiment 2
    a. received estradiol 48 hours before being tested.
    b. received progesterone 4 hours before being tested.
    c. were ovariectomized.
    d. all of the above
    e. both a and b

5. In both experiments, the alcohol was administered
    a. with a bit of soda water and a twist of lemon.
    b. intraperitoneally.
    c. subcutaneously.
    d. intravenuously.
    e. intracranially.

6. In Experiment 2, which of the following tests was a noninhibitory test?
    a. the tenth test
    b. the thirteenth test
    c. the eleventh test
    d. the ninth test
    e. the first test

The answers to the preceding questions are on page 290.

Food-For-Thought-Questions

1. Since the publication of these experiments in 1989, the importance of basic research on the disinhibitory effects of drugs has been dramatically illustrated. It was recently shown that people who have learned to practice safe sex tend to engage in risky sexual practices when they are under the influence of alcohol or other addictive drugs—even those who are at great risk (e.g., homosexual drug users). Discuss with respect to Experiment 2 and the prospects for future research on sexual disinhibition.

2. Many people assume that human behavior is unique and that the study of other species has little to tell us about human biopsychological processes. Discuss with respect to the present experiments.

3. In view of the fact that most human rapes are committed under the influence of alcohol, Experiment 2 may provide an animal model of some aspects of this serious social problem. Discuss.

ARTICLE 20

Buprenorphine Suppresses Cocaine Self-Administration by Rhesus Monkeys

N.K. Mello, J.H. Mendelson, M.P. Bree, and S.E. Lukas
Reprinted from Science, 1989, volume 245, 859-862.

Animal models provide a means of bringing the analytic power of laboratory experimentation to bear on human disorders. Disorders are created in laboratory animals that mimic human disorders in key respects, and then these model disorders are subjected to experimental analysis. The following article describes how Mello and her collaborators have used the drug self-administration model to demonstrate the potential of a new approach to the treatment of human cocaine abuse.

Monkeys were trained to press a button for food pellets or to administer small doses of cocaine through an implanted intravenous catheter. They received eight 1-hour test sessions a day—food pellets were the reinforcers during four of the sessions, and cocaine injections were the reinforcers during the other four. After several months of cocaine self-administration, the monkeys received 30 daily infusions of the mixed opioid agonist-antagonist buprenorphine, 0.40 mg/kg for the first 15 days and 0.70 mg/kg for the second 15 days. By the end of the 30-day regimen of buprenorphine injections, the average number of cocaine self-injections per day was reduced from over 50 to about 2. The fact that there was only a minor reduction in responding for food during this same period suggested that buprenorphine reduced cocaine self-administration specifically by reducing the reinforcing properties of cocaine. Following the cessation of the buprenorphine injections, the monkeys returned to their original rate of cocaine self-administration in an average of 30.5 days.

These results suggest that buprenorphine might prove to be the basis of an effective pharmacotherapy for the treatment of cocaine abuse. If a controlled study in human cocaine users confirms this hypothesis, this laboratory experiment by Mello and her collaborators will have played a key role in helping free tens of thousands of people from the dire economic, social, and medical consequences of cocaine abuse.

Buprenorphine Suppresses Cocaine Self-Administration by Rhesus Monkeys

Nancy K. Mello, Jack H. Mendelson, Mark P. Bree, Scott E. Lukas

Cocaine abuse has reached epidemic proportions in the United States, and the search for an effective pharmacotherapy continues. Because primates self-administer most of the drugs abused by humans, they can be used to predict the abuse liability of new drugs and for preclinical evaluation of new pharmacotherapies for drug abuse treatment. Daily administration of buprenorphine (an opioid mixed agonist-antagonist) significantly suppressed cocaine self-administration by rhesus monkeys for 30 consecutive days. The effects of buprenorphine were dose-dependent. The suppression of cocaine self-administration by buprenorphine did not reflect a generalized suppression of behavior. These data suggest that buprenorphine would be a useful pharmacotherapy for treatment of cocaine abuse. Because buprenorphine is a safe and effective pharmacotherapy for heroin dependence, buprenorphine treatment may also attenuate dual abuse of cocaine and heroin.

COCAINE ABUSE IS WIDESPREAD IN the general population (1) and has also increased among heroin-dependent persons, including those in methadone maintenance treatment programs (2). The many adverse medical consequences of cocaine abuse (3) are augmented by the combined use of cocaine and heroin (4). For example, dual addiction to intravenous

Alcohol and Drug Abuse Research Center, Harvard Medical School–McLean Hospital, Belmont, MA 02178.

cocaine and heroin may increase the risk of acquired immunodeficiency syndrome (AIDS), both through needle sharing and through the combined immunosuppressive effects of both drugs (5). Intravenous drug abuse was estimated to account for more than 30% of AIDS victims in the United States in 1988 (1).

At present, there is no uniformly effective pharmacotherapy for cocaine abuse (6), and the dual abuse of cocaine plus heroin is an even more difficult treatment challenge. Heroin abuse can be treated with opiate agonists [methadone and α-l-acetylmethadol (LAAM)] (7) and the opiate antagonist naltrexone (8), but these pharmacotherapies are not useful for combined cocaine and heroin abuse (9). Although desipramine (a tricyclic antidepressant) reduces cocaine abuse in some patients (6, 10), it can stimulate relapse to cocaine abuse in abstinent patients (11). Treatment with methadone and desipramine has yielded inconsistent effects on cocaine use by heroin abusers (12).

An ideal pharmacotherapy would be one that antagonized the reinforcing effects of cocaine and that had minimal adverse side effects or potential for abuse liability. The opioid mixed agonist-antagonist buprenorphine (13) meets these criteria for the treatment of opiate abuse. Buprenorphine effectively suppressed heroin self-administration by heroin-dependent men during inpatient studies (14) and blocked opiate effects for more than 24 hours (15). Cessation of buprenorphine treatment does not produce severe and protracted withdrawal signs and symptoms in man (14, 16, 17). Buprenorphine is safer than methadone because its antagonist component appears to prevent lethal overdose, even at approximately ten times the analgesic therapeutic dose (18). Buprenorphine is also effective for the outpatient detoxification of heroin-dependent persons (19). The opioid agonist effects of buprenorphine make it acceptable to heroin abusers (14, 16), but illicit diversion has been minimal in comparison to heroin (20). Preclinical studies indicate that buprenorphine is less reinforcing than other opioids in rhesus monkey and baboon (21, 22).

Here we describe the effect of buprenorphine treatment on cocaine self-administration by rhesus monkeys. Cocaine effectively maintains operant responding, leading to its intravenous administration in primates, and it is well established that primates self-administer most drugs abused by man (23). The primate model of drug self-administration is a useful method for the prediction of drug abuse liability and can be used to evaluate new pharmacotherapies for drug abuse disorders (24).

Two male and three female adult rhesus monkeys (*Macaca mulatta*) (25) with a 262 ± 79 day history of cocaine self-administration were studied. Each monkey was implanted with a double-lumen silicone rubber intravenous catheter under aseptic conditions to permit administration of buprenorphine or saline during cocaine self-administration. The intravenous catheter was protected by a custom-designed tether system (Spaulding Medical Products) that permits monkeys to move freely. Monkeys worked for food (1-g banana pellets) and for intravenous cocaine (0.05 or 0.10 mg per kilogram of body weight per injection) on the same operant schedule of reinforcement. An average of 64 responses was required for each food pellet or cocaine injection under a second-order schedule of reinforcement (26). Food and cocaine each were available during four 1-hour sessions each day. Food sessions began at 11 a.m., 3 p.m., 7 p.m., and 7 a.m.; cocaine sessions at 12 noon, 4 p.m., 8 p.m., and 8 a.m. Each food or drug session lasted for 1 hour or until 20 drug injections or 65 food pellets were delivered. The total number of cocaine injections was limited to 80 per day to minimize the possibility of adverse drug effects. The nutritionally fortified diet of banana pellets was supplemented with fresh fruit, vegetables, biscuits, and multiple vitamins each day.

Buprenorphine treatment was administered at two doses (0.40 and 0.70 mg/kg per day) that effectively suppressed opiate self-administration in our previous studies in primates (24). Buprenorphine (or an equal volume of saline solution) was administered daily beginning at 9:30 a.m. Buprenorphine and saline were gradually infused at a rate of 1 ml of solution every 12 min and flushed through the catheter with sterile saline in a volume that exceeded the estimated catheter dead space. Each dose of buprenorphine and saline was studied for 15 consecutive days (60 sessions). After 30 days of treatment, buprenorphine was abruptly discontinued and daily saline treatment was resumed.

We measured cocaine and food self-administration during 15 days of saline treatment and six successive 5-day periods of buprenorphine treatment (Fig. 1). During base-line saline treatment, each of the five monkeys self-administered 2.1 to 4 mg/kg per day of cocaine [group average (± SEM) of 3.07 ± 0.17 mg/kg per day]. This level of cocaine self-administration corresponds to that commonly reported by cocaine abusers (1 to 2 g per week is equivalent to 2.04 to 4.08 mg/kg per day in man) (27). All animals reduced their cocaine self-administration significantly during buprenorphine treatment ($P < 0.0001$) (Fig. 1). On the first day of buprenorphine treatment, cocaine self-administration decreased by 50% or more in four of the five subjects (range 50 to 67%). Average cocaine self-administration decreased by 49 ± 15.5% to 1.60 ± 0.25 mg/kg per day during the first 5 days of buprenorphine treatment ($P < 0.01$). Average cocaine self-administration then decreased to 77 ± 7.4% and 83 ± 8.2% below base-line levels during buprenorphine treatment days 6 to 10 and 11 to 15, respectively. Cocaine self-administration averaged 0.98 ± 0.11 mg/kg per day over the first 15 days of buprenorphine treatment at 0.40 mg/kg per day (Fig. 1).

During the second 15 days of buprenorphine treatment at 0.70 mg/kg per day, cocaine self-administration decreased to between 91 ± 2.7% and 97 ± 0.9% below base-line levels (Fig. 1). Monkeys self-administered an average of 0.19 ± 0.03 mg/kg per day of cocaine. Analysis of individual subject data showed that the rate and degree of suppression by buprenorphine of cocaine-maintained responding was equivalent

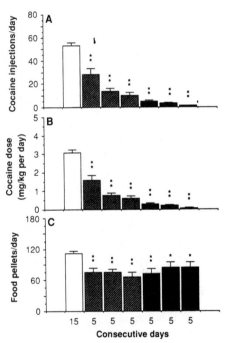

Fig. 1. The effects of single daily infusions of buprenorphine or a control saline solution on cocaine and food self-administration. Saline treatment is shown as an open bar and buprenorphine treatment as a striped bar (0.40 mg/kg per day) and a solid bar (0.70 mg/kg per day). The number of days that each treatment condition was in effect is shown on the abscissa. Each data point is the mean ± SEM of five subjects. (**A**) The average number of cocaine injections self-administered; (**B**) the average dose of cocaine (milligrams per kilogram per day) self-administered; (**C**) the average number of food pellets self-administered. The statistical significance of each change from the saline treatment base line as determined by analysis of variance for repeated measures and Dunnett's tests for multiple comparisons (39) is shown (*$P < 0.05$; **$P < 0.01$).

in animals that self-administered relatively high (4 mg/kg per day) and low (2.1 mg/kg per day) doses of cocaine during the saline base-line treatment period. Cocaine-maintained operant responding remained suppressed for at least 15 days after cessation of buprenorphine treatment. This time course is similar to clinical reports of abstinence signs and symptoms 15 to 21 days after abrupt withdrawal of buprenorphine (16) and probably reflects the slow dissociation of buprenorphine from the opiate receptor (13). Individual monkeys returned to baseline levels of cocaine self-administration at different rates ranging from 15 to 58 days (mean, 30.5 ± 10 days).

In contrast to its dose-dependent effects on cocaine self-administration, buprenorphine administration (0.40 mg/kg per day) suppressed food-maintained responding by 31 ± 8.3% during the first 15 days of treatment. Then food self-administration gradually recovered to average 20 ± 12.5% below base line during the second 15 days of treatment with a higher dose of buprenorphine (Fig. 1). Although these changes were statistically significant ($P < 0.05$ to 0.01), it is unlikely that they were biologically significant. There were no correlated changes in body weight, and animals continued to eat daily fruit and vegetable supplements. Moreover, food self-administration during the first daily session after buprenorphine treatment was not suppressed in comparison to saline treatment. The distribution of food intake across the four daily food sessions was equivalent during saline and buprenorphine treatment conditions. Four of five animals returned to base-line levels of food-maintained operant responding within 3 to 17 days after cessation of buprenorphine treatment (8.5 ± 2.9 days). Animals were not sedated during buprenorphine treatment and activity levels appeared normal. These data suggest that buprenorphine treatment suppressed cocaine-maintained responding but did not produce a generalized suppression of behavior.

These data are consistent with our previous observations that chronic buprenorphine self-administration (0.01 to 1.0 mg/kg per injection) did not significantly suppress food self-administration by rhesus monkeys (28, 29). Although administration of single doses of buprenorphine (0.10 to 0.30 mg/kg) significantly suppressed food-maintained responding (28) during chronic buprenorphine administration (1.0 mg/kg), recovery of food-maintained responding occurred rapidly (29). After an initial suppression, food self-administration increased significantly above control levels during 25 days of buprenorphine treatment (1.0 mg/kg) (29). The fact that buprenorphine is not an appetite suppressant in primates (28, 29) or in man (14) is another indicator of its relative safety during chronic use.

These preclinical data suggest that buprenorphine may be an effective pharmacotherapy for the treatment of cocaine abuse. However, clinical evaluation of buprenorphine treatment will require double-blind (buprenorphine versus placebo) trials with randomized patient assignment and independent indices of compliance with the treatment regimen (for example, buprenorphine blood levels) and objective measures of drug use (frequent drug urine screens). One advantage of the primate model for preclinical evaluation of pharmacotherapies is that compliance and multiple drug use are not at issue. It is important to emphasize that if buprenorphine treatment of cocaine abuse were to prove clinically efficacious, this would not be a "substitute addiction" with a less toxic cocaine-like stimulant drug analogous to methadone treatment of heroin dependence. Buprenorphine is an opioid mixed agonist-antagonist (13), whereas cocaine is a stimulant drug (3). Moreover, buprenorphine does not substitute for cocaine in primate drug self-administration studies (21).

We do not yet understand the mechanisms accounting for the suppression of cocaine self-administration by buprenorphine. The relative contribution of buprenorphine's opioid agonist and antagonist components to its effects on the reinforcing properties of cocaine are unknown. However, since opioid antagonists such as naloxone and naltrexone do not suppress cocaine self-administration in primates (30) or in rodents (31), we postulate that either the opioid agonist component or the opioid agonist-antagonist combination is critical for the effects of buprenorphine on cocaine self-administration. Clinical and primate studies of opioid agonist effects on cocaine self-administration are inconsistent. Methadone treatment did not reduce the incidence of cocaine-positive urines in heroin-dependent patients (9), but morphine treatment suppressed cocaine self-administration in a dose-dependent manner in squirrel monkeys (32).

There is a consensus that dopaminergic neural systems play a critical role in cocaine reinforcement (33), and our data suggest that buprenorphine modifies the reinforcing properties of cocaine. This interpretation is consistent with several lines of evidence indicating comodulatory interactions between endogenous opioid and dopaminergic systems in brain (34–36). Neuroendocrine (34), neuropharmacological (35), and behavioral studies (36) suggest that dopaminergic systems modulate endogenous opioid system activity and the converse. Attenuation of cocaine self-administration by buprenorphine further illustrates an interrelationship between opioid and dopamine systems. Our findings also suggest the importance of examining commonalities in the way in which abused drugs maintain behavior leading to their self-administration (37).

Buprenorphine is potentially valuable for the treatment of dual addiction to cocaine and heroin because it suppresses heroin use by heroin addicts (14). Empirical support for this prospect comes from a report of an open clinical trial (38). Opioid-dependent patients treated with methadone had a significantly higher incidence of cocaine-positive urines than patients treated for 1 month with daily sublingual doses of buprenorphine (average 3.2 mg/day; range 2 to 8 mg) (38). If buprenorphine reduces cocaine abuse, as well as dual cocaine and heroin abuse, it could be very beneficial to society in reducing drug abuse problems and the associated risks for human immunodeficiency virus infection.

REFERENCES AND NOTES

1. N. J. Kozel and E. H. Adams, *Science* **234**, 970 (1986); *National Institute on Drug Abuse, ADAMHA* (Alcohol, Drug Abuse, and Mental Health Administration), *Request for Applications DA-89-01* (December 1988).
2. T. R. Kosten, B. J. Rounsaville, F. H. Gawin, H. D. Kleber, *Am. J. Drug Alcohol Abuse* **12**, 1 (1986); B. Kaul and B. Davidow, *ibid.* **8**, 27 (1981).
3. L. L. Cregler and H. Mark, *N. Engl. J. Med.* **315**, 1495 (1986); J. H. Mendelson and N. K. Mello, in *Harrison's Principles of Internal Medicine*, E. Braunwald *et al.*, Eds. (McGraw-Hill, New York, ed. 11, 1986), pp. 2115–2118.
4. M. J. Kreek, in *Psychopharmacology, The Third Generation of Progress*, H. Y. Meltzer, Ed. (Raven, New York, 1987), pp. 1597–1604.
5. R. M. Donahoe and A. Falek, in *Psychological, Neuropsychiatric and Substance Abuse Aspects of AIDS*, T. P. Bridge *et al.*, Eds. (Raven, New York, 1988), pp. 145–158; T. W. Klein, C. A. Newton, H. Friedman, *ibid.*, pp. 139–143.
6. H. D. Kleber and F. H. Gawin, *J. Clin. Psychiatry* **45**, 18 (1984); F. H. Gawin and E. H. Ellinwood, *N. Engl. J. Med.* **318**, 1173 (1988).
7. V. P. Dole and M. Nyswander, *J. Am. Med. Assoc.* **193**, 646 (1965); J. B. Blaine *et al.*, *Ann. N.Y. Acad. Sci.* **311**, 214 (1978); J. B. Blaine *et al.*, *ibid.* **362**, 101 (1981).
8. R. E. Meyer and S. M. Mirin, *The Heroin Stimulus* (Plenum, New York, 1979); W. R. Martin *et al.*, *Arch. Gen. Psychiatry* **28**, 784 (1973); N. K. Mello *et al.*, *J. Pharmacol. Exp. Ther.* **216**, 45 (1981).
9. T. R. Kosten, B. J. Rounsaville, H. D. Kleber, *Arch. Gen. Psychiatry* **44**, 281 (1987).
10. F. H. Gawin and H. E. Kleber, *ibid.* **41**, 903 (1984); F. S. Tennant, Jr., and R. A. Rawson, in *Problems of Drug Dependence, 1982*, L. S. Harris, Ed. (Committee on Problems of Drug Dependence, Washington, DC, 1983).
11. R. E. Weiss, *J. Am. Med. Assoc.* **260**, 2545 (1988).
12. C. P. O'Brien *et al.*, *J. Clin. Psychiatry* **49**, 17 (1988); T. R. Kosten *et al.*, *ibid.* **48**, 442 (1987).
13. The opioid mixed agonist-antagonist buprenorphine is an oripavine derivative of thebaine with partial μ opioid agonist activity. It is a congener of a potent opioid agonist, etorphine, and an opioid antagonist, diprenorphine. The structure and chemical derivation of this opioid mixed agonist-antagonist have been described by J. W. Lewis, in *Narcotic*

Antagonists: Advances in Biochemical Pharmacology, M. Braude et al., Eds. (Raven, New York, 1974), vol. 8, pp. 123–136. The pharmacology of buprenorphine has been described by J. W. Lewis, M. J. Rance, and D. J. Sanger [in *Advances in Substance Abuse, Behavioral and Biological Research*, N. K. Mello, Ed. (JAI, Greenwich, CT, 1983), vol. 3, pp. 103–154].

14. N. K. Mello and J. H. Mendelson, *Science* **207**, 657 (1980); _____, J. C. Kuehnle, *J. Pharmacol. Exp. Ther.* **223**, 30 (1982); N. K. Mello and J. H. Mendelson, *Drug Alcohol Depend.* **14**, 282 (1985).

15. W. K. Bickel et al., *J. Pharmacol. Exp. Ther.* **247**, 47 (1988).

16. D. R. Jasinski, J. S. Pevnick, J. D. Griffith, *Arch. Gen. Psychiatry* **35**, 601 (1978).

17. S. E. Lukas, D. R. Jasinski, R. E. Johnson, *Clin. Pharmacol. Ther.* **36**, 127 (1984); P. J. Fudala, R. E. Johnson, E. Bunker, ibid. **45**, 186 (1989).

18. C. D. Banks, *N.Z. Med. J.* **89**, 255 (1979).

19. W. K. Bickel et al., *Clin. Pharmacol. Ther.* **43**, 72 (1988); T. R. Kosten and H. D. Kleber, *Life Sci.* **42**, 635 (1988).

20. J. J. O'Connor et al., *Br. J. Addict.* **83**, 1085 (1988).

21. S. E. Lukas, J. V. Brady, R. R. Griffiths, *J. Pharmacol. Exp. Ther.* **238**, 924 (1986).

22. N. K. Mello, S. E. Lukas, M. P. Bree, *Drug Alcohol Depend.* **21**, 81 (1988); J. H. Woods, in *Proceedings* (Committee on Problems of Drug Dependence, Washington, DC, 1977); A. M. Young et al., *J. Pharmacol. Exp. Ther.* **229**, 118 (1984).

23. T. Thompson and K. R. Unna, Eds., *Predicting Dependence Liability of Stimulant and Depressant Drugs* (University Park Press, Baltimore, 1977); R. R. Griffiths, G. E. Bigelow, J. E. Henningfield, in *Advances in Substance Abuse, Behavioral and Biological Research*, N. K. Mello, Ed. (JAI, Greenwich, CT, 1980), vol. 1, pp. 1–90; R. R. Griffiths and R. L. Balster, *Clin. Pharmacol. Ther.* **25**, 611 (1979).

24. N. K. Mello, M. P. Bree, J. H. Mendelson, *J. Pharmacol. Exp. Ther.* **225**, 378 (1983).

25. Animal maintenance and research were conducted in accordance with guidelines provided by the Committee on Laboratory Animal Resources. The facility is licensed by the U.S. Department of Agriculture and protocols were approved by the Institutional Animal Care and Use Committee. The health of the animals was periodically monitored by a consultant veterinarian from the New England Regional Primate Research Center. Surgical implantation of an intravenous catheter for drug infusion was performed under aseptic conditions. A surgical level of anesthesia was induced with ketamine (25 mg/kg, intramuscular) or pentobarbital (30 mg/kg, intravenous). Because the procedure usually takes 25 min, supplemental doses of anesthetic were seldom required. A mild analgesic (Tylenol) was administered every 4 to 6 hours for the first 24 hours after surgery.

26. Completion of a fixed ratio (FR) of four consecutive variable ratio (VR) components, in which an average of 16 responses produced a brief stimulus light (S+), was required for cocaine or food delivery. This is a second-order FR 4 schedule with VR 16 components [FR 4 (VR 16:S)].

27. J. H. Mendelson et al., *Am. J. Psychiatry* **145**, 1094 (1988); J. H. Mendelson et al., in preparation.

28. N. K. Mello et al., *Pharmacol. Biochem. Behav.* **23**, 1037 (1985).

29. S. E. Lukas et al., ibid. **30**, 977 (1988).

30. J. H. Woods and C. R. Schuster, in *Stimulus Properties of Drugs*, T. Thompson and R. Pickens, Eds. (Appleton-Century-Crofts, New York, 1971), pp. 163–175; A. K. Killian et al., *Drug Alcohol Depend.* **3**, 243 (1978); S. R. Goldberg et al., *Pharmacol. Exp. Ther.* **176**, 464 (1971).

31. A. Ettenberg, H. O. Pettit, F. E. Bloom, G. F. Koob, *Psychopharmacol. Ser. (Berlin)* **78**, 204 (1982); M. E. Carroll, S. T. Lac, M. J. Walker, R. Kragh, T. Newman, *J. Pharmacol. Exp. Ther.* **238**, 1 (1986).

32. R. Stretch, *Can. J. Physiol. Pharmacol.* **55**, 778 (1977).

33. M. W. Fischman, in *Psychopharmacology: The Third Generation of Progress*, H. Y. Meltzer, Ed. (Raven, New York, 1987), pp. 1543–1553; C. A. Dackis and M. S. Gold, *Neurosci. Biobehav. Rev.* **9**, 469 (1985); M. J. Kuhar, M. C. Ritz, J. Sharkey, *Natl. Inst. Drug Abuse Research Monogr. Ser.* **88** (1988), pp. 14–22; M. C. Ritz, R. J. Lamb, S. R. Goldberg, M. J. Kuhar, *Science* **237**, 1219 (1987); W. L. Woolverton, L. I. Goldberg, J. Z. Ginos, *J. Pharmacol. Exp. Ther.* **230**, 118 (1984); W. L. Woolverton, *Pharmacol. Biochem. Behav.* **24**, 531 (1986).

34. J. H. Mendelson, N. K. Mello, P. Cristofaro, A. Skupny, J. Ellingboe, *Pharmacol. Biochem. Behav.* **24**, 309 (1986); R. O. Kuljis and J. P. Advis, *Endocrinology* **124**, 1579 (1989); N. K. Mello, J. H. Mendelson, M. P. Bree, M. Kelly, in *Problems of Drug Dependence 1989*, L. S. Harris, Ed. (Committee on Problems of Drug Dependence, Washington, DC, 1989).

35. Y. Ishizuka et al., *Life Sci.* **43**, 2275 (1988); G. Di Chiara and A. Imperato, *J. Pharmacol. Exp. Ther.* **244**, 1067 (1988).

36. M. A. Bozarth and R. A. Wise, *Life Sci.* **19**, 1881 (1981); H. Blumberg and C. Ikeda, *J. Pharmacol. Exp. Ther.* **206**, 303 (1978); T. S. Shippenberg and A. Herz, *Brain Res.* **436**, 169 (1987).

37. N. K. Mello, in *The Pathogenesis of Alcoholism, Biological Factors*, B. Kissin and H. Begleiter, Eds. (Plenum, New York, 1983), vol. 7, pp. 133–198; G. Di Chiara and A. Imperato, *Ann. N. Y. Acad. Sci.* **473**, 367 (1986).

38. T. R. Kosten et al., *Life Sci.* **44**, 887 (1989).

39. B. J. Winer, *Statistical Principles in Experimental Design* (McGraw-Hill, New York, 1971).

40. Supported in part by grants DA-02519, DA-04059, DA-00101, DA-00064, and DA-00115 from the National Institute on Drug Abuse, ADAMHA, and grant RR-05484 from NIH. We thank N. Diaz-Migoyo, M. Kaviani, and K. Hunt for technical assistance, and J. Drieze for data analysis and graphics preparation.

7 March 1989; accepted 12 May 1989

Glossary

analgesic - pain killing

aseptic - not sterile; not antiseptic

buprenorphine - a mixed opioid agonist-antagonist, which has been used successfully in the treatment of opiate addiction

catheter - an implanted tube through which fluids can be removed or injected into the body

dopamine - one of the neurotransmitter chemicals; dopaminergic neurons are thought to play a role in mediating the rewarding effects of cocaine and heroin

double-blind - a study in which the patient does not know which treatment that she or he is receiving (drug or placebo) and neither does the experimenter

endogenous - occurring naturally in the body

heroin - a semisynthetic opiate; heroin is produced by a minor modification in the morphine molecule; heroin penetrates the blood-brain barrier more readily than morphine, and thus it is a more potent psychoactive agent

methadone - a synthetic opiate that is given to opiate addicts as a replacement for their morphine or heroin; addicts are maintained on controlled doses of methadone to keep them from experiencing the illness associated with morphine or heroin withdrawal

mixed opioid agonist-antagonist - any drug that can increase some of the effects of opiates and decrease others; for example, buprenorphine

naloxone - an opiate antagonist

naltrexone - an opiate antagonist

opioid - pertaining to opiates; an opiate is any psychoactive chemical derived from the opium poppy (e.g., morphine) or any synthetic (e.g., methadone) or semisynthetic (e.g., heroin) drug that has effects similar to those of naturally occurring opiates

opioid agonist - any drug that increases the effects of opiates or produces similar effects

opioid antagonist - any drug that decreases the effects of opiates; for example, naloxone and naltrexone

pharmacotherapy - the treatment of a disorder by administering drugs

second-order schedule - a schedule of reinforcement that is the product of two different schedules; for example, in this experiment, a light came on after an average of every 16th response, and every fourth light was accompanied by a reinforcement (see footnote 26 in the article)

self-administration - an animal model of human drug addiction; laboratory animals readily learn to perform an operant response (e.g., a lever press) to inject drugs into themselves; drugs to which humans tend to become addicted tend to be the same ones that are self-administered by laboratory animals

Essay Study Questions

1. What would be an ideal pharmacotherapy for cocaine abuse?

2. What was known about buprenorphine prior to this experiment?

3. Why did the authors include the food-pellet condition in this experiment? How would the authors have had to change their interpretation if the monkeys had stopped responding for food as well as cocaine?

4. What evidence suggests that buprenorphine is a safe drug?

5. What evidence suggests that buprenorphine may be useful for treating combined cocaine and heroin abuse?

Multiple-Choice Study Questions

1. Which drug is commonly used in the treatment of heroin abusers?
    a. desipramine
    b. methadone
    c. dopamine
    d. aseptic
    e. morphine

2. Clinical studies have indicated that buprenorphine is
    a. effective against heroin self-administration in inpatient studies.
    b. effective in blocking some opiate effects for more than 24 hours.
    c. less reinforcing than other opiates.
    d. not lethal even at doses 10 times greater than an analgesic dose.
    e. all of the above

3. Which drug is an opiate antagonist?
    a. naloxone
    b. methadone
    c. naltrexone
    d. all of the above
    e. both a and c

4. Buprenorphine
    a. increased responding for food pellets.
    b. produced a substantial weight loss.
    c. eliminated responding for food pellets although it did not influence the consumption of other available foods.
    d. both b and c
    e. none of the above

5. Which of the following is not a factor in the results of those experiments on the treatment of drug abuse that employ the animal drug self-administration model?
    a. compliance
    b. multiple drug use
    c. side effects
    d. both a and b
    e. both b and c

6. Mello et al. think that buprenorphine reduces the rewarding effects of cocaine by its action on
    a. catecholaminergic neurons.
    b. dopaminergic neurons.
    c. noradrenergic neurons.
    d. adrenergic neurons.
    e. both b and c

The answers to the preceding questions are on page 290.

Food-For-Thought Questions

1. What experiment do you recommend that these authors do next? What procedure do you recommend?

2. Methadone is an opioid agonist that has many of the same effects as heroin, except that it does not produce the same degree of euphoria. Why do you think that it is so widely prescribed for opiate addicts? Do you think that there is any advantage in being a methadone addict rather than a heroin addict?

ARTICLE 21

Lesions of the Nucleus Accumbens in Rats Reduce Opiate Reward but Do Not Alter Context-Specific Opiate Tolerance

J.E. Kelsey, W.A. Carlezon, Jr., and W.A. Falls
Reprinted from Behavioral Neuroscience, 1989, volume 103, 1327-1334.

It has been hypothesized that the dopaminergic neurons that have their cell bodies in the ventral tegmental area (VTA) of the midbrain and their axon terminals in the nucleus accumbens of the forebrain mediate the rewarding effects of opiate drugs (e.g., morphine and heroin). In Experiment 1 of the following article, Kelsey, Carlezon, and Falls used the conditioned place-preference paradigm to test this hypothesis. They observed that control rats that experienced the effects of a morphine injection in one compartment of a two-compartment box tended to prefer that compartment when they were tested the next day drug free, but that rats with bilateral nucleus accumbens lesions did not.

Kelsey, Carlezon, and Falls pointed out that although the results of their first experiment were consistent with the hypothesis that rats with nucleus accumbens lesions are less capable of experiencing the rewarding effects of morphine, there is another equally tenable explanation for the results: perhaps the rats with nucleus accumbens lesions were capable of experiencing the rewarding effects of morphine but were incapable of learning the association between the effects of morphine and the environmental context. To choose between these two possible interpretations of Experiment 1, Kelsey, Carlezon, and Falls assessed the degree to which nucleus accumbens lesions blocked the development of context-specific tolerance to morphine. They found that intact rats that received a tolerance test in the same environment in which they had previously experienced four morphine injections were more tolerant to the analgesic effects of morphine than were intact rats that had experienced the four morphine injections in a different environment, and that nucleus accumbens lesions did not diminish the magnitude of this context-specific tolerance. Accordingly, nucleus accumbens lesions do not appear to eliminate the ability of rats to associate morphine effects with their environmental context.

The results of these two experiments, when considered together, support the hypothesis that lesions of the nucleus accumbens reduce an animal's ability to experience the rewarding effects of opiates.

Lesions of the Nucleus Accumbens in Rats Reduce Opiate Reward but Do Not Alter Context-Specific Opiate Tolerance

John E. Kelsey, William A. Carlezon, Jr., and William A. Falls
Bates College

Bilateral electrolytic lesions of the nucleus accumbens in rats eliminated the capacity of 10 mg/kg morphine to produce a conditioned place preference (Experiment 1). However, these lesions did not alter the capacity to establish context-specific tolerance to the analgesic effects of 5 mg/kg of morphine (Experiment 2). This latter finding indicates that rats with nucleus accumbens lesions are not impaired in associating the effects of morphine with a particular location. Thus, the failure of morphine to produce a conditioned place preference in these lesioned rats probably cannot be attributed to an inability to associate the effects of morphine with a particular chamber, i.e., the initially nonpreferred chamber. Rather, morphine may fail to establish a conditioned place preference in these rats because nucleus accumbens lesions disrupt a pathway that is critical in mediating the rewarding effects of opiates.

Wise and Bozarth (1984) have argued that opiates, such as morphine, are rewarding, partly because they stimulate the dopamine-releasing neurons in the ventral tegmental area (VTA) that project rostrally to the nucleus accumbens of the forebrain (Lindvall & Björklund, 1974). Implicating the VTA, Bozarth and Wise (1981) found that rats learn to press a lever for injections of morphine directly into the VTA but not for injections into other areas that also contain high densities of opiate receptors, such as the periventricular gray and caudate nucleus (Bozarth & Wise, 1982). Similarly, Phillips and LePiane (1980) demonstrated that injections of morphine directly into the VTA were sufficient to establish a conditioned place preference, whereas injections dorsal to the VTA were not.

Evidence implicating the dopamine projection from the VTA to the nucleus accumbens, however, is not as compelling. For example, damage to the cell bodies of the nucleus accumbens by intra-accumbens injections of kainic acid has been reported to reduce intravenous morphine and heroin self-administration (Dworkin, Guerin, Goeders, & Smith, 1988; Zito, Vickers, & Roberts, 1985). However, Pettit, Ettenberg, Bloom, and Koob (1984) reported that more specific damage to the dopamine neurons that terminate in the nucleus accumbens by intra-accumbens injections of 6-hydroxydopamine (6-OHDA) did not markedly alter intravenous heroin self-administration. On the other hand, Spyraki, Fibiger, and Phillips (1983) reported that intra-accumbens injections of 6-OHDA reduced the ability of systemic heroin to produce a conditioned place preference. However, this reduction was modest, and in the absence of a lesioned group not injected with heroin, it is difficult to interpret.

To further examine the hypothesis that the nucleus accumbens is involved in mediating some of the rewarding effects of opiates, in Experiment 1 we compared the ability of both systemic morphine and saline to produce a conditioned place preference in rats with lesions of the nucleus accumbens. In Experiment 2, we examined the alternative hypothesis that, rather than specifically reducing opiate reward, nucleus accumbens lesions reduce the ability to associate the consequences of morphine with a particular environment.

Experiment 1: Conditioned Place Preference

In this experiment, we examined morphine's capacity to produce a conditioned place preference as an assay of the drug's rewarding effects so that we could avoid some of the practical and theoretical problems associated with other methods, such as self-administration (e.g., Mucha, van der Kooy, O'Shaughnessy, & Bucenieks, 1982). To develop a conditioned place preference, rats are typically allowed to establish a preference for one of two or three interconnecting chambers. The rats are then injected with a presumably rewarding drug, such as morphine, while they are confined to their nonpreferred chamber and then are retested for preference after the effect of the drug has worn off. The assumption is that the association of the rewarding properties of the drug with the nonpreferred chamber will cause the rats to increase their preference for that chamber. Thus, if the nucleus accumbens is involved in mediating some of the rewarding effects of morphine, then lesions of the nucleus accumbens should reduce the capacity of morphine to act as a reward and, thus, to produce a conditioned place preference.

Experiment 1 was based on research submitted to Bates College by William A. Carlezon, Jr., in partial fulfillment of the requirements of a BS degree in biopsychology. The data from both experiments were presented at the Society for Neuroscience Convention in New Orleans, Louisiana, 1987.

William A. Carlezon, Jr., is now at the Department of Biological Research, Hoechst-Roussel Pharmaceuticals, Somerville, New Jersey. William A. Falls is now at Yale University, New Haven, Connecticut.

Correspondence concerning this article should be addressed to John E. Kelsey, Department of Psychology, Bates College, Lewiston, Maine 04240.

To be certain that changes in place preference were due to the morphine, some rats were injected with saline, instead of morphine, while being confined to the nonpreferred chamber.

Method

Subjects. The 48 naive male Sprague-Dawley rats weighed 320–440 g at the time of surgery and were housed individually in clear Plexiglas cages with wood shavings for bedding in a colony that was lighted from 7 a.m. to 7 p.m. each day. Food and water were available ad lib except during testing.

Surgery. The rats were anesthetized with an injection of sodium pentobarbital (65 mg/kg i.p.) supplemented with ethyl ether. Bilateral lesions of the nucleus accumbens were performed in 24 rats by passing 1.0 mA of anodal current for 15 s through stereotaxically positioned electrodes (AP = 3.2 mm anterior to bregma, H = 7.2 mm below the dura mater, and L = ±1.7; Pellegrino & Cushman, 1967). The electrodes were no. 1 insect pins insulated with Epoxylite except for the flattened tip. The remaining 24 rats received one of three types of sham operations. Ten rats underwent surgery as described for the lesions, except that the electrodes were lowered only 6.2 mm below the dura mater and no current was passed through the electrodes. Nine rats received identical operations, except that the electrodes were not lowered into the brain, and 5 rats were simply anesthetized.

Apparatus. The 20 × 49 × 31 cm shuttle box was made of galvanized steel except for the clear Plexiglas top. The box was divided into two 20 × 24.3 cm interconnecting chambers by a partition that had a 7 × 8 cm opening in the center flush with the grid floor. Except for the Plexiglas top and grid floor, one chamber was painted black, and the other chamber was left the natural silver-gray tone of the metal. The silver-gray side was illuminated by a 6-W bulb housed on the center of the Plexiglas top, and a steel mesh screen with 1.0 × 1.0 cm openings was placed over the grid floor on that side. The grid floor was made of 0.6-cm-diameter stainless steel rods separated by 2.0 cm center to center and running across the width of the two chambers. This floor pivoted around a bearing located under the partition and was supported on both ends by adjustable weights and switches. The weights were adjusted for each rat so that when the center of the rat was between the fifth and sixth grids from the center of the box, the floor pivoted and tripped the switch on the opposite end, signalling a crossing response. Speakers located at both ends of the box supplied white noise of 75 dB (SPL). The room containing the shuttle box was illuminated by a 25-W red bulb.

Procedure. One to 4 weeks after surgery ($M \pm SE$ = 15.8 ± 1.0 days), each rat was weighed and transferred to the experimental room at approximately the same time every day (1–4 p.m.). After a 5-min adaptation period during which the shuttle box was cleaned with 0.003 M acetic acid, each rat was placed into the silver-gray, lighted chamber facing toward the passage into the black chamber. A microprocessor automatically recorded the number of crossing responses and the cumulative time (to the nearest 0.1 s) spent in each chamber during each 15-min (900-s) session. Preference testing was continued daily in this fashion until each rat established a stable preference for one chamber for 2 consecutive days. Rats were assumed to have a preference when they spent at least 67% of the time (more than 600 s) in one chamber (black or silver gray). *Stable preference* was defined as less than 200 s of variation between the 2 days.

On the following day, approximately half of the rats in each group were injected with 10 mg/kg i.p. morphine sulfate (0.67 cc/kg), and the other half were injected with an equal volume of isotonic saline. After 15 min, each rat was transferred to the experimental room for a 5-min adaptation period. Each rat was then confined to its nonpreferred chamber for 30 min. On the next day, all rats were injected with 0.67 cc/kg i.p. isotonic saline, moved to the experimental room after 15 min, and confined to the previously preferred chamber for 30 min after a 5-min adaptation period.

Changes in preference were assessed on the next (and final) day by following the procedure used during initial assessment of place preference. Each rat was taken from the colony, and after a 5-min adaptation period in the experimental room, each was placed in the middle of the silver-gray, lighted chamber facing the black chamber and allowed to cross freely between the two chambers for 15 min.

Histology. After testing, the lesioned rats were killed with ethyl ether and perfused intracardially with isotonic saline followed by a 10% formol–saline solution. Every fourth 64-μm-thick frozen section in the area of the lesion was saved, stained with cresyl violet, and examined under a microscope.

Data analysis. Because there were no differences in the behavior of the three sham-operated groups, their data were combined for subsequent analyses. The stability of the initial place preference as measured by the time spent in the nonpreferred chamber during the last 2 criterion days was analyzed by a Lesion (control and nucleus accumbens lesion) × Drug (saline and morphine) × Days (2) analysis of variance (ANOVA) with repeated measures on days. Subsequent analyses used the average for these 2 criterion days for each group. Changes in place preference as measured by the time spent in the nonpreferred chamber were analyzed by a Lesion (2) × Drug (2) × Days (initial and final preference) ANOVA with repeated measures on days.

Results

Histology. The lesions produced moderate damage to the anterior portion of the nucleus accumbens surrounding the anterior commissure (see Figure 1), and thus were likely to have damaged many of the dopamine terminals arising from the VTA (Fallon & Moore, 1978; Lindvall & Björklund, 1974). The lesions were centered at AP = 9.4, H = −1.0, and L = ±2.0 (Pellegrino & Cushman, 1967). Damage to the overlying striatum and the underlying preoptic area was minimal. The data from 6 rats were eliminated from the analyses because the lesions of these rats extended well outside the nucleus accumbens.

Behavior. The rats established a stable preference (mean difference of 74.7 ± 8.8 s between the 2 criterion days) for a chamber of the shuttle box in 3.2 ± 0.2 days ($M \pm SE$). There were no significant differences among the groups in the number of days required to establish a preference or in the stability of the preference. Also, there were no significant differences among the four groups in their initial preference for the black chamber (see Figure 2). In fact, none of the four groups had an average initial preference that differed from chance (450 s; $ps > .20$), indicating that approximately half of the rats in each group preferred the black chamber, whereas the other half preferred the silver-gray chamber (see Figure 2). However, when the preference for each rat was expressed in terms of the time spent in the nonpreferred chamber, whichever chamber that was, the rats in all four groups had strong and statistically similar initial preferences, spending approximately 22% of the time in their nonpreferred chamber (see Figure 3).

Further analysis of these place preference data (see Figure

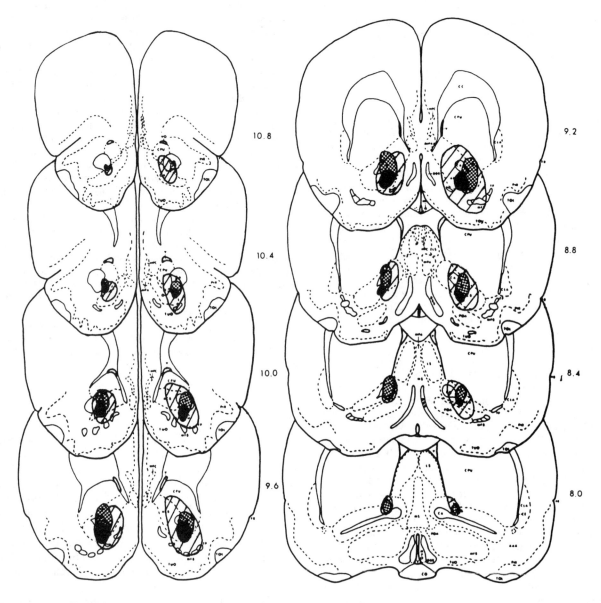

Figure 1. Reconstructions of the smallest (solid), typical (dotted), and largest (hatched) lesions of the nucleus accumbens. (From *A Stereotaxic Atlas of the Rat Brain* (pp. 7, 9, 11, 13, 15, 17, 19, and 21) by L. J. Pellegrino and A. J. Cushman, 1967, New York: Appleton-Century-Crofts. Copyright 1967 by Appleton-Century-Crofts. Adapted by permission.)

3), revealed a marginally significant Lesion × Drug × Days interaction, $F(1, 38) = 3.27$, $p < .08$. A subsequent simple Drug × Days interaction, $F(1, 38) = 11.51$, $p < .01$, indicated that the sham-operated control rats injected with morphine while being confined to the nonpreferred chamber increased their preference for the nonpreferred chamber more than did the control rats injected with saline in the nonpreferred chamber (see Figure 3). In contrast, subsequent simple interactions indicated that the rats with nucleus accumbens lesions injected with morphine while being confined to their nonpreferred chamber failed to increase their preference for that chamber more than did the lesioned rats injected with only saline, $F(1, 38) = 0.70$, and increased their preference less than did the control rats injected with morphine, $F(1, 38) = 7.74$, $p < .01$ (see Figure 3). There were no significant differences among the four groups in the number of times they crossed from one chamber to another during initial or final preference testing (see Table 1).

There were no significant differences in body weight among any of the four groups before surgery (see Table 2). However, a Lesion × Days interaction, $F(1, 38) = 6.81$, $p < .05$, indicated that the rats with nucleus accumbens lesions gained less weight following surgery (see Table 2). A Lesion × Drug × Days interaction, $F(1, 38) = 4.74$, $p < .05$, and subsequent Newman-Keuls tests indicated that whereas the sham-operated control rats injected with saline and both groups of

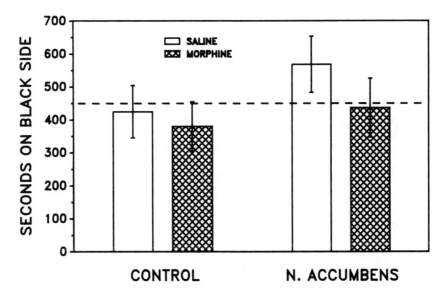

Figure 2. Preference of all four groups for the black chamber during initial testing. (The dotted line represents no preference. The error bars reflect ±1 *SE*. N. Accumbens = nucleus accumbens.)

rats with nucleus accumbens lesions gained equivalent amounts of weight during testing (*p*s < .05), the control rats injected with morphine lost weight (see Table 2).

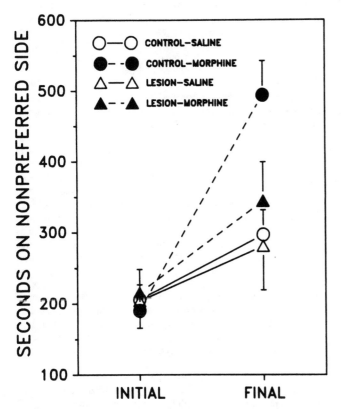

Figure 3. Preference of all four groups for the nonpreferred chamber during initial and final testing. (Initial represents the mean preference on the last 2 criterion days before drugs were administered, and Final represents the preference on the test day. The error bars reflect ±1 *SE*.)

Discussion

As predicted, electrolytic lesions of the nucleus accumbens eliminated the capacity of morphine to produce a conditioned place preference. Injections of morphine in the nonpreferred chamber increased the preference of the control rats for the nonpreferred chamber more than did injections of saline. However, similar injections of morphine into rats with nucleus accumbens lesions failed to increase their preference more than did injections of saline, and these injections increased the preference less than did morphine injections into control rats. Thus, this finding is consistent with and extends the finding of Spyraki et al. (1983) that 6-OHDA lesions of the nucleus accumbens reduced the ability of systemic heroin to produce a conditioned place preference.

The failure of morphine to alter the place preference of the rats with nucleus accumbens lesions more than saline altered it cannot be attributed to differences in initial preference, activity, or body weight because all four groups were statistically equivalent on these measures. Consequently, nucleus accumbens lesions may eliminate the opiate-induced conditioned place preference because, as hypothesized by Spyraki et al. (1983) and Wise and Bozarth (1984), the lesions disrupt a crucial component of a pathway mediating the rewarding effects of opiates.

However, the nucleus accumbens lesions may have blocked the formation of the conditioned place preference not by reducing opiate reward but by interfering with learning. In particular, these lesions may have interfered with the ability to associate the effects of morphine with a distinct environment, in this case the initially nonpreferred chamber. Indicating that nucleus accumbens lesions can interfere with learning in various settings, Taghzouti, Louilot, Herman, LeMoal, and Simon (1985) found that rats with bilateral 6-OHDA lesions of the nucleus accumbens showed impaired spontaneous alternation, deficient acquisition and reversal of a spatial discrimination, and perseveration during extinction.

Experiment 2: Context-Specific Tolerance

To examine the hypothesis that lesions of the nucleus accumbens reduce the ability of rats to associate the effects of morphine with a distinct environment, we examined the effects of the lesions on the ability to acquire context-specific tolerance to the analgesic effects of morphine. Siegel and his co-workers (e.g., Hinson & Siegel, 1982; Siegel, 1979, 1983), and others have shown that tolerance to morphine is greater if the rats are tested for tolerance in the environment in which they had been injected with morphine than if they are tested in a different environment. For example, Siegel, Hinson, and Krank (1978) found that morphine produced less analgesia in rats previously injected with morphine in the test environment than in rats previously injected with the same dosages of morphine in another environment or in rats being injected with morphine for the first time.

Although the precise conditioning mechanisms underlying this context-specific tolerance are in dispute (e.g., Baker & Tiffany, 1985; Falls & Kelsey, 1989; Paletta & Wagner, 1986; Siegel, 1979, 1983), there is general agreement that this context specificity is the result of learning to pair the effects of morphine with a specific environment, precisely the same kind of learning that was essential to produce a conditioned place preference in Experiment 1. However, there is one critical difference: Acquisition of context-specific tolerance to the analgesic effects of morphine presumably does not require that morphine be perceived as rewarding. Consequently, if nucleus accumbens lesions specifically interfere with the rewarding properties of morphine, as postulated by Spyraki et al. (1983) and Wise and Bozarth (1984), then these lesions should not interfere with the ability to acquire context-specific tolerance to morphine. If, on the other hand, nucleus accumbens lesions interfere with the formation of a conditioned place preference by reducing the ability to associate the consequences of morphine with a distinct environment, then nucleus accumbens lesions should also interfere with the ability to acquire context-specific morphine tolerance. This experiment was designed to evaluate these hypotheses by examining the effects of nucleus accumbens lesions on development of context-specific tolerance to the analgesic effects of morphine.

Table 1
Number of Crossing Responses During Initial and Final Preference Tests

Group	n	Initial test	Final test
Control			
Saline	11	18.0 ± 3.6	22.4 ± 3.9
Morphine	13	17.1 ± 2.4	19.5 ± 2.8
Lesion			
Saline	9	16.9 ± 3.1	15.7 ± 3.5
Morphine	9	17.7 ± 2.4	21.0 ± 3.5

Note. Values are $M \pm SE$.

Table 2
Body Weight on the Day of Surgery and on the First and Last Days of Testing

Group	n	Day of surgery	First day	Last day
Control				
Saline	11	360.6 ± 13.4	395.6 ± 10.7	406.9 ± 10.5
Morphine	13	381.3 ± 8.0	418.1 ± 9.5	415.8 ± 10.2
Lesion				
Saline	9	375.4 ± 9.4	399.3 ± 9.8	407.8 ± 8.5
Morphine	9	368.0 ± 7.7	372.2 ± 7.6	379.3 ± 7.6

Note. Values are in grams ($M \pm SE$).

Method

Subjects. Eight of the sham-operated control rats and 10 of the rats with nucleus accumbens lesions had served as subjects in Experiment 1. The remaining 17 rats were similar naive male Sprague-Dawley rats housed and treated as in Experiment 1.

Surgery. Nine of the naive rats received nucleus accumbens lesions, and 8 underwent sham operations as described in Experiment 1.

Apparatus. During testing for pain reactivity, each rat was placed in a 6.5-cm-diameter wire mesh restraining tube that had a 7.5 × 33.5 cm steel bottom. This tube was placed in a sliding Plexiglas base with a 0.6-cm-deep longitudinal depression trough in which the rat's tail was laid during testing. Radiant heat was supplied by a Sylvania EHL 300-W spotlight focused through a condenser lens. The light was adjusted with a rheostat to an intensity that reliably elicited tail flicks in approximately 5 s in undrugged rats (see Figure 4). Pain reactivity was tested in an illuminated room containing speakers that supplied white noise of 75 dB (SPL).

Procedure. Testing began 28.5 ± 6.1 days after the completion of Experiment 1 for the 18 rats used in that experiment and 19.9 ± 0.5 days after surgery for the 17 naive rats. The rats were run in pairs consisting of a control and a lesioned rat. Pain reactivity was measured by the latency of a rat to flick its tail away from a heat source (D'Amour & Smith, 1941). On the 1st day, the rats were placed in the restraining tubes and taken to the testing environment. A towel was placed over them to reduce distractions. After approximately 5 min of adaptation, a rat's tail was placed under the heat lamp. With a stopwatch, we measured pain reactivity as the latency to the nearest 0.1 s for the rat to flick its tail away from the heat source after the spotlight was turned on. A *tail flick* was defined as the rat moving the portion of its tail directly under the heat source more than 1 cm in any direction. If the rat did not move its tail within 10 s, the heat source was turned off to reduce damage to the tail. To further reduce damage to any one area of the tail, the point of focus of the heat source was varied along the posterior 7 cm of the tail by moving the restraining tube along the Plexiglas base before each trial. Trials occurred at 1-min intervals until consistent tail-flick latencies were obtained for four consecutive trials. The average latency for these four trials was recorded as the baseline latency for that rat.

Immediately after the test for pain reactivity, the rats were removed from the restraining tubes, and approximately half of the rats of each group were injected with 5 mg/kg s.c. morphine sulfate (0.33 cc/kg; paired animals), and the other half were injected with an equal volume of saline (unpaired animals). The rats were returned to the restraining tubes, covered with a towel, and left in the test environment for 1 hr. They were then returned to their home cages in the colony. Three hours later, the rats were injected in the colony with

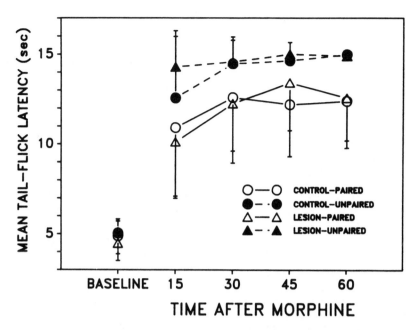

Figure 4. Tolerance to the analgesic effects of 5 mg/kg morphine in all four groups as measured by pain reactivity. (The error bars reflect ±1 *SE*.)

either 0.33 cc/kg s.c. saline (paired animals) or 5 mg/kg s.c. morphine in an equal volume (unpaired animals) and returned to their home cages. This procedure was followed for 4 consecutive days, with the exception that pain reactivity was measured only on the 1st day.

On the 5th day, all of the rats were placed in the restraining tubes and taken to the test environment as on the previous days. All rats were then injected with 5 mg/kg s.c. morphine sulfate (0.33 cc/kg) and tested for pain reactivity 15, 30, 45, and 60 min after the injection. At each interval, each rat received three trials spaced 1 min apart, and the latencies for the three trials were averaged.

Data analysis. Because there were no differences in behavior between the rats run in Experiment 1 and the rats run only in this experiment, we combined the data to form four groups. Tail-flick latencies after morphine treatment on Day 5 were analyzed by a Lesion (control and nucleus accumbens lesion) × Pairing (paired and unpaired) × Time After Morphine (15, 30, 45, and 60 min) ANOVA with repeated measures on time.

Results

Histology. The lesions were similar to those described in Experiment 1 (see Figure 1). The data from 3 rats were eliminated from the analyses because the lesions of these rats extended below the nucleus accumbens. Two of the rats already had been eliminated from Experiment 1.

Behavior. There were no differences among the four groups in baseline reactivity to pain before the drug injections (see Figure 4). However, after 4 days of morphine injections, the rats that had received morphine in the testing environment (paired animals) were more tolerant to morphine in that environment than were the rats that had received morphine in the colony (unpaired animals). In particular, a significant main effect of pairing, $F(1, 28) = 11.62$, $p < .01$, indicated that 5 mg/kg morphine produced less analgesia in the paired rats than it did in the unpaired rats (see Figure 4).

The lack of a significant main effect of the lesion or a significant Lesion × Pairing interaction ($Fs < 1.0$) indicated that this context-specific tolerance to morphine was not altered by lesions of the nucleus accumbens. The paired rats with nucleus accumbens lesions were as tolerant to the analgesic properties of morphine as were the paired sham-operated control animals, and the rats of both paired groups were equally more tolerant than were the rats of both unpaired groups (see Figure 4).

A significant main effect of pairing, $F(1, 28) = 7.24$, $p < .05$, indicated that, by chance, the paired rats weighed less than the unpaired rats (see Table 3). A significant Lesion × Days interaction, $F(1, 28) = 10.78$, $p < .01$, indicated that the control rats lost more weight than did the rats with nucleus accumbens lesions during the 5 days of this experiment (see Table 3).

Discussion

Clearly, the procedure used in the experiment was sufficient to produce context-specific tolerance to the analgesic effects of morphine. The rats tested for pain reactivity in the environment in which they had received morphine (paired animals) were more tolerant to the analgesic effects of morphine than were the rats that had received the same injections of morphine in a different environment (unpaired animals). To examine the possibility that this context-specific tolerance was due to the coincidental differences in body weight between the paired and unpaired rats, we eliminated the lightest rat from each of the two paired groups and the heaviest rat from each of the two unpaired groups. This selection eliminated the differences in weight between the paired and unpaired rats but did not alter the significant main effect of pairing on tail-flick latencies after morphine treatment, $F(1,$

Table 3
Body Weight on the First and Last Days of Testing for Context-Specific Tolerance

Group	First day	Last day
Control		
Paired	431.6 ± 12.4	406.6 ± 10.2
Unpaired	466.0 ± 9.8	428.5 ± 14.7
Lesion		
Paired	402.2 ± 15.9	395.4 ± 11.3
Unpaired	443.0 ± 18.6	430.0 ± 13.7

Note. Values are in grams ($M ± SE$); $n = 8$ for each group.

24) = 12.18, $p < .01$. Thus, the occurrence of context-specific tolerance clearly did not depend on initial differences in body weight.

More important is the finding that nucleus accumbens lesions had no effect on the ability to acquire context-specific tolerance. Insofar as the ability to acquire context-specific tolerance requires the ability to associate the effects of morphine with a particular context (Baker & Tiffany, 1985; Paletta & Wagner, 1986; Siegel, 1979, 1983), these data indicate that rats with nucleus accumbens lesions are as capable as control rats of learning to associate the consequences of morphine with a particular environment.

The additional finding that the rats with nucleus accumbens lesions lost less weight than did the controls during tolerance development suggests that morphine has less of a weight- or appetite-suppressing effect on rats with nucleus accumbens lesions. Thus, these data are consistent with the data of Experiment 1, which indicated that rats with nucleus accumbens lesions were less affected by the weight-reducing effects of morphine.

General Discussion

Experiment 1 demonstrated that moderately sized bilateral lesions of the nucleus accumbens eliminated the capacity of morphine to produce a conditioned place preference. Experiment 2 indicated that these lesions did not alter the capacity to produce context-specific tolerance to the analgesic effects of morphine. Insofar as the acquisition of context-specific tolerance depends on the ability to associate the effects of morphine with a particular environment, the latter finding indicates that nucleus accumbens lesions do not interfere with the formation of context–morphine associations. Therefore, the failure of morphine to establish a conditioned place preference in the rats with nucleus accumbens lesions in Experiment 1 probably was not due to a more general deficit in the ability to associate the effects of morphine with a particular environment, that is, the initially nonpreferred chamber. However, note that the conditioning procedures used to produce context-specific tolerance in the second experiment differed in several ways from those used to produce a conditioned place preference in the first experiment. Thus, for example, it is possible that nucleus accumbens lesions interfered with context–morphine associations after one conditioning trial (Experiment 1) and this associative deficit was overcome after three additional trials (Experiment 2). However, the finding that 6-OHDA lesions of the nucleus accumbens reduced the capacity of 2.0 mg/kg heroin to induce a place preference even after four conditioning trials (Spyraki et al., 1983) suggests that this possibility is unlikely. Consequently, by indicating that lesions of the nucleus accumbens apparently do not disrupt the formation of context–morphine associations, this study indirectly strengthens the hypothesis that nucleus accumbens lesions reduce the ability of morphine to establish a conditioned place preference because they destroy neurons that are essential for the rewarding effects of opiates. If this interpretation is correct, our study adds to the growing literature that one of the roles of the nucleus accumbens is to mediate the rewarding effects of opiates (Spyraki et al., 1983; Wise & Bozarth, 1984).

Further research is needed to determine whether these effects of nucleus accumbens lesions are due to damage to the dopaminergic fibers that project from the VTA (Spyraki et al., 1983; Wise & Bozarth, 1984), to damage to opiate-sensitive cell bodies in the nucleus accumbens (e.g., Olds, 1982; Vaccarino, Bloom, & Koob, 1985), or to damage to other systems (Spyraki, Nomikos, Galanopoulou, & Daïfotis, 1988). Further research is also needed to determine whether the nucleus accumbens is involved in mediating other effects of opiates, such as analgesia and withdrawal. The failure of nucleus accumbens lesions to alter baseline pain reactivity or the analgesic effects of morphine in this experiment suggests that the nucleus accumbens is not involved in mediating the analgesic effects of opiates. The additional finding that nucleus accumbens lesions did not alter the development of context-specific or context-independent tolerance suggests that this area is not involved in mediating opiate tolerance and therefore may not be involved in mediating opiate withdrawal. These data suggest that the nucleus accumbens may be rather specifically involved in mediating the rewarding (and weight-reducing) effects of opiates.

References

Baker, T. B., & Tiffany, S. T. (1985). Morphine tolerance as habituation. *Psychological Review, 92,* 78–108.

Bozarth, M. A., & Wise, R. A. (1981). Intracranial self-administration of morphine into the ventral tegmental area in rats. *Life Sciences, 28,* 551–555.

Bozarth, M. A., & Wise, R. A. (1982). Localization of the reward-relevant opiate receptors. In L. S. Harris (Ed.), *Problems of drug dependence* (Research Monograph No. 41). Rockville, MD: National Institute on Drug Abuse.

D'Amour, F. E., & Smith, D. L. (1941). A method for determining loss of pain sensation. *Journal of Pharmacology and Experimental Therapeutics, 72,* 74–79.

Dworkin, S. I., Guerin, G. F., Goeders, N. E., & Smith, J. E. (1988). Kainic acid lesions of the nucleus accumbens selectively attenuate morphine self-administration. *Pharmacology Biochemistry and Behavior, 29,* 175–181.

Fallon, J. H., & Moore, R. Y. (1978). Catecholamine innervation of the basal forebrain. IV. Topography of the dopamine projection to the basal forebrain and neostriatum. *Journal of Comparative Neurology, 180,* 545–580.

Falls, W. A., & Kelsey, J. E. (1989). Procedures that produce context-specific tolerance to morphine in rats also produce context-specific withdrawal. *Behavioral Neuroscience, 103,* 842–849.

Hinson, R. E., & Siegel, S. (1982). Nonpharmacological bases of drug tolerance and dependence. *Journal of Psychosomatic Research, 26,* 495–503.

Lindvall, O., & Björklund, A. (1974). The organization of the ascending catecholamine neuron systems in the rat brain as revealed by the glyoxylic acid fluorescence method. *Acta Physiologica Scandinavica Supplement, 412,* 1–48.

Mucha, R. F., van der Kooy, D., O'Shaughnessy, M., & Bucenieks, P. (1982). Drug reinforcement studied by the use of place conditioning in rat. *Brain Research, 243,* 91–105.

Olds, M. E. (1982). Reinforcing effects of morphine in the nucleus accumbens. *Brain Research, 237,* 429–440.

Paletta, M. S., & Wagner, A. R. (1986). Development of context-specific tolerance to morphine: Support for a dual-process interpretation. *Behavioral Neuroscience, 100,* 611–623.

Pellegrino, L. J., & Cushman, A. J. (1967). *A stereotaxic atlas of the rat brain.* New York: Appleton-Century-Crofts.

Pettit, H. O., Ettenberg, A., Bloom, F. E., & Koob, G. F. (1984). Destruction of dopamine in the nucleus accumbens selectively attenuates cocaine but not heroin self-administration in rats. *Psychopharmacology, 84,* 167–173.

Phillips, A. G., & LePiane, F. G. (1980). Reinforcing effects of morphine microinjection into the ventral tegmental area. *Pharmacology Biochemistry and Behavior, 12,* 965–968.

Siegel, S. (1979). The role of conditioning in drug tolerance and addiction. In J. D. Keehn (Ed.), *Psychopathology in animals: Research and clinical implications* (pp. 143–168). New York: Academic Press.

Siegel, S. (1983). Classical conditioning, drug tolerance, and drug dependence. In R. G. Smart, F. B. Glaser, Y. Israel, H. Kalant, R. E. Popham, & W. Schmidt (Eds.), *Research advances in alcohol and drug problems* (pp. 207–246). New York: Plenum Press.

Siegel, S., Hinson, R. E., & Krank, M. D. (1978). The role of predrug signals in morphine analgesic tolerance: Support for a Pavlovian conditioning model of tolerance. *Journal of Experimental Psychology: Animal Behavior Processes, 4,* 188–196.

Spyraki, C., Fibiger, H. C., & Phillips, A. G. (1983). Attenuation of heroin reward in rats by disruption of the mesolimbic dopamine system. *Psychopharmacology, 79,* 278–283.

Spyraki, C., Nomikos, G. G., Galanopoulou, P., & Daïfotis, Z. (1988). Drug-induced place preference in rats with 5,7-dihydroxytryptamine lesions of the nucleus accumbens. *Behavioural Brain Research, 29,* 127–134.

Taghzouti, K., Louilot, A., Herman, J. P., LeMoal, M., & Simon, H. (1985). Alternation behavior, spatial discrimination, and reversal disturbances following 6-hydroxydopamine lesions in the nucleus accumbens of the rat. *Behavioral and Neural Biology, 44,* 354–363.

Vaccarino, F. J., Bloom, F. E., & Koob, G. F. (1985). Blockade of nucleus accumbens opiate receptors attenuates intravenous heroin reward in the rat. *Psychopharmacology, 86,* 37–42.

Wise, R. A., & Bozarth, M. A. (1984). Brain reward circuitry: Four circuit elements "wired" in apparent series. *Brain Research Bulletin, 12,* 203–208.

Zito, K. A., Vickers, G., & Roberts, D. C. S. (1985). Disruption of cocaine and heroin self-administration following kainic acid lesions of the nucleus accumbens. *Pharmacology Biochemistry and Behavior, 23,* 1029–1036.

Received September 29, 1988
Revision received May 9, 1989
Accepted June 14, 1989 ∎

Glossary

6-hydroxydopamine (6-OHDA) - a neurotoxin that is used to lesion dopaminergic neurons (and noradrenergic neurons); when 6-OHDA is injected into an area of the brain, it is taken up by any dopaminergic neurons (and noradrenergic neurons) in the vicinity and selectively destroys them

analgesic - pain reducing

bregma - the intersection between two major skull sutures (i.e., seams), which is often used as a reference point (i.e., a point from which measurements are taken) for stereotaxic surgery

caudate nucleus - a forebrain nucleus that has a high density of opiate receptors

conditioned place-preference test - a method of measuring the rewarding properties of drugs; on the training day or days the subject experiences the effects of a drug in one chamber of a two- or three-chamber box; later, the degree to which the subject's tendency to prefer the drug chamber is assessed when the subject is drug free

context-specific tolerance - drug tolerance that is expressed only when the subject receives the drug in the same context (i.e., the same environment) in which it previously experienced the drug's effects

electrolytic lesions - lesions that are produced by the passage of electric current

heroin - a powerful semisynthetic opiate; heroin is manufactured by making a slight chemical modification to the morphine molecule, which increases its ability to penetrate the blood-brain barrier

intracardially - into the heart

isotonic - a solution that is of the same concentration as the bodily fluids; when an isotonic solution is injected into an organism, it does not cause a net flow of water into or out of cells

kainic acid - a neurotoxin that is used to lesion nuclei (i.e., clusters of cell bodies) in the central nervous system; its main advantage is that when it is injected into a nucleus, it destroys neurons with cell bodies in the area without destroying neural fibers that are merely passing through it

morphine - the major psychoactive ingredient of the opium poppy

nucleus accumbens - a forebrain nucleus that receives dopaminergic projections from the ventral tegmental area; the nucleus accumbens is situated just below the striatum and just above the preoptic area

opiates - drugs derived from the opium poppy; or drugs with similar structures or effects

periventricular gray - an area of gray matter located in the midbrain around the third ventricle; the periventricular gray contains a high density of opiate receptors

rheostat - a variable resistor, which can be used to adjust the current flow in an electrical circuit

rostrally - toward the front end of a vertebrate

sham-operated subjects - subjects that are subjected to the same surgical procedures as the experimental subjects, except that the surgical manipulation that is the focus of the investigation is not performed; for example, in the experiments of Kelsey et al. sham-operated subjects were anesthetized and "opened up", but they received no lesions

sodium pentobarbital - a barbiturate drug that is commonly used as a general anesthetic for laboratory animals

tail-flick test - a test of analgesia in rats; the tip of the rat's tail is put under a heat lamp (or in a hot water bath) and the time that it takes for the rat to move its tail (i.e, to flick its tail) is taken as a measure of the rat's insensitivity to pain; analgesics such as morphine increase the tail-flick latency

tolerance - any reduction in a drug's effects that is caused by previous exposure to the drug

Essay Study Questions

1. Describe the neural pathway that has been hypothesized to mediate the rewarding effects of opiates. What evidence supports this hypothesis?

2. Describe the place-preference paradigm. Its main advantage is that the subjects are tested while drug free; explain.

3. What effects have nucleus accumbens lesions been found to have on morphine-conditioned place preference? Two different explanations of these results have been proposed; what are they?

4. What is context-specific tolerance?

5. What effect did lesions of the nucleus accumbens have on context-specific tolerance to the analgesic effect of morphine?

6. Describe the tail-flick test.

7. What were the major conclusions of Kelsey, Carlezon, and Falls? What further research do they recommend?

Multiple-Choice Study Questions

1. The nucleus accumbens receives major dopaminergic input from the
    a. caudate nucleus.
    b. periventricular gray.
    c. ventral tegmental area.
    d. preoptic area.
    e. both a and b

2. Which of the following agents is used by biopsychologists and other neuroscientists to produce experimental brain lesions.
    a. kainic acid injections
    b. electricity
    c. 6-OHDA injections
    d. all of the above
    e. both a and c

3. The conditioned place-preference test is commonly used as a measure of
    a. analgesia.
    b. conditionability.
    c. drug tolerance.
    d. dopamine function.
    e. none of the above

4. An increase in tail-flick latency indicates
    a. decreased sensitivity to pain.
    b. analgesia.
    c. decreased reward.
    d. increased sensitivity to pain.
    e. both a and b

5. Which of the following drugs is an opiate?
    a. sodium pentobarbital
    b. heroin
    c. morphine sulphate
    d. all of the above
    e. both b and c

6. In Experiment 2, the experiment on context-specific tolerance, the rats in the control-paired group displayed significantly more tolerance to the analgesic effects of morphine than did the rats in the
    a. lesion-unpaired group.
    b. control-unpaired group.
    c. lesion-paired group.
    d. both a and b
    e. none of the above

The answers to the preceding questions are on page 290.

Food-For-Thought Questions

1. Considering the serious problems created by opiate addicts in many societies, do you think that voluntary surgical treatment would be warranted in some cases? Based on the results of Kelsey, Carlezon, and Falls, what surgical treatment would you recommend? Discuss the ethics of your proposed surgical treatment for opiate addiction.

2. Research on context-specific tolerance suggests that an addict who always injects his drug in the same context would be more susceptible than usual to death from overdose if he injected the drug in a different context. Explain and discuss.

ARTICLE 22

The Corpus Callosum Is Larger With Right-Hemisphere Cerebral Speech Dominance

J. O'Kusky, E. Strauss, B. Kosaka, J. Wada, D. Li, M. Druhan, and J. Petrie
Reprinted from Annals of Neurology, 1988, volume 24, 379-383.

In the midnineteenth century, the study of the effects of unilateral brain damage (i.e., brain damage restricted to one side of the brain) revealed that the two hemispheres do not contribute equally to the performance of certain kinds of cognitive tasks. Of particular interest was the discovery that the left hemisphere usually plays a greater role than the right hemisphere in language-related tasks. In 1949, the development of the sodium amytal test greatly facilitated the study of the cerebral asymmetry of speech. Injections of the anesthetic, sodium amytal into the carotid artery on the same side of the neck as the hemisphere dominant for language-related activities render the patient temporarily mute, whereas intracarotid injections to the other side produce only minor, momentary speech disturbances. The results of sodium amytal tests indicate that the left hemisphere is dominant for speech in more than 95% of right-handers and less than 70% of left-handers.

In the present study, O'Kusky and his collaborators measured the midsagittal cross-sectional area of the corpus callosums of 100 subjects: 50 epileptic subjects and 50 neurologically normal control subjects—the corpus callosum is the major fiber bundle connecting the left and right hemispheres. Its area was determined from magnetic resonance images. All 100 subjects had their speech laterality assessed by the verbal dichotic listening test, and because 44 of the epileptic subjects were candidates for brain surgery, they had their cerebral speech laterality assessed by the sodium amytal test, as well.

The total midsagittal area of the corpus callosums did not differ between the epileptic and control subjects, between left-handers and right-handers, or between females and males. However, the corpus callosums of subjects whose right hemispheres were dominant for speech were found to be larger than those of subjects whose left hemispheres were dominant for speech and those of subjects with bilateral speech representation. The mean callosal area in patients with right-hemisphere speech dominance was 18% greater than that of patients with left-hemisphere speech dominance and 28% greater than that of patients with bilateral speech representation. This suggests that persons with right-hemisphere speech dominance tend to have many more interhemispheric axonal fibers.

The Corpus Callosum Is Larger with Right-Hemisphere Cerebral Speech Dominance

John O'Kusky, PhD,* Esther Strauss, PhD,# Brenda Kosaka, MA,# Juhn Wada, MD,†‡ David Li, MD,§
Margaret Druhan, RN,†‖ and Julie Petrie, BA#

Variations in the size of the human corpus callosum were examined as a possible morphological substrate of functional asymmetries of the cerebral hemispheres, such as cerebral speech dominance. The midsagittal surface area of the corpus callosum, obtained by magnetic resonance imaging, was measured in 50 patients with epilepsy and 50 neurologically normal control subjects. The mean callosal area did not differ significantly between patients and control subjects, between left-handed and right-handed subjects, or between men and women. When measurements were compared among 44 patients, whose cerebral speech dominance had been determined by the intracarotid injection of sodium amytal, the area of the corpus callosum was significantly greater in patients with right-hemisphere cerebral speech dominance. The mean callosal area was greater by 109 to 159 square millimeters (18–28%) when compared to that of patients with either left-hemisphere speech dominance or bilateral speech representation. This difference in midsagittal surface area could represent as many as 37 to 54 million additional callosal axons in subjects with right-hemisphere cerebral speech dominance.

O'Kusky J, Strauss E, Kosaka B, Wada J, Li D, Druhan M, Petrie J. The corpus callosum is larger with right-hemisphere cerebral speech dominance.
Ann Neurol 1988;24:379–383

Functional asymmetries of the human cerebral hemispheres in the performance of certain cognitive tasks have been known since the mid-nineteenth century [1]. The most frequent pattern of hemisphere specialization, with linguistic-sequential tasks processed in the left hemisphere and spatial tasks processed in the right hemisphere, has been extensively reviewed in recent literature [2–4]. Patterns of functional lateralization that differ from the statistical norm, most often associated with handedness, gender, brain injury, or various learning disabilities, have become the subject of much research and speculation [1, 5–7].

Morphological asymmetries of the cerebral hemispheres exist in the posterior sylvian region and the frontal operculum, regions that coincide with the classic language areas of Wernicke and Broca, respectively. Studies of both gross anatomy and histology in human autopsy material indicate that these regions are larger in the left hemisphere than in the right hemisphere of most adult brains [8–10]. These observations have led to the hypothesis that such asymmetries are the neuroanatomical substrate of left-hemisphere cerebral speech dominance.

In a recent study of postmortem human brain, the midsagittal area of the corpus callosum was reported to be larger in left-handed and ambidextrous subjects than in consistently right-handed subjects [11]. It was concluded that the greater bihemispheric representation of cognitive functions in left-handers and mixed-handers [1] is associated with an increase in the number of callosal axons connecting the two hemispheres. However, these differences were not confirmed in subsequent studies [12, 13], where measurements of callosal area were made on midsagittal slices of human brain visualized by magnetic resonance imaging (MRI).

The extent to which callosal area varies as a function of cerebral speech dominance could be a confounding variable in studies in which subjects were classified according to handedness. Using intracarotid injection of sodium amytal to determine cerebral speech dominance, previous investigators have reported that more than 95% of right-handed persons have speech represented in the left hemisphere, as opposed to only 50 to 70% of left-handed persons [6, 14, 15]. In the present study the surface area of the corpus callosum was measured on midsagittal sections of human brain obtained

From the *Department of Pathology, Division of Medical Microbiology, the †Divisions of Neurological Sciences and ‡Neurology, and the Departments of §Diagnostic Radiology and ‖Psychiatry, University of British Columbia, Vancouver, BC; and the #Department of Psychology, University of Victoria, Victoria, BC, Canada.

Received Aug 18, 1987, and in revised form Dec 15, 1987, and Mar 16, 1988. Accepted for publication Mar 24, 1988.

Address correspondence to Dr O'Kusky, Division of Medical Microbiology, Department of Pathology, Room 364, C-Floor, Heather Pavilion, 2733 Heather St, Vancouver, BC, Canada V5Z 1M9.

Table 1. Comparison of the Total Midsagittal Area of Corpus Callosum in Epileptic Patients and Control Subjects in the Five Laterality Groups

	Laterality Groups[a]				
	LL	MLL	MRL	RL	MBL
Normal control subjects					
Callosal area (mm^2)	627.5 ± 27.6	700.2 ± 20.7	619.8 ± 25.0	615.2 ± 32.8	668.6 ± 30.8
Range	517.1–758.4	565.2–785.9	474.6–716.7	454.1–745.0	497.8–814.1
n	10	10	11	10	9
Male/female	5/5	5/5	6/5	5/5	5/4
Age	30.7 ± 2.8	27.5 ± 2.9	26.2 ± 1.2	27.8 ± 3.7	30.1 ± 3.8
Epileptic patients					
Callosal area (mm^2)	820.3	668.7 ± 81.1	614.2 ± 25.2	656.6 ± 31.2	608.9 ± 31.4
Range		466.4–838.0	349.1–796.6	488.4–822.9	464.1–794.1
n	1	4	20	13	12
Male/female	1/0	3/1	10/10	4/9	5/7
Age	46	29.8 ± 5.6	28.7 ± 1.9	25.4 ± 3.5	30.8 ± 3.8

[a]Values presented are the mean ± standard error.

LL = left-lateral; MLL = mixed left-lateral; MRL = mixed right-lateral; RL = right-lateral; MBL = mixed bilateral.

by MRI from subjects of known cerebral speech dominance and handedness. Differences in the size of the corpus callosum, which presumably reflect differences in the number of callosal axons projecting between the cerebral hemispheres, were investigated as a possible neuroanatomical substrate of functional asymmetries of the human cerebral hemispheres.

Materials and Methods

Subjects included 50 patients with medically refractory seizures (ranging in age from 11–57 years) and 50 neurologically normal control subjects (ranging in age from 17–60 years). Patients were excluded from the study if they had undergone previous neurosurgical treatment. Control subjects were selected after initial screening to exclude individuals with past episodes of head injury, hypoxic insult, or serious systemic illnesses. Control subjects were matched to the patient population for age, sex, and handedness as far as possible. Signed informed consent was obtained from each subject before participation in the study. Although patients with seizure disorders are not considered at risk for MRI [16], medical staff were in attendance for both patients and control subjects. Of the 50 patients with epilepsy, 44 had been investigated for possible surgical intervention with cerebral speech dominance determined by intracarotid injection of sodium amytal [17, 18]. Speech was mediated exclusively by the left hemisphere in 30 patients, by the right hemisphere in 5 patients, and by both hemispheres in 9 patients.

All subjects completed a standardized questionnaire to determine hand, foot, eye, and ear preference [19]. For sample questions such as "With which hand would you throw a ball? and "With which eye would you look through a telescope?" responses of "right," "left," or "both" were scored +1, −1, or 0, respectively. The criteria for classification were based on the average score in each of four categories of hand, foot, eye, and ear preference: *left-lateral*, average scores of −1; *mixed left-lateral*, average scores between 0 and −1; *mixed right-lateral*, average scores between 0 and +1; and *right-lateral*, average scores of +1. Of the 100 subjects, 11 were classified as consistently left-lateral, 14 as mixed left-lateral, 31 as mixed right-lateral, and 23 as consistently right-lateral (Table 1). The remaining 21 subjects were classified as *mixed bilateral*, for example, demonstrating consistent left-hand preference with consistent right-eye preference.

All subjects completed a free-recall verbal dichotic listening test [20]. Briefly, the test consisted of 22 trials, each containing three dichotic pairs of monosyllabic words. Subjects were instructed to report as many of the six words as possible on each trial. A laterality index was computed from left-ear (L) and right-ear (R) scores: $(R-L)/(R+L)$. This test was validated on a population of patients whose cerebral speech dominance had been determined using the carotid amytal test [20]. A negative index, indicating a left-ear advantage, suggests right-hemisphere speech representation. A positive index, indicating a right-ear advantage, suggests left-hemisphere speech dominance.

MRI was performed with a superconducting whole-body scanner (Picker International Cryogenic 2000, 0.15 Tesla). Midsagittal slices were 8.5-mm-thick inversion recovery sequences (TR, 1,666 msec; TI, 400 msec). Image reconstruction was performed with a two-dimensional Fourier transformation method on a 256 × 256 matrix, with a pixel size of 1.1 mm. MRI scans were enlarged to a final magnification of ×1, and the outline of the corpus callosum was traced from the midsagittal section. The MRI films were coded to prevent experimenter bias during the tracing and measurement protocol. The total midsagittal surface area of the corpus callosum was measured using an image analysis system (IBM/BIOQUANT System IV, R&M Biometrics, Nashville, TN). In addition, the areas of five individual sections and the callosal thickness through four tangents were measured, as illustrated in Figure 1. Tracings of the corpus callosum and measurements of the midsagittal surface area were repeated by a second independent examiner. Correlation between the two sets of measurements for total midsagittal area was $r = 0.96$ ($p < 0.001$).

The statistical significance of differences between patients and control subjects and among the five laterality groups was

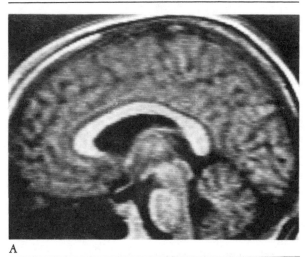

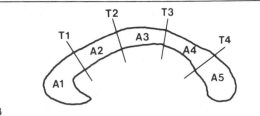

Fig 1. (A) Magnetic resonance image showing the midsagittal plane of the human corpus callosum. The genu (anterior) is oriented on the left with the splenium (posterior) on the right. (B) The tracing of this corpus callosum illustrates the five sections (A1–A5) employed in the measurement protocol. The length of the ventral surface was measured from the tip of the genu to the end of the splenium and divided into five equal segments. Tangents were drawn from the ventral surface at the points of junction for these five segments. The lengths of these tangents (T1–T4) were measured as estimates of callosal thickness.

tested using individual 2 × 5 (subject groups × laterality) analyses of variance [21]. The statistical significance of differences among the patients with documented cerebral speech dominance was determined using single-factor analysis of variance [21].

Results

Mean values for the total midsagittal area of the corpus callosum for patients and control subjects in the five laterality groups are presented in Table 1. Despite the wide range of callosal areas (349–838 mm^2), the within-group variability was reasonably small. The standard error expressed as a percentage of the mean varied only from 3 to 12% in the individual groups. Analyses of variance (subject groups × laterality) were performed for total callosal area, the areas of sections A1 to A5, and the lengths of tangents T1 to T4. The total area of the corpus callosum (see Table 1) did not differ significantly between epileptic patients and control subjects ($F = 0.92$; $df = 1,90$; $p = 0.34$) or among the five laterality groups ($F = 1.52$; $df = 4,90$; $p = 0.20$). No interactions of groups by laterality were significant ($F = 1.52$; $df = 4,90$; $p = 0.19$). No significant differences were found for any of the individual area or length measurements. A 2 × 5 (sex × laterality) analysis of variance for total callosal area revealed no significant differences between men and women ($F = 1.30$; $df = 1,90$; $p = 0.26$) or among the five laterality groups ($F = 1.00$; $df = 4,90$; $p = 0.41$). No interactions of gender by laterality were significant ($F = 0.79$; $df = 4,90$; $p = 0.54$). No significant differences were found between men and women for any of the individual area or length measurements.

When callosal measurements among the 44 patients with epilepsy of known cerebral speech dominance were compared (Table 2), the total area of the corpus callosum was found to be significantly greater in subjects having right-hemisphere speech dominance. The mean callosal area in patients with right-hemisphere speech dominance (RH) was 18% greater than that of patients with left-hemisphere speech (LH) and 28% greater than that of patients with bilateral speech representation (BL). The difference between LH and BL subjects was not statistically significant. Upon examination of the areas of sections A1 to A5, similar results were obtained for sections A2, A3, and A4, corresponding approximately to the body of the corpus callosum (Fig 2). In RH subjects the mean areas of these sections were 20 to 27% greater than in LH subjects and 30 to 55% greater than in BL subjects. The differences between LH and BL subjects were not statistically significant. The mean area of section A1, corresponding to the genu of the corpus callosum, tended to be larger in RH subjects than in LH (17%) or BL (29%) subjects, but these differences were of only borderline significance ($F = 3.09$; $df = 2,41$; $p = 0.055$). The thickness of the callosum through tangents T2 and T3 was also greater in RH subjects (see Table 2).

These results indicate that callosal area, particularly for the body of the corpus callosum, is significantly greater in epileptic patients with right-hemisphere cerebral speech dominance. Results of the dichotic listening test suggest that similar differences exist among the control subjects. There was a significant negative correlation between total callosal area and dichotic listening index ($r = -0.31$; $df = 48$; $p = 0.03$), demonstrating that a left-ear advantage, suggesting right-hemisphere speech representation, tends to be associated with a greater callosal area. Similarly, significant correlations were found between dichotic listening indices and the areas of sections A1 ($r = -0.28$; $df = 48$; $p = 0.04$) and A2 ($r = -0.30$; $df = 48$; $p = 0.03$).

The correlation between handedness scores and callosal area was not statistically significant ($r = -0.13$; $df = 48$; $p = 0.35$). When measurements of callosal

Table 2. Comparison of Callosal Area and Thickness Among Epileptic Patients with Left-Hemisphere, Right-Hemisphere, and Bilateral Cerebral Speech Dominance as Determined by Intracarotid Injection of Sodium Amytal

	Cerebral Speech Dominance[a]			ANOVA	
	Left-Hemisphere	Right-Hemisphere	Bilateral	$F(df = 2,41)$	p
Callosal area (mm²)					
Total area	622.2 ± 22.1[b]	731.1 ± 48.4[c]	571.7 ± 33.2	3.25	.048
A1	170.8 ± 6.5	199.8 ± 11.6	155.3 ± 8.6	3.09	.055
A2	100.0 ± 3.7[b]	119.7 ± 7.4[c]	92.1 ± 4.8	3.48	.039
A3	94.8 ± 3.6[b]	118.7 ± 5.6[d]	82.6 ± 5.7	5.85	.006
A4	88.8 ± 5.0[b]	113.1 ± 6.5[c]	73.1 ± 9.4	3.64	.034
A5	167.8 ± 8.0	179.8 ± 27.2	168.6 ± 13.4	0.19	.833
Callosal thickness (mm)					
T1	6.19 ± 0.27	7.40 ± 0.38	5.81 ± 0.39	2.32	.109
T2	5.72 ± 0.20[b]	6.80 ± 0.48[d]	5.13 ± 0.15	4.64	.015
T3	5.04 ± 0.27	5.36 ± 0.19[d]	3.78 ± 0.19	4.12	.023
T4	7.37 ± 0.34	8.04 ± 1.19	7.14 ± 0.88	0.32	.723
Age	30.1 ± 1.9	29.0 ± 4.9	25.9 ± 3.0	0.61	.552
n	30	5	9		
Male/female	14/16	2/3	3/6		

[a]Values presented are the mean ± standard error.
[b]$p < 0.05$.
[c]$p < 0.01$.
[d]$p < 0.001$.

ANOVA = analysis of variance.

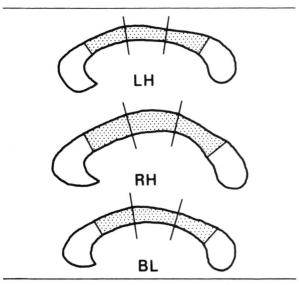

Fig 2. Tracings of the corpus callosum from 3 subjects whose cerebral speech dominance had been determined by the intracarotid injection of sodium amytal: left-hemisphere speech dominance (LH), right-hemisphere speech dominance (RH), and bilateral speech representation (BL). Callosal areas on these traces approximate the mean value for each group. The shaded portion corresponds to sections A2 through A4, and the tangents T2 and T3 are indicated.

area and thickness were regrouped according to handedness scores, using a more traditional classification scheme ([11, 22] e.g., right-, mixed-, and left-handed), the results were unchanged. No significant differences were found among right-handed, mixed-handed, and left-handed subjects for any of the individual area or length measurements.

Discussion

The mean callosal area in patients with right-hemisphere cerebral speech dominance was significantly greater by 109 to 159 mm² when compared to that of patients with either left-hemisphere or bilateral speech representation. Using published estimates of the number of axons per square millimeter of midsagittal callosal area in human postmortem brain [23], this difference may represent as many as 37 to 54 million additional callosal axons.

In animal studies of postnatal development of the corpus callosum, the number of callosal axons in neonates exceeds that of young adults, suggesting that normal development involves the remodeling of axonal projections between the two hemispheres with a subsequent elimination of callosal axons [24–27]. This reduction in the number of callosal axons, which reflects the selective elimination of axon collaterals or callosal neurons during the early postnatal period, can be manipulated experimentally by altering sensory or motor experience during early development [28–31].

It would be tempting to suggest that either trauma

to the developing human brain or environmental influences during the early postnatal period, which would be sufficient to retard the elimination of callosal axons, would influence the localization of speech in the right hemisphere. The greater incidence of right-hemisphere speech following early left-hemisphere lesions, which encroach upon the frontal and parietal speech areas [6], would tend to support this hypothesis. However, it remains to be determined that the increased callosal area, viewed on MRI scans in patients with right-hemisphere speech, actually corresponds to a greater number of callosal axons connecting the two hemispheres. A possible cause-effect relationship between developmental changes in the number of callosal axons and the incidence of right-hemisphere speech dominance remains an intriguing subject for future developmental studies.

We are grateful to the Medical Research Council of Canada, the Vancouver Society for Epilepsy Research, the Vancouver Foundation, Dr W. Koerner, and the University of British Columbia Health Sciences Centre Hospital for financial support of this project, and to Dr P. L. McGeer for helpful discussions. Dr O'Kusky is a Research Scholar of the Medical Research Council of Canada.

References

1. Bryden MP. Laterality, functional asymmetry in the intact brain. New York: Academic, 1982
2. Corballis MC. Human laterality. New York: Academic, 1983
3. Geschwind N, Galaburda AM. Cerebral dominance, the biological foundations. Cambridge, MA: Harvard University Press, 1984
4. Heilman KM, Valenstein E. Clinical neuropsychology. Oxford, UK: Oxford University Press, 1985
5. Hynd GW, Cohen M, eds. Dyslexia: neuropsychological theory, research and clinical differentiation. New York: Grune and Stratton, 1983
6. Rasmussen T, Milner B. The role of early left-brain injury in determining lateralization of cerebral speech functions. Ann NY Acad Sci 1977;299:355–369
7. Witelson SF. On hemispheric specialization and cerebral plasticity from birth. In: Best C, ed. Hemispheric function and collaboration in the child. New York: Academic, 1985
8. Geschwind N. The anatomical basis of hemisphere differentiation. In: Diamond SJ, Beaumont JG, eds. Hemisphere function in the human brain. London: Elek, 1974
9. Strauss E, Kosaka B, Wada J. The neurological basis of lateralized cerebral function: a review. Human Neurobiol 1983;2:115–127
10. Witelson S. Anatomic asymmetry in the temporal lobes: its documentation, phylogenesis, and relationship to functional asymmetry. Ann NY Acad Sci 1977;299:328–354
11. Witelson SF. The brain connection: the corpus callosum is larger in left-handers. Science 1985;229:665–668
12. Kertesz A, Polk M, Howell J, Black SE. Cerebral dominance, sex, and callosal size in MRI. Neurology 1987;37:1385–1388
13. Nasrallah HA, Andreasen NC, Coffman JA, et al. The corpus callosum is not larger in left-handers. Soc Neurosci Abstr 1986;12:720
14. Branch C, Milner B, Rasmussen T. Intracarotid sodium amytal for the lateralization of cerebral speech dominance. J Neurosurg 1964;21:399–405
15. Strauss E, Wada J. Lateral preferences and cerebral speech dominance. Cortex 1983;19:165–177
16. National Radiological Protection Board Ad Hoc Advisory Group on Nuclear Magnetic Resonance Clinical Imaging. Revised guide on acceptable limits to exposure during nuclear magnetic resonance clinical imaging. Br J Radiol 1983;56:974–977
17. Wada J. A new method for the determination of the side of cerebral speech dominance. Igaku to Seibutsugaku [Medicine and Biology] 1949;14:221–222
18. Wada J, Rasmussen T. Intracarotid injection of sodium amytal for the lateralization of cerebral speech dominance. J Neurosurg 1960;17:226–282
19. Porac C, Coren S. Lateral preferences and human behavior. New York: Springer, 1981
20. Strauss E, Gaddes WH, Wada J. Performance on a free-recall verbal dichotic listening task and cerebral speech dominance determined by the carotid amytal test. Neuropsychologia 1987;25:747–753
21. Winer BJ. Statistical principles in experimental design. New York: McGraw-Hill, 1971
22. Annett M. The binomial distribution of right, mixed and left handedness. Q J Exp Psychol 1967;19:327–333
23. Blinkov SM, Glezer II. The human brain in figures and tables. New York: Plenum, 1968
24. Feng JZ, Brugge JF. Postnatal development of auditory callosal connections in the kitten. J Comp Neurol 1983;214:416–426
25. Innocenti GM, Fiore L, Caminiti R. Exuberant projection into the corpus callosum from the visual cortex of newborn cats. Neurosci Lett 1977;4:237–242
26. Ivy GO, Killackey HP. The ontogeny of the distribution of callosal projection neurons in the rat parietal cortex. J Comp Neurol 1981;195:367–389
27. LaMantia AS, Rakic P. The number, size, myelination, and regional variation of axons in the corpus callosum and anterior commissure of the developing rhesus monkey. Soc Neurosci Abstr 1984;10:1081
28. Innocenti GM, Frost DO. Abnormal visual experience stabilizes juvenile patterns of interhemispheric connections. Nature 1979;280:231–234
29. Juraska JM, Meyer M. Environmental, but not sex, differences exist in the gross size of the rat corpus callosum. Soc Neurosci Abstr 1985;11:528
30. Kopcik JR, Juraska JM, Washburne DL. Sex and environmental effects on the ultrastructure of the rat corpus callosum. Soc Neurosci Abstr 1986;12:1218
31. Lund RD, Mitchell DE, Henry GH. Squint induced modification of callosal connections in cats. Brain Res 1978;144:169–172

Glossary

axon - the single long fiber extending from the cell body of a neuron; not all neurons have axons

axon collaterals - branches extending from the main axon

Broca's area - a loosely defined area in the left frontal lobe of humans; one theory suggests that it is the center for the programming of speech production

cognitive - pertaining to complex mental processes such as thinking, imagining, and remembering

dichotic listening test - on each trial of this test, three pairs of words are presented in sequence through headphones; the words in each pair are presented simultaneously, one to each ear; at the end of each sequence, each subject reports any of the six words that she or he recalls; subjects who are left-hemisphere dominant for speech tend to recall more of the words presented to the right ear, and subjects who are right hemisphere dominant for speech tend to recall more of the words presented to the left ear

frontal operculum - an area of human frontal cortex; in the left hemisphere, it roughly corresponds to Broca's area

genu - the front, curved portion of the corpus callosum

hypoxic - pertaining to hypoxia or lack of oxygen

intracarotid - into a carotid artery; there are two carotid arteries in the neck, one feeding the left hemisphere and one feeding the right hemisphere

magnetic resonance imaging (MNI) - a technique for visualizing the living human brain; it provides clearer images than computerized axial tomography (CAT)

midsagittal - refers to the central plane that divides the brain into two equal halves, left and right

neonates - newborn organisms

posterior - towards the back of the brain

postmortem - after death

sodium amytal - an anesthetic drug

sodium amytal test - a test in which sodium amytal is injected first into one carotid artery and then, after a period of time, into the other; each of the two sodium amytal injections temporarily anesthetizes the hemisphere on the side of the injection, thus the degree of speech disruption following each of the two injections provides a basis for determining the cerebral lateralization of speech; this test is not normally given to healthy subjects; it is used to determine the speech lateralization of people who are candidates for brain surgery

Sylvian - pertaining to the lateral fissure of the brain, which is also called the Sylvian fissure

Wernicke's area - a loosely defined area in the temporal lobe of the left hemisphere just below the Sylvian fissure; one theory suggests that it is the center for language comprehension

Essay Study Questions

1. Describe the relation between handedness and cerebral speech lateralization in the general population.

2. Compare the sodium amytal test and the dichotic listening test. In what situations are these respective tests typically used?

3. The authors speculate that early brain trauma causes larger callosums. Explain.

4. How did the authors assess hand, foot, eye, and ear preference? Would direct observation of the behavior of their subjects been better?

5. How did the authors determine the size of the corpus callosum of each of their subjects?

Multiple-Choice Study Questions

1. The first influential study of the functional asymmetry of the human cerebral hemispheres was published in the
    a. 1940s.
    b. 1950s.
    c. 1980s.
    d. mid 1200s.
    e. none of the above

2. Frontal operculum is to posterior Sylvian cortex as
    a. left hemisphere is to right hemisphere.
    b. Broca is to Wernicke.
    c. O'Kusky is to Wada.
    d. Pinel is to a mollusk.
    e. temporal lobe is to frontal lobe.

3. Which of the following is a test of cerebral speech laterality that is commonly given to patients who are candidates for brain surgery but is not normally given to healthy subjects?
    a. magnetic resonance imaging
    b. dichotic listening test
    c. sodium amytal test
    d. both a and c
    e. both b and c

4. Which of the following is a method of viewing the living human brain?
    a. sodium amytal test
    b. magnetic resonance imaging
    c. dichotic listening test
    d. postmortem examination
    e. linguistic-sequential test

5. Which of the following groups is likely to contain more people who are right-hemisphere dominant for speech than people who are left-hemisphere dominant for speech?
    a. right-handers
    b. left-handers
    c. ambidextrous females
    d. left-handed females
    e. none of the above

6. The size of the corpus callosum was estimated in this study
    a. by postmortem examination.
    b. by direct measurement of midsagittal tissue slices.
    c. from midsagittal MRI scans.
    d. both a and c
    e. none of the above

The answers to the preceding questions are on page 290.

Food-For-Thought Questions

1. O'Kusky and his collaborators were appropriately cautious in the interpretation of their results. They correctly pointed out that a larger corpus callosum does not necessarily mean more callosal axons. What are the other possibilities?

2. If you were a researcher interested in taking up the challenge of O'Kusky et al.'s results, how would you try to identify the factor that leads to the development of larger corpus callosums in people who are right hemisphere dominant for speech?

3. The results of this study are consistent with the inconsistency of previous reports of a relation between handedness and corpus callosum size. Explain.

ARTICLE
23

Reading With One Hemisphere

K. Patterson, F. Vargha-Khadem, and C.E. Polkey

Reprinted from Brain, 1989, volume 112, 39-63.

The left hemispheres of most left-handers and almost all right handers play the dominant role reading and other language-related tasks. What language-related skills are possessed by the right hemisphere? There are two ways of approaching this question: one is to assess the language abilities of the right hemisphere performing alone to determine what abilities it has; the other is to assess the language deficits of the left hemisphere performing alone to see what abilities it lacks. In their intriguing neuropsychological study, Patterson, Vargha-Khadem, and Polkey assessed the reading skills of two teenaged girls following the surgical removal of one entire hemisphere (hemispherectomy): N.I. was left with only a right hemisphere; H.P. was left with only a left hemisphere.

Not surprisingly, N.I. had difficulty reading after the removal of her language-dominant left hemisphere, but more important for understanding the capacities of her right hemisphere were the reading skills that she retained. N.I. retained the ability to perform some reading tests but not others: Her ability to match upper case and lower case letters was perfect, she was capable, with some difficulty, of naming individual numbers and letters, and she could recognize, and in some cases read, highly familiar concrete words. However, she was totally incapable of performing any task that required the translation of printed letters into sounds. For example, she could not give the sounds represented by individual letters, pronounce simple nonwords (e.g., "neg"), or combine individual speech sounds to form a word.

In contrast to N.I.'s meager postoperative reading abilities, H.P.'s reading abilities, although rather poor for her age, were not abnormally poor given the circumstances pathological. Patterson, Vargha-Khadem, and Polkey believe that H.P.'s poor reading performance was entirely attributable to the fact that she had suffered from an incapacitating disorder for 5 years prior to her operation and thus had had little chance to practice her reading skills during this period (i.e., when she was between 10 and 15 years old), when most children are consolidating and extending their reading skills. They also believe that the right hemisphere participates in the initial acquisition of reading skills and in the rapid performance of difficult reading tasks. Accordingly, removal of the right hemisphere from an individual (e.g., H.P.) who has already learned to read would be expected to produce few deficits on the tests of basic reading skill.

READING WITH ONE HEMISPHERE

by KARALYN PATTERSON, FARANEH VARGHA-KHADEM and
CHARLES E. POLKEY

(From the MRC Applied Psychology Unit, Cambridge, the Department of Developmental Paediatrics, Institute of Child Health and Department of Psychological Medicine, Hospital for Sick Children, London, and the Neurosurgical Unit, Maudsley Hospital, London)

SUMMARY

The subjects of this study were 2 originally right-handed teen-aged girls who had undergone complete hemispherectomy (1 left, 1 right) for intractable epilepsy. Both subjects had developed normal language and reading capacities before the onset of their illness. The reading performance of H.P. (whose right hemisphere had been removed), while not as advanced in level as that of a normal 17-yr-old, showed no abnormality in any subcomponent of reading skill. The reading performance of N.I. (whose left hemisphere had been removed) was poor, but with a pattern of retained and impaired subskills strikingly similar to adult deep dyslexic patients and to split-brain patients given reading tasks lateralized to the left visual field (right hemisphere). The results are discussed with regard to implications for the reading capacity of the nondominant right hemisphere and also its putative contribution to normal reading.

INTRODUCTION

The vast majority of right-handed people have left hemisphere (LH) dominance for language skills such as reading. It is an unresolved question, however, whether the right hemisphere (RH) of a LH-dominant individual also develops significant capacity for reading. There are two approaches to this question, and several sources of evidence within each. The first approach assesses reading performance when the RH must function predominantly on its own. Pertinent sources of evidence include: (1) patients with extensive LH lesions resulting from stroke or other cerebral lesion; (2) patients with surgical separation of the hemispheres who are given language input restricted to the RH; and (3) patients who have had the LH surgically removed for intractable epilepsy and/or progressive brain disease.

The second approach assesses the contribution of the RH to normal reading performance. One possible source of evidence here involves split-field presentation of written language to normal subjects; and many interesting results have emerged from the use of this paradigm (e.g., Hardyck *et al.*, 1985; Young and Ellis, 1985). However, while adequately controlled studies can ensure that material presented in the left visual field arrives initially in the RH, it is difficult both to control and

Correspondence to: Dr K. Patterson, MRC Applied Psychology Unit, 15 Chaucer Road, Cambridge CB2 2EF, UK.

© Oxford University Press 1989

to ascertain the locus of subsequent processing in subjects with two intact and connected hemispheres. Therefore one turns to neuropsychological evidence derived from the same three types of clinical populations, but now on the right side: the reading performance of adults with RH lesions, of split-brain patients given right visual field input, and of patients with right hemispherectomy. The logic of this approach is that, if the RH normally makes a significant contribution to skilled reading, then a person forced to rely exclusively or primarily on the LH should read abnormally.

With so many sources of evidence germane to this issue, why does it remain controversial? The answer, of course, is that none of the sources is unimpeachable. With both split-brain and hemispherectomy cases, the majority of patients have had congenital or early brain abnormalities, probably resulting in atypical lateralization of language functions (Rasmussen and Milner, 1977). This problem does not afflict interpretation of adult stroke patients. On the other hand, such patients with LH lesions virtually never have total LH destruction, leaving the question as to whether their residual language abilities derive primarily from the damaged LH or from the intact but linguistically inadequate RH (*see* Lambert, 1982, for further discussion of the difficulties inherent in these various sources of evidence).

The most compelling source of evidence on the question of RH reading would be provided by cases of hemispherectomy for brain disease with a late onset, after normal lateralization would be expected to have occurred. Such cases (e.g., Smith, 1966) are extremely rare, and were mainly studied before current standards for detailed and theoretically informed assessment had emerged. The exceptions are the hemispherectomy cases, one right and one left, extensively and elegantly studied by Zaidel (1977; 1978*a*, *b*; 1980; 1982). However, as age at onset of symptoms in these 2 cases was 6:7 and 7:8, respectively, these patients may still have been within a period of relative plasticity (although there is considerable controversy surrounding this notion, which will be addressed further in the Discussion). Furthermore, while Zaidel's case of left hemispherectomy is assuredly relevant to the question of language processing in the RH, she had essentially no reading ability (e.g., Zaidel, 1978*b*). Thus, apart from marking the lower end of its range, she cannot inform us about the particular linguistic skill of interest here.

The present study reports the reading performance of 2 hemispherectomy patients, one left and one right, with normal early development followed by onset of brain disease at approximate ages of 13 and 10 yrs. The framework guiding this investigation is the cognitive neuropsychological approach applied in recent years to the study of acquired reading disorders such as surface dyslexia (Patterson *et al.*, 1985) and deep dyslexia (Coltheart *et al.*, 1980). Indeed, deep dyslexia, first investigated in detail by Marshall and Newcombe (1966, 1973), is particularly germane to the current investigation. Deep dyslexic patients (most of whom have suffered large infarcts in the territory of the left middle cerebral artery) show a complex pattern of reading impairments including a total failure to pronounce even the simplest printed nonsense words (e.g., CAG); a considerable advantage in oral word reading for concrete, imageable nouns (e.g., CHICKEN) relative to either abstract nouns (e.g., CONTEXT) or grammatical words such as prepositions (e.g., AT, OF, BY); and a tendency, in single word oral reading, to produce error responses with a semantic (but no orthographic or phonological) resemblance to the target word (e.g., HERMIT—'recluse').

This pattern of acquired dyslexia is relevant in view of a suggestion by Coltheart (1980, 1983) and Saffran *et al.*, (1980) that deep dyslexic reading is most plausibly and economically understood as reading with the RH. This hypothesis has been subject both to support (e.g., Zaidel and Schweiger, 1984) and criticism (e.g., Warrington, 1981). Since neither the RH hypothesis of deep dyslexia nor any contending hypothesis has been able to give a wholly adequate account of the observed pattern of performance (*see* Coltheart *et al.*, 1987), the issue remains unresolved. This theoretical debate adds another focus of interest to our case of indisputable RH reading.

CASE REPORT

Case 1 (N.I.)

Until the onset of her illness at the age of 13 yrs, N.I. (born 1969) was a normal, right-handed girl, attending secondary school with performance in the average range, and with no personal or family history of neurological abnormality. In September 1982, she presented with a generalized convulsion preceded by lethargy and episodes of dysphasia for 6 weeks. Over the next 4 months she had further generalized attacks, one of these followed by transient weakness of the right-sided limbs. In May 1983, some 8 months after the first generalized convulsion, focal motor attacks involving the right arm and leg appeared, and within weeks these became epilepsia partialis continua. Attacks lasted 30–40 s and were generally preceded by an unpleasant taste in the mouth. On examination 1 yr after the onset of the illness, N.I. had a right hemiparesis (grade 4), an extensor right plantar response, a right homonymous hemianopia, and some dyspraxia. She had become left handed and was sufficiently disabled that she had to hold on to furniture to move around.

From the outset the attacks included dysphasia. Although speech and language functions were frequently interfered with as a result of the seizures and postictal states, N.I.'s speech only deteriorated radically some 14 months after the onset of her illness.

Diagnostic tests had consistently implicated N.I.'s left hemisphere. Electroencephalography (EEG) showed widespread LH abnormalities. Although CT scans initially did not indicate any gross changes, within 18 months postonset they reflected evidence of LH atrophy. The preoperative scan showed moderate left hemiatrophy with an area of low density in the left posterior region of an apparent ischaemic origin.

Preoperative psychometry showed scaled scores of (1) on both the vocabulary and comprehension subtests and scaled scores of (4) and (1) on block design and object assembly subtests of the WISC-R (Wechsler, 1976). Administration of the Schonell Graded Word Reading Test (Schonell, 1942) indicated a reading age of 6:5. Auditory comprehension for single words as measured by the British Picture Vocabulary Scale (Dunn and Dunn, 1982) was at the age level of 5:5.

A sodium amytal test (Branch *et al.*, 1964) was carried out preoperatively to determine whether N.I.'s RH was capable of subserving gross speech and memory functions. Injections were given on two separate days, and the LH was injected three times, the RH once. The evidence derived from these tests was inconclusive, but speech arrest coincident with the onset of hemiparesis was only evident when the right cerebral hemisphere was inactivated.

Before operation N.I. was having continuous seizures. She had a moderate expressive dysphasia, a complete right homonymous hemianopia, a right facial palsy, grade 2 cortical weakness of the right upper limb with no useful movement in the hand, and similar weakness of her right lower limb. She had cortical sensory loss involving the whole right side of the body, and increased tendon jerks and an extensor plantar response on that side.

A total left hemispherectomy using Adam's modification (Adams, 1983) was performed in 1985, 30 months after the onset of N.I.'s illness. The postoperative diagnosis was 'Rasmussen's encephalitis' (Rasmussen, 1978, 1983). This is a syndrome whose aetiology is obscure. It invariably begins in childhood or early adolescence, and is characterized by the onset of unilateral focal motor seizures followed by a progressive hemiparesis. The neuropathological features include widespread cortical lesions of an 'ischaemic' type, of variable size, and perivascular cuffing with inflammatory cells. It is this latter feature that has led to the use of the term 'encephalitis', although no aetiological agent has ever been implicated and the CSF during the illness shows no inflammatory changes. Histopathology in N.I. showed widespread abnormalities affecting the left neocortex. Major structural landmarks were either shrunken because of neuronal cell loss or grossly distorted as a result of the disease process.

Seizures were totally arrested following the operation and N.I. has remained seizure-free to date. She is still on a moderate dose of anticonvulsants. Her neurological state has remained unchanged. She has been attending a nonacademic college where she spends her time learning to cook and to do handicrafts.

N.I.'s spontaneous speech is very limited both in content and grammatical form, perhaps more anomic than classically agrammatic. Here is a sample of her speech, in conversation with one of the authors:

K.P.: What do you particularly like to eat? N.I.: I like ... er ... you know ... the ... I can't say it now ... well, I like ... I don't like chips a lot but ... I like Bolognaise. K.P.: Do you cook? N.I.: Yes. K.P.: What do you cook? N.I.: Different things. I made ... er ... I made ... these (pointing to the apple that she was eating). But not that ... these. ... K.P.: What did you do with them? N.I.: Crust over the top. K.P.: What's that called? N.I.: An apple. K.P.: Yes, that's an apple, but with the crust over the top, what do we call it? N.I.: I can't remember. K.P.: An apple ... N.I.: Pie!

Case 2 (H.P.)

H.P. was the product of a normal pregnancy and delivery. Shortly after her birth in 1970, she was diagnosed as suffering from Turner's syndrome. This chromosomal abnormality is not generally associated with intellectual deficit (Menkes, 1980), and indeed H.P. went on to develop normally and to achieve appropriate developmental milestones. She was right handed, and her early years were medically unremarkable. She attended a normal school locally and was performing in the above average range in all academic subjects until her epilepsy commenced. When she was nearly 10 yrs old, she was found totally unresponsive, lying on her left side with her eyes closed. A month later, a similar episode occurred, and over the ensuing months a pattern of minor seizures changed to focal attacks involving the left limbs. The severity of these increased such that within months she was admitted to hospital with epilepsia partialis continua and a moderate Todd's paralysis.

A CT scan carried out at this stage indicated some right cerebral atrophy. The EEG showed multifocal discharges maximal in the right rolandic region. The sleep EEG showed continuous epileptic discharges over the RH accompanied by movements of the left arm and leg. Neurological examination revealed increased tone on the left side with decreased power and coordination. Major neurosurgery was contemplated at this stage but the decision was postponed in favour of drug therapy. Over the ensuing 5 yrs, many attempts were made to bring the seizures under control using different regimes of anticonvulsant medications and corticosteroids, but without enduring success.

Preoperatively, H.P. was suffering from frequent seizures, up to six per hour, involving the left arm and leg with deviation of the head to the left. Neurological examination indicated an extensive RH lesion. She had a left homonymous hemianopia, cortical weakness of the left upper limb (grade

3) with no useful movement in the hand and similar weakness of her left leg, and left hemisensory loss with impaired joint position sense. There were increased tendon jerks on the left with a left extensor plantar response and clonus at the left ankle. Preoperative CT scanning had shown increasing right hemiatrophy. EEGs revealed gross abnormality with runs of high voltage sharp-slow wave complexes and no normal background activity.

Psychometry before operation yielded a verbal IQ of 82 and a performance IQ of 67. Verbal memory was mildly impaired and spatial memory skills were severely deficient.

A total right hemispherectomy (Adams' modification) was performed in 1985 and, on the basis of medical history and histopathological features, a diagnosis of 'Rasmussen's encephalitis' was made. Although macroscopic examination revealed a curious absence of focal lesions, detailed histology of the cortex showed varied and extensive degrees of damage and atrophy. In the most severely affected areas, there was narrowing of the full thickness of the cortex with neuronal loss, and astrocytic and vascular proliferation.

Following the operation H.P.'s seizures were totally arrested and she was weaned off anticonvulsant medication within the same year as the operation. She has been attending a local college of further education pursuing courses in English, mathematics and domestic science. After a year's placement in the special needs group, she has now been integrated into the regular stream.

RESULTS

Appendix 1 summarizes background information for both cases and provides a general description of the 2 patients. All the basic language assessments in Appendix 1 reveal poor performance for N.I. and substantially better abilities for H.P. H.P., however, does not perform at the level of a normal 17-yr-old, but more like an 11-yr-old child. This is perhaps not surprising in view of the fact that, over a 5-yr period, she increasingly suffered epileptic seizures (as frequently as six per hour shortly before her operation) which completely disrupted normal life.

Presentation of the results of our detailed testing on the reading performance of N.I. and H.P. will be organized into three main sections: (1) orthography; (2) from orthography to phonology; and (3) from orthography to semantics. Wherever possible, the same tests have been given to the 2 subjects, but in some specific instances this was either not necessary or not feasible. In particular, assessments of some skills (such as letter identification), relevant for N.I., were superfluous for H.P. given her success at more complex tasks that require this subskill; and some of the lists of words that H.P. was asked to read were so far beyond the reading capability of N.I. that their administration would have been unkind to her and uninformative to us.

Orthography

Letter naming, sounding and recognition (N.I. only)

Method. Individual upper case letters were presented in random order and N.I. was asked either to give each letter's name or the sound that it has in a word. Letter recognition was further tested with upper-lower case matching of pairs of letters (n = 24), words (n = 40) and nonwords (n = 40), printed on cards with one item above the other. One member of each pair was in lower case and one in upper case. Half of the pairs in each condition were composed of the same item (e.g., G—g, ROAD—road, FERB—ferb), for which the response should be 'yes', and half were composed of different items (e.g., G—y, ROAD—roar, FERB—ferd), for which the response should be 'no'.

Results. N.I. was very poor at naming randomly presented single letters. She usually performed the task by starting with the letter A and speaking her way through the alphabet, often getting lost and starting again at the beginning. As this strategy made the test very slow to administer, only 17 of the 26 letters were presented. She was correct (i.e., knew where to stop in the alphabet) on 7/17. Her 10 errors consisted of 4 'close' errors of the type that this strategy might be expected to produce (e.g., naming F as 'E' and H as 'J'), 1 other error (P—'H'), and 5 instances where she simply said 'I don't know'. She was worse at sounding than naming letters, producing the sound for S but for no other letter. It is quite clear, however, that these poor performances reflect difficulties in knowing the appropriate sounds, not in recognizing or identifying the letters. In the tests of upper-lower case matching, she was 100% correct with both letters and words, and 95% correct with nonwords.

Number reading

Method. Individual numbers from 1-4 digits in length, written in Arabic numerals, were presented for reading aloud.

Results. N.I. was much more accurate, although no faster, at naming numbers than letters. In fact, she made no errors in number reading; but no number higher than 23 was tested because she used the same serial strategy as with letters, starting at 1 and counting up. In contrast, H.P. was quick and accurate at reading not only 1 and 2-digit numbers but also multidigit numbers, dates, etc.

Word recognition

Method. Recognition of printed words was tested using the lexical decision paradigm, in which subjects are instructed to respond 'yes' if they recognize a letter string as a real word and 'no' if they do not. The first test consisted of 76 high frequency content and function words (all > 350/million, Kučera and Francis, 1967), 3-5 letters in length, plus 76 nonwords created by changing a single letter in each of the test words (e.g., SILCE from SINCE). The second list was derived from a lexical decision test designed by Rickard (1987) to evaluate the influences of word familiarity and word imageability on lexical decision performance. Both of these variables were determined by rating scores on 7-point scales (from the MRC Data Base, Coltheart, 1981); therefore familiarity in this case refers to subjective familiarity rather than objective frequency of word usage as in word frequency norms such as those of Kučera and Francis (1967). The complete test consists of 12 blocks, where three levels of Familiarity (High, Medium and Low) are crossed with four levels of Imageability (High, Medium, Low and Very Low). In view of the restricted educational level of our subjects, we administered only the eight blocks formed from High and Medium Familiarity words at the four levels of Imageability. Examples of these words are shown in Table 1. The nonwords in this test, like those in the first test, were created by changing a single letter in a test word; however the word and nonword members of an orthographic pair (e.g., NERVOUS and NERDOUS) never appeared in the same testing block. The number of real words per block varied between 15 and 30, with a total of 165 words (and an equal number of nonwords) over the eight blocks.

Results. Results for both subjects appear in Table 1. H.P.'s performance on Test 2 showed a significant influence of familiarity, with an average hit rate of 1.0

TABLE 1. LEXICAL DECISION PERFORMANCE*

	N.I.				H.P.			
	Hit rate	FP rate	d'	% correct	Hit rate	FP rate	d'	% correct
Test 1	0.93	0.16	2.46	89%	0.99	0.04	4.07	97%
Test 2	0.88	0.43	1.36	72%	0.90	0.08	2.68	92%

Word sets in Test 2	(n words)	Example	Hit rate N.I.	Hit rate H.P.
Hi Fam Hi Im	(20)	money	1.0	1.0
Hi Fam Med Im	(20)	brake	0.95	1.0
Hi Fam Lo Im	(30)	proof	0.97	1.0
Hi Fam V Lo Im	(20)	choose	1.0	1.0
Med Fam Hi Im	(20)	knight	0.75	0.95
Med Fam Med Im	(20)	curve	0.70	0.75
Med Fam Lo Im	(20)	eager	0.85	0.75
Med Fam V Lo Im	(15)	vague	0.73	0.67

The words in Test 1 were very high frequency content and function words. Those in Test 2 systematically varied in both Familiarity and Imageability. Hit Rate = the proportion of 'yes' responses to words; FP = False Positive Rate the proportion of 'yes' responses to nonwords; d' = the measure of sensitivity (basically a Z-score) derived from signal detection theory.

* Hi = high, Med = medium, Lo = Low, V Lo = very low, Familiarity (Fam) or Imageability (Im).

on High Familiarity words and of 0.79 on Medium Familiarity words, χ^2 (1 df) = 21.3, $P < 0.001$. Obviously no imageability effect was detectable within the High Familiarity words due to ceiling effects. Within Medium Familiarity words, although an unordered χ^2 test was not significant (χ^2, 3 df = 4.79, $P = 0.19$), an ordered trend test for the effect of imageability ($\bar{\chi}^2$) just reached significance ($P = 0.04$). Normal adult subjects performing this test (Rickard, 1987) have essentially perfect accuracy on the eight blocks used here; they begin to show small but reliable effects of both variables only in the four Low Familiarity blocks. Therefore, H.P. can be characterized as having a normal pattern of performance on this test; but, due to her restricted reading vocabulary, she shows both overall lower accuracy and 'earlier' sensitivity to the variables of familiarity and imageability than a normal adult. An interaction between frequency and concreteness in lexical decision was also shown to be characteristic of normal adults by James (1975).

In contrast, N.I. achieved reasonable (though never normal) discrimination between words and nonwords only given words of the highest possible frequency, as in the first test. In the second test, like H.P., N.I. showed a significant influence of familiarity: average hit rates on High and Medium Familiarity words were 0.98 and 0.76, respectively; χ^2 (1 df) = 18.2, $P < 0.001$. There was no evidence of an imageability effect. Given N.I.'s high acceptance rate on nonwords (FP rate = 0.43), it must be asked whether she showed statistically significant discrimination between Medium Familiarity words and nonwords. A comparison of hit rates on the four Medium Familiarity sets of words to the upper band of the 95% confidence limits on the FP rate (0.49) showed reliable discrimination for the High and Low Imageability sets ($P = 0.016$ and 0.001, respectively) and marginally significant discrimination for the Medium and Very Low Imageability sets ($P = 0.048$ and 0.051, respectively).

From orthography to phonology

Single word reading

Method. For all tests of single word reading, words were printed in lower case letters on individual cards; the patient was given a stack of cards and asked to go through them at her own pace reading each word aloud. Self-corrections were accepted as correct responses; this occurred rarely for H.P. but more often for N.I., at least on words that she had some success in reading (e.g., FRUIT—'juice . . . it's apples and pears and . . . fruit'; RABBIT—'ears and . . . not a hamster . . . a rabbit'). Four different sets of words were used in the assessment of oral reading: (1) (both N.I. and H.P.) 60 content and 60 function words, 3–6 letters in length, all of very high frequency (> 350/million, Kučera and Francis, 1967); (2) (H.P. only: too difficult for N.I.) 40 high and 40 low-imageability words matched for length, frequency and age of acquisition (from V. Coltheart *et al.*, 1988); (3) (H.P. only: too difficult for N.I.) 78 words, 39 with regular and 39 with exceptional correspondences between orthography and phonology (e.g., PINE and PINT, respectively); the two sets are matched for length, frequency and part of speech (from Coltheart *et al.*, 1979); (4) (N.I. only: too easy for H.P.) 235 common concrete nouns selected from the MRC Data Base (Coltheart, 1981), each word with mean values greater than 5.5 for all three of the variables of familiarity, concreteness and imageability; plus 20 common animal names (e.g., COW, TIGER, DONKEY), 10 colour names (e.g., BLUE, YELLOW, PINK) and the number words from ONE to TEN. This total list of 275 words was administered to N.I. over two sessions.

TABLE 2. PERCENTAGE CORRECT ORAL READING

Test no.	Type of word	(n)	N.I.	H.P.
1	High frequency content words	(60)	20%	98%
	High frequency function words	(60)	07%	97%
2	High imageability nouns/verbs	(40)	—	88%
	Low imageability nouns/verbs	(40)	—	90%
3	Spelling-to-sound regular nouns	(39)	—	95%
	Spelling-to-sound irregular nouns	(39)	—	79%
4	Common concrete nouns	(235)	35%	—
	Common animal names	(20)	70%	—
	Colour names	(10)	90%	—
	Number words (one to ten)	(10)	100%	—

Results. Performance on all tests of single word reading is shown in Table 2. H.P. read the 120 words in (1) quickly and fluently, making a total of 3 errors: 2 visual errors (QUITE—'quiet' and THOUGH—'through') and 1 error that could be described either as visual or as a regularization of a word with an irregular spelling-to-sound correspondence (GREAT—'greet'). On list (2), normal 10-yr-old children with average reading ability show no effect of imageability (High Imageability words: 84% correct, Low Imageability: 83% correct) whereas age-matched poor readers are sensitive to imageability (54% vs 42%) (V. Coltheart *et al.*, 1988). Table 2 shows that H.P., whose word naming performance is somewhat better than that of an average 10-yr-old reader, is similar in being uninfluenced by imageability. On list (3), H.P. misnamed only 2 of the regular words (CHECK—'cheek'; PROTEIN, a word that she did not know—/prouteni:/) but 8 of the exception words. Every one of these errors is an example of a standard type of children's reading error: sounding a silent letter (for example, SUBTLE—/sʌbtail/), regularizing an irregular correspondence (e.g., SEW—/su:/) or confusing one word with another highly similar in orthography (e.g., THOROUGH—'through'). Consistent with her performance on the standardized reading tests presented in the Appendix, the pattern of H.P.'s oral reading performance is precisely like that of a normal child somewhat younger than her own age of 17 yrs.

N.I.'s attempt to read the words in list (1) reveals a slight advantage for content over function words, but even her performance on the content words was extremely poor. This is probably due to the fact that word concreteness or imageability is a major determinant of her oral reading performance, and these very high frequency content words (e.g., END, PART, PLACE) are mostly rather low in concreteness. N.I.'s performance on this test, together with her performance on the first lexical decision test presented earlier, demonstrates a major dissociation between recognition and oral reading of function words. Her hit rate for recognizing high-frequency function words as real words was 95%, but she was virtually incapable of reading any of these words aloud.

N.I.'s success rate on list (4) is shown in Table 2, and Table 3 presents the results of an error analysis based on the classification scheme typically used in analysing the reading errors of adult deep dyslexic patients (Coltheart *et al.*, 1980). N.I.'s error pattern is strikingly similar to that of acquired deep dyslexic patients,

TABLE 3. DETAILED DESCRIPTION OF N.I.'S READING RESPONSES TO COMMON CONCRETE NOUNS (n = 235)

Response Category	Examples	Number	%
Correct	SHOP THUMB	81	35
Omission	POND RUG	79	34
Circumlocution	HOTEL→ 'It's a flat and . . a room and you stay a couple of days' THIEF→ 'The bank, and lots of money but I can't say it'	26	11
Semantic error	ARM→ finger; PIGEON→ cockatoo	17	7
Morphological error	DUCK→ ducks; SMOKE→ smoking	12	5
Part response	NEWSPAPER→ paper; SAW→ hacksaw	7	3
Visual error	BUSH→ brush; FROST→ forest	7	3
Visual and/or Semantic	MUD→ muck; SWALLOW→ swan	4	2
Other	JERSEY→ diamonds, sapphires	2	—
	Total	235	100

although there are also some differences, quantitative if not qualitative. First, although published cases of adult deep dyslexia show considerable variation in reading skill, a number of them would succeed in reading the majority of words selected to maximize the performance of a patient sensitive to frequency, concreteness and imageability. In comparison, N.I.'s overall success rate was unimpressive, although she was better on the special categories of animals, colours and numbers (the last of which she read, just as with Arabic numerals, by counting up from one). Secondly, whilst there is also substantial variability in proportions of the various error types for different deep dyslexic patients (Shallice and Warrington, 1980), N.I. has a higher rate of omissions and a lower rate of visual errors than most of the reported cases.

Of more note than the differences listed above, however, are the resemblances: in particular, in response to concrete nouns, N.I. produced a number of single word semantic paralexias and circumlocutions, both of which reflect (at least partial) comprehension of a printed word despite failure to retrieve its correct phonology. Single word semantic paralexias have occupied a central role in considerations of deep dyslexia (Coltheart *et al.*, 1980). These errors suggest a patient who, lacking any direct mechanism(s) for translating an orthographic string into a phonological code, is forced to do so entirely with reference to the word's meaning.

Comparison of reading with naming (N.I. only)

Whenever poor oral reading is observed in a patient with limited speech output, the hypothesis of a single deficit at the level of phonological output, responsible for both problems, should be considered. One method of assessing this hypothesis is to compare the patient's success at producing a set of names, first in response to pictures typically named by those words and secondly in response to the printed words corresponding to the names. Such comparisons have proved informative with anomic patients who display the characteristics of either surface dyslexia (e.g., Goldblum, 1985; Kremin, 1985; Margolin *et al.*, 1985) or deep dyslexia (Howard, 1985). This issue does not arise for H.P., whose naming and oral reading performance was normal in pattern albeit poorer than average for her age; but it is germane to N.I. Her scores on naming tests and her spontaneous speech indicate substantial anomia; is it possible that many of her failures in oral reading have the same cause, that is, a limit on either the size of or the accessibility of her speech output lexicon? It is already clear that the entirety of her reading impairment cannot be explained in this way, since her spontaneous speech is well supplied with function words, virtually none of which she can read aloud. The production of function words in spontaneous speech may however be supported by other processes (syntactic, prosodic) which do not contribute significantly to single-word oral reading. Therefore, with regard to content words, it is still pertinent to compare N.I.'s naming and reading.

Method. The stimulus materials for naming consisted of 54 line drawings of people engaged in common actions (e.g., a girl climbing a ladder, a boy combing his hair, a man reading a book, etc.). For reading, each of these drawings was converted into a simple written sentence describing the picture, with the verb underlined. The 54 line drawings and their corresponding sentences were divided into two sets, A and B. On one day, set A was given for naming and set B for reading, and on another day, the reverse. N.I. was asked to produce the verb from both picture and sentence, with the instruction (and given examples) that it was only necessary to say the action word rather than the whole sentence. The main reason for presenting the written verbs in a sentence context was to balance as much as possible the amount of potential information in the two conditions. The picture for the action of climbing shows not only the person climbing but also a ladder, which may facilitate retrieval of 'climbing'. The presence, therefore, of the written word LADDER in the sentence might provide the same assistance with regard to oral reading of CLIMBING.

TABLE 4. COMPARISON OF READING AND NAMING FOR N.I.:
54 COMMON ACTION WORDS

		Naming		
		Correct	Incorrect	Total
Reading	Correct	15	2	17
	Incorrect	26	11	37
	Total	41	13	54

Results. N.I.'s performance, shown in Table 4, reveals a highly significant advantage for naming over reading ($P < 0.001$ by a two-tailed McNemar test). N.I.'s level of success at naming actions demonstrates that her failure to read the corresponding words cannot be attributed to their absence from or inaccessibility in her speech lexicon. Interestingly, a number of adult deep dyslexic patients who have been assessed on this same comparison (Howard, 1985) show a significant advantage in the opposite direction, that is, better reading than naming.

Nonword reading

Method. Thirty simple nonsense words (such as NEG, LOAT, FOD) were printed on individual cards and presented to both H.P. and N.I. for oral reading.

Results. H.P.'s performance, shown at the top of Table 5, is certainly normal for her reading age and probably even normal for her chronological age (Masterson,

TABLE 5. PERCENTAGE CORRECT PERFORMANCE ON
VARIOUS TESTS OF PHONOLOGICAL PROCESSING

	n	N.I.	H.P.
Nonword reading	30		
Strict criterion		0	80
Lenient criterion		0	97
Nonword repetition	30	100	100
Auditory rhyme judgements	60	93	95
Rhyme production	20	0	100
Phoneme deletion	24	0	96
Phoneme addition	24	0	100

1985). N.I., on the other hand, could not read any of the nonwords. Under conditions of extreme encouragement to say what a particular nonword looked like or what she thought it might sound like, she produced visually similar real words to a few of the items (e.g., NOKE—'coke', FOD—'it's supposed to be a fog'), but her preferred response was always 'I don't know, I'm so sorry'. It should be clear that this is a failure specifically in the ability to translate orthography to phonology: N.I. has no impairment either in recognition of the component letters of a nonword (as demonstrated earlier by her virtually perfect matching of upper and lower case nonwords) or in the ability to produce nonwords orally. Asked to say the same 30 nonwords in an immediate repetition task, she repeated every one perfectly.

Other phonological tasks

It is now thoroughly established that explicit awareness of the phonological structure of words is intimately related to the acquisition of reading skill (*see*, e.g., Liberman *et al.*, 1974; Rozin and Gleitman, 1977; Morais *et al.*, 1979; Bradley and Bryant, 1983). This issue, which has been at the forefront of research on developmental reading disorders (Bertelson and De Gelder, 1988), has been largely ignored in work on acquired disorders. Acquired dyslexic patients are typically tested on nonword reading; but phonological segmentation and assembly, undoubted prerequisites for nonword reading, are rarely assessed (but *see* Derouesné and Beauvois, 1985, for an exception). The fact that deep dyslexic patients like D.E. and P.W. (studied extensively by Patterson, 1978, 1979, and Morton and Patterson, 1980*a,b*) are completely incapable of either deleting a phoneme from or adding a phoneme to a spoken word appears not to have been reported in the literature.

Method. Four relevant tasks in the auditory/phonological domain were administered. (1) 60 pairs of monosyllabic words (30 of which rhyme) were spoken to the subject who was instructed to say 'yes' if the 2 words rhymed and 'no' if they did not. (2) The subjects were requested to produce a rhyme for each of 20 single-syllable spoken words. (3) The subjects were instructed to delete the first sound from each of 24 spoken words and say what remained. All stimulus items yielded another real word if the deletion was done properly (e.g., 'mother', 'gold', 'witch', 'meat', 'steam', etc.). (4) H.P. and N.I. were asked to add a sound on to the front of a spoken word and produce the result; as with the phoneme deletion task, correct responses to all 24 items were real words (e.g., /m/ plus 'add'—'mad', /g/ plus 'aim'—'game', /h/ plus 'ill'—'hill', /k/ plus 'lock'—'clock', etc.).

Results. As Table 5 demonstrates, each patient made a few errors on the auditory rhyme judgement test, but in fact they were both reasonably (and equally) good at it. N.I., however, failed totally on all three of the other tasks. In rhyme production, N.I. could only repeat back an example just given to her. Offered 'cat' and 'bat' as a sample rhyme, if she was immediately asked to give a rhyming word for 'cat' she would say 'bat'; but she could never produce either a different rhyme for 'cat' or a rhyme for a test word not used in an example. Asked to provide a rhyming word for 'five' (and note her near-perfect performance at judging that 'five' and 'hive' rhyme whilst 'five' and 'give' do not), she responded 'ten'. With phoneme deletion, even after numerous examples, N.I. had no idea how to perform the task. Despite attempts to facilitate her performance with an easier version of the test in which the first sound of the word was identified for her (e.g., the first sound of 'meat' is 'mmm'; if you take 'mmm' away from 'meat', what is left?), the task remained beyond her ability. Finally, for phoneme addition, N.I. could repeat the 2 individual elements of each stimulus presentation but never succeeded in blending them into a single response word.

Apart from recognition of spoken rhymes, N.I. is utterly lacking in any awareness of or ability to manipulate the phonological structure of spoken words. Given what is known about the relationship of these skills to reading acquisition, N.I. (who apparently learned to read in a perfectly normal manner at a perfectly average age) would almost certainly have been capable of performing these tasks before her illness. This suggests that phonological manipulation is an ability completely lateralized to the LH. H.P.'s success on these tasks is certainly consonant with this view. She made one error in phoneme deletion (asked to delete the first sound of 'stall', she replied 'all') but her performance on these tasks was essentially normal.

From orthography to semantics

Precision of comprehension

Method. Some indication of the basic ability of both N.I. and H.P. to comprehend single printed words is offered by their performance at matching words to pictures (*see* the visual BPVS scores in Appendix 1). This test does not, however, provide an adequate basis for conclusions regarding precision of comprehension. Success in matching the printed word HAND to a picture of a hand where the alternative choices are pictures of a duck, a ball and a shoe can assuredly be achieved on the basis of very gross or even partial comprehension of the word. To obtain a more demanding test of word comprehension, Howard and Orchard-Lisle (1984) developed the Pyramids and Palmtrees test in which 1 of 2 response words (e.g., BUTTERFLY and DRAGONFLY) must be selected to go with a third word (CATERPILLAR). The test was designed such that the 2 response words in any triad are semantically related (usually they are coordinates); a correct choice between them thus requires a reasonably precise understanding of the meaning of the 3 component words composing each item of the test. An additional advantage of the Pyramids and Palmtrees test is that it can be administered with either words or pictures. Performance by N.I. and H.P. on this test using words as stimuli will provide a better indication of their detailed comprehension of written words, and a comparison of this with their scores on the same test using pictures as stimuli will provide some perspective on their semantic knowledge as accessed through these different modalities. The two versions of the test were administered to each patient in different sessions with several months intervening between them.

Results. Performance is shown in Table 6. H.P. made 3 errors with the word version and 2 with the picture version; although these scores are slightly below the means for control groups of normal adult subjects (99%; D. Howard and K. Patterson, unpublished observations), they are just within the normal range as the maximum number of errors made by any control subject was 3. N.I. showed subnormal performance even on the picture version but a significant advantage on pictures over words; χ^2 (1 df) = 9.01, $P < 0.01$. By a binomial test, her

TABLE 6. PERCENTAGE CORRECT PERFORMANCE
ON TWO COMPREHENSION TESTS*

	n	N.I.	H.P.
Pyramids and Palmtrees			
Words	50	62	94
Pictures	50	88	96
LUVS			
Printed words	40	85	100
Spoken words	40	98	100

* Chance performance on Pyramids and Palmtrees = 50% and on LUVS = 20%.

performance on the word version is only marginally ($P = 0.06$) above the chance level of 50%.

The nature of comprehension errors

A much discussed issue with regard to deep dyslexic patients (*see*, e.g., Morton and Patterson, 1980*a*; Newcombe and Marshall, 1980; Patterson and Besner, 1984) concerns the locus of their semantic errors in oral reading (e.g., SWAN—'duck'). If such errors reflect only a problem in the procedures whereby semantic features specify a phonological response, then a patient who reads SWAN as 'duck' ought to succeed in selecting a picture of a swan (rejecting the duck) in response to the printed word SWAN. If, on the other hand, the reading error reflects a lack of the precise semantic knowledge that enables a person to know the difference between swans and ducks, then one would expect the patient to make errors in the picture selection task as well. In fact, amongst a set of adult deep dyslexic patients who have been given this sort of test (*see* Bishop and Byng, 1984), some do and some do not make semantic errors in word/picture matching, suggesting two different sources of the reading errors. N.I. makes semantic reading errors; will she also make semantic comprehension errors?

Method. Kay *et al.*, (1989) have designed a word/picture matching test, based on the principle of Bishop's LUVS test (Lexical Understanding with Visual and Semantic Distractors; *see* Bishop and Byng, 1984). Each of the 40 items comprises 1 word such as NAIL (either spoken or printed) and 5 line drawings: the correct item (for this item, obviously, a nail), a visually similar distractor (a pencil), a close semantic distractor (a screw), a distant semantic distractor (a pair of pliers) and an unrelated distractor (a letter: the sort that you post). The two versions of the test (once with printed words and once with spoken words) were administered to both N.I. and H.P., in separate sessions with several months intervening.

Results. H.P.'s performance was perfect in both modalities. N.I.'s single error in the auditory modality was choosing the unrelated alternative to a target word that she did not know, 'syringe'. Her 6 errors in the visual modality comprised 3 choices of a close semantic distractor, 1 distant semantic alternative, 1 visually similar alternative and 1 unrelated distractor. Even though two-thirds of her errors with printed words involved choice of a semantically related alternative, the small total number of errors clearly prevents statistical support for a claim of semantic errors in this task. It should be noted, however, that claims for semantic errors in word-picture matching both for split-brain patients (patients L.B. and N.G., Zaidel, 1982) and for some deep dyslexic patients (e.g., patients V.S. and R.P., Bishop and Byng, 1984) have been based on total numbers of errors and proportions of semantic errors no greater than those shown by N.I.

DISCUSSION

The Introduction to this paper described two neuropsychological approaches to the question of RH representation of reading skills. In one, with the focus on a patient with primary or sole use of the RH, the basic approach is to examine the extent and nature of reading performance possible. In the other, where a patient with primary or sole use of the LH is studied, the question is whether any reading deficit is found which might be attributable to the absence of a contribution from the RH. It was argued for both approaches that the most convincing source of evidence would be cases of hemispherectomy for neurological disease of late onset. Now that 2 such cases have been presented, what can be concluded about reading with one hemisphere?

N.I., whose symptoms of LH abnormality commenced at the age of 13 yrs and whose LH was removed at the age of 15, is poor at virtually all aspects of reading; but her pattern of reading performance is by no means one of undifferentiated impairment. She is essentially perfect at recognizing letters, although not very successful at naming them and totally unable to give their sound equivalents. She is reasonably good at discriminating very common words from orthographically similar nonwords, but her lexical decision performance falls off quickly as word frequency declines. She can comprehend printed words (as measured by matching to pictures) when these correspond to common concrete objects, but she occasionally makes semantic errors in this task, and her success is dependent upon maximal degrees of both word familiarity and concreteness. She also has some degree of success in oral reading of the most familiar and highly imageable words, but here, too, she is prone to semantic paralexic errors. That she only succeeds in oral reading with words of the type that she can understand is no coincidence: instead, it reflects the fact that she has no means of translating a printed word into a phonological response except on the basis of its meaning. She cannot read aloud nonsense words at all (although she repeats them perfectly). She cannot pronounce subcomponents of written words and, indeed, within the domain of the spoken word, she has no awareness that words are divisible into subcomponents.

Before further discussion of N.I.'s reading, it is necessary to consider the justification for generalizing from her reading performance to the representation of reading abilities in the RH of right-handed adults, either those who are neurologically normal or those who were neurologically normal prior to a stroke or other incident. The potential obstacle to such generalization is N.I.'s speech. Although she is very aphasic, her speech is rather better than and quite different from the speech of globally aphasic stroke cases with massive LH lesions, most of whom are either severely agrammatic, or restricted to single word utterances, or even restricted to the same recurring utterance. N.I.'s relatively good speech raises two questions. (1) Might she have had either RH or bilateral representation of language before the onset of her brain disease? (2) If the answer to the first question is negative, with the consequent interpretation that her limited speech represents interhemispheric reorganization of speech in response to the onset of her brain disease, then might the same be true of her limited reading ability?

Most easily rejected is the notion that N.I. might originally have had RH dominance for language: the fact that all her language abilities are now so impoverished would be inexplicable. Almost as easy to reject, on statistical grounds, is the notion that N.I. had bilateral representation of speech. In a sample of 140 right handers given sodium amytal tests (Milner, 1975), the procedure indicated LH speech representation for 134 individuals (96% of the sample) and RH speech representation for 6 (4%). Not a single case had bilateral speech. Amytal testing is necessarily time limited and therefore inadequate for assessing a range of language functions. Nevertheless, the most likely conclusion is that N.I., completely right handed and completely normal until age 13, had normal LH dominance for language.

Resolution of the remaining query is less straightforward. In the first instance, the conclusion that N.I.'s speech represents a late reorganization of hemisphere dominance could be surprising in the light of prevalent assumptions about complete lateralization of language by about the age of 5 yrs (*see* Searleman, 1977, Zaidel, 1979, and Hahn, 1987, for discussion). However, there has also been an alternative view that some degree of plasticity remains until puberty (Basser, 1962; Lenneberg, 1967) or even into adulthood (Teuber, 1975). Furthermore, an extensive review of the hemispherectomy literature by St James-Roberts (1981) revealed considerable variability of language function in left hemispherectomy cases independent of age at onset. Finally, it must be remembered that N.I. is severely impaired in every aspect of language. The debate as to whether left hemispherectomy can fully spare certain aspects of language processing only arises for cases with early onset of disease (e.g., Dennis and Whitaker, 1976; Dennis, 1980; Bishop, 1983; Strauss and Verity, 1983).

If N.I.'s speech reflects hemisphere reorganization, then presumably no one would attempt to generalize from her spoken language to the probable representation of speech abilities in the RH of a right-handed adult who is either normal or has just suffered a stroke. Why is it then legitimate to draw inferences about probable RH reading abilities in such individuals on the basis of N.I.'s reading performance? Relevant arguments offer plausibility but no proof. The first point concerns the ecological importance to an individual of basic speech comprehension and production, without which he or she must live in genuine isolation. The same cannot be said of reading skills. In response to a deteriorating dominant hemisphere, therefore, it seems plausible that the brain will concentrate on reestablishing as much basic communication skill as possible in the other hemisphere. The second point is that, from the onset of N.I.'s illness to the present, and including the time at which brain reorganization was presumably occurring, N.I. had constant exposure to and experience of comprehending and producing speech but virtually no reading experience. The final point is that her reading skill is dramatically inferior to her speech comprehension and production. In assessments pertaining to the present study, she comprehended many spoken sentences but no written ones; she repeated sentences up to about 5 words in length without error but never correctly read aloud a sentence; she easily repeated spoken nonsense words but failed entirely to read them aloud. If N.I.'s RH did take over reading skill as a result of reorganization, it did so rather unsuccessfully. The most plausible assumption is that N.I.'s current reading abilities are essentially like (and in particular no better than) those of the RH of a normal right-handed 13-yr-old girl.

If this interpretation is accepted, then the data for N.I. probably constitute the best available source of evidence on the reading capacities of the nondominant hemisphere. These data can now be compared with evidence from other sources regarding this question.

N.I.'s reading performance is highly similar to the pattern demonstrated[1] by Zaidel and his colleagues (e.g., Zaidel and Peters, 1981; Zaidel, 1982) for the split-brain patients L.B. and N.G. in reading tasks with stimuli lateralized to the left visual field. The major difference is that N.I. (with her limited speech represented in the RH) can accomplish some oral reading whereas L.B. and N.G. (with their normal speech represented in the LH) cannot read aloud words presented in the left visual field. The main similarities are an inability to convert orthography to phonology by any direct translation procedures, sensitivity to the familiarity and especially to the concreteness or imageability of words, and the occurrence of at least some semantic errors in matching printed words to pictures. Many writers have expressed worries about generalizing from split-brain patients to normal subjects (c.g., Lambert, 1982; Gazzaniga, 1983; Patterson and Besner, 1984; Coltheart, 1985). Furthermore, there is extensive debate regarding interpretation of the RH language skills shown by L.B. and N.G., not only with regard to normal subjects but also in relation to other split-brain cases (Gazzaniga, 1983; Levy, 1983; Zaidel, 1983; Myers, 1984). Although such generalizations are inherently risky, some of the results are impressively consistent. In particular, Zaidel's results for L.B. and N.G. and the present results for N.I. suggest the same conclusions about the nondominant RH's capacity for reading.

Secondly, N.I.'s reading performance is identical in pattern, though not in level, to most cases of acquired deep dyslexia described in the literature. As indicated in the Introduction, a strong case for the equation of deep dyslexic reading with RH reading has been made by Coltheart (1980, 1983) and Saffran et al. (1980), and supported by Landis et al. (1983), Zaidel and Schweiger (1984), Jones and Martin (1985) and others. One of the authors of this paper has been associated with grave reservations regarding this notion (Patterson and Besner, 1984; Marshall and Patterson, 1985), but it is clear that the hypothesis receives a considerable boost from the current results on N.I.'s reading. The case is not (because it cannot be) proven. First of all, there is the important point made by Rabinowicz and Moscovitch (1984) that some types of LH damage may result in residual LH language abilities resembling the abilities of the isolated RH. Thus acquired deep dyslexic reading performance could still be LH reading, at least in some patients. Furthermore, the quantitative differences between N.I.'s reading skill and that of most deep dyslexic patients should not be ignored; some possible accounts of this discrepancy are mentioned below. Despite these reservations and some unresolved issues (such as alexia without agraphia: *see* discussions by Landis *et al.*, 1980; Patterson and Besner, 1984; Zaidel and Schweiger, 1984; Shallice and Saffran, 1986), the empirical conclusion is clear: adult deep dyslexics, who may be reading with the RH, and N.I., who must be reading with the RH, are strikingly similar.

There would seem to be three potential explanations for N.I.'s quantitatively poor performance (e.g., in oral reading of familiar concrete words) relative to many published cases of acquired deep dyslexia resulting from cerebrovascular lesions, head injury or missile wounds. The first possibility relates to the fact, noted earlier, that N.I.'s reading experience came to a virtually complete halt at the age of 13. If a normal LH-dominant individual has a RH reading lexicon, and if learning within already established subcomponents of a skill continues throughout an individual's life, then perhaps the reading lexicon in N.I.'s RH would be expected to underestimate the RH reading lexicon of a 50-yr-old (a more typical age for onset of acquired deep dyslexia). It may be worth noting that one of the few relatively young cases of acquired deep dyslexia, W.S. (Saffran, 1980), apparently had extensive reading instruction after his missile wound at the age of 11 yrs (*see* Saffran *et al.*, 1980). This is not, however, true of the deep dyslexic patient D.E. (Patterson, 1978, 1979) with a brain injury at the age of 16 yrs, who had left school at 15 and was probably no great reading enthusiast at any age. While by no means the best reader of the various well-studied cases of acquired deep dyslexia, D.E.'s performance is substantially better than N.I.'s.

The second possibility concerns the neurological and cognitive status of N.I.'s RH: while this evinces no gross damage, it did suffer the effects of extremely severe LH seizures over a period of about $2\frac{1}{2}$ yrs. The majority of adult deep dyslexic patients sustained an abrupt insult (such as a cerebrovascular lesion), confined to the LH. Perhaps their RHs are more intact than N.I.'s RH, and that explains their superior reading. To the extent that this argument has force, then the RH reading performance of split-brain patients (with a history of seizures) may constitute a more satisfactory basis for comparison with N.I. And indeed, as emphasized by Patterson and Besner (1984), the split-brain cases extensively tested on RH reading (*see* references by Zaidel) are, like N.I., qualitatively similar but quantitatively quite inferior to deep dyslexic patients.

Both of the first two explanations are predicated on the assumption that deep dyslexic reading arises primarily or exclusively from the RH. The remaining explanation acknowledges that, however damaged it may be, these patients do have a LH whereas N.I. does not. The evidence may converge convincingly on a major RH *contribution* to the pattern of performance in deep dyslexic reading; but it remains possible that most deep dyslexic patients read at a higher level than N.I. because their reading can rely both on an intact RH and some limited LH resources.

Turning much more briefly to H.P. and the other approach to the question of RH reading, the present assessment indicates that H.P.'s reading abilities are delayed but in no way deviant. Her reading lexicon is smaller than that of a normal 17-yr-old, and thus she is naturally more sensitive to variables such as familiarity and spelling-to-sound regularity. These are, however, quantitative differences only; there were no qualitatively abnormal features in H.P.'s reading, nor was there any subskill of a normal person's reading profile that was lacking in H.P.'s abilities. The discrepancy between her chronological and reading ages may be entirely explicable in terms of the 5 yr history of preoperative illness. H.P.'s performance suggests that the RH plays no necessary role in supporting reading skills, at least of the type assessed in this study (mainly single word recognition, comprehension and pronunciation).

Zaidel and Schweiger addressed themselves to the question of the role of the RH in normal reading: 'Perhaps it is important both for reading acquisition and for mature, efficient reading. In particular, the RH may be important for quick pattern recognition during speed reading, and for semantic-thematic orientation to a situation or narrative. On this account, loss of RH contribution in adults would lead to subtle deficits in higher-order reading which are rarely assessed in the neurological clinic and have yet to be formally tested in split-brain patients.' (Zaidel and Schweiger, 1984, p. 362).

These three points all seem thoroughly sound. Normal acquisition of any skill as complex as reading undoubtedly requires two functioning hemispheres, and there is indeed evidence that children with one hemisphere have difficulty learning to read whether it is the LH or the RH that has been removed. Although rather little is known on the subject of speed reading, there is an intriguing report by Andreewsky *et al.* (1980) of speed-reading ability in a case of deep dyslexia. Finally, although assessments of possible higher order language or reading deficits in patients with RH lesions are rare, a few are beginning to appear (e.g., Cavalli *et al.*, 1981; Wapner *et al.*, 1981; Chiarello and Church, 1986; Grossman and Haberman, 1987). Most of these studies have revealed significant (if sometimes small) deficits, relative to control subjects, in patients with RH lesions, especially (e.g., in the Grossman and Haberman study) for anterior RH lesions. There is, however, no indication that such deficits are specifically concerned with linguistic abilities as opposed to more general higher order cognitive skills.

ACKNOWLEDGEMENTS

We express our gratitude to Dr E. M. Brett for making it possible for us to investigate his patient. We would also like to thank the families of N.I. and H.P. who graciously tolerated the long travelling distances necessary for their visits. Last but not least we thank N.I. and H.P. for their unlimited patience and cooperation. The preoperative psychometric results were kindly provided by the clinical psychology service of the Maudsley Hospital. This study was supported by Project Grants CFM2 and CFM3 from the British Medical Research Council to F.V.-K.

REFERENCES

ADAMS CBT (1983) Hemispherectomy—a modification. *Journal of Neurology, Neurosurgery and Psychiatry*, **46**, 617–619.

ANDREEWSKY E, DELOCHE G, KOSSANYI P (1980) Analogies between speed-reading and deep dyslexia: towards a procedural understanding of reading. In: *Deep Dyslexia*. Edited by M. Coltheart, K. Patterson and J. C. Marshall. London: Routledge and Kegan Paul, pp. 307–325.

BASSER LS (1962) Hemiplegia of early onset and the faculty of speech with special reference to the effects of hemispherectomy. *Brain*, **85**, 427–460.

BERTELSON P, DE GELDER B (1988) Learning about reading from illiterates. In: *From Reading to Neurons. Toward Theory and Methods for Research on Developmental Dyslexia*. Edited by A. M. Galaburda. Cambridge, MA: MIT Press. In press.

BISHOP DVM (1982) *TROG: Test for Reception of Grammar*. Abingdon, Oxon: Thomas Leach (for the Medical Research Council).

BISHOP DVM (1983) Linguistic impairment after left hemidecortication for infantile hemiplegia? A reappraisal. *Quarterly Journal of Experimental Psychology*, **35A**, 199–207.

BISHOP DVM, BYNG S (1984) Assessing semantic comprehension: methodological considerations, and a new clinical test. *Cognitive Neuropsychology*, **1**, 233–244.

BRADLEY L, BRYANT PE (1983) Categorizing sounds and learning to read—a causal connection. *Nature, London*, **301**, 419–421.

BRANCH C, MILNER B, RASMUSSEN T (1964) Intracarotid sodium amytal for the lateralization of cerebral speech dominance: observations in 123 patients. *Journal of Neurosurgery*, **21**, 399–405.

CAVALLI M, DE RENZI E, FAGLIONI P, VITALE A (1981) Impairment of right brain-damaged patients on a linguistic cognitive task. *Cortex*, **17**, 545–555.

CHIARELLO C, CHURCH KL (1986) Lexical judgments after right- or left-hemisphere injury. *Neuropsychologia*, **24**, 623–630.

COLTHEART M (1980) Deep dyslexia: a right hemisphere hypothesis. In: *Deep Dyslexia*. Edited by M. Coltheart, K. Patterson and J. C. Marshall. London: Routledge and Kegan Paul, pp. 326–380.

COLTHEART M (1981) *MRC Psycholinguistic Database User Manual: Version 1*. London: Medical Research Council.

COLTHEART M (1983) The right hemisphere and disorders of reading. In: *Functions of the Right Cerebral Hemisphere*. Edited by A. W. Young. London: Academic Press, pp. 171–201.

COLTHEART M (1985) Right-hemisphere reading revisited. *Behavioral and Brain Sciences*, **8**, 363–365.

COLTHEART M, BESNER D, JONASSON JT, DAVELAAR E (1979) Phonological encoding in the lexical decision task. *Quarterly Journal of Experimental Psychology*, **31**, 489–507.

COLTHEART M, PATTERSON K, MARSHALL JC (editors) (1980) *Deep Dyslexia*. London: Routledge and Kegan Paul.

COLTHEART M, PATTERSON K, MARSHALL JC (1987) Deep dyslexia since 1980. In: *Deep Dyslexia*. Second edition. Edited by M. Coltheart, K. Patterson and J. C. Marshall. London: Routledge and Kegan Paul, pp. 407–451.

COLTHEART V, LAXON VJ, KEATING C (1988) Effects of word imageability and age of acquisition on children's reading. *British Journal of Psychology*, **79**, 1–12.

DENNIS M (1980) Capacity and strategy for syntactic comprehension after left or right hemidecortication. *Brain and Language*, **10**, 287–317.

DENNIS M, WHITAKER HA (1976) Language acquisition following hemidecortication: linguistic superiority of the left over the right hemisphere. *Brain and Language*, **3**, 404–433.

DE RENZI E, VIGNOLO LA (1962) The Token Test: a sensitive test to detect receptive disturbances in aphasics. *Brain*, **85**, 665–678.

DEROUESNÉ J, BEAUVOIS M-F (1985) The 'phonemic' stage in the non-lexical reading process: evidence from a case of phonological alexia. In: *Surface Dyslexia: Neuropsychological and Cognitive Studies of Phonological Reading*. Edited by K. E. Patterson, J. C. Marshall and M. Coltheart. London: Lawrence Erlbaum, pp. 399–457.

DUNN LM, DUNN LM, WHETTON C, PINTILLIE D (1982) *British Picture Vocabulary Scales*. Windsor: NFER-Nelson.

GAZZANIGA MS (1983) Right hemisphere language following brain bisection: a 20-year perspective. *American Psychologist*, **38**, 525–537.

GOLDBLUM M-C (1985) Word comprehension in surface dyslexia. In: *Surface Dyslexia: Neuropsychological and Cognitive Studies of Phonological Reading*. Edited by K. E. Patterson, J. C. Marshall and M. Coltheart. London: Lawrence Erlbaum, pp. 175–205.

GROSSMAN M, HABERMAN S (1987) The detection of errors in sentences after right hemisphere brain damage. *Neuropsychologia*, **25**, 163–172.

HAHN WK (1987) Cerebral lateralization of function: from infancy through childhood. *Psychological Bulletin*, **101**, 376–392.

HARDYCK C, CHIARELLO C, DRONKERS NF, SIMPSON GV (1985) Orienting attention within visual fields: how efficient is interhemispheric transfer? *Journal of Experimental Psychology: Human Perception and Performance*, **11**, 650–666.

HOWARD D (1985) *The Organisation of the Lexicon: Evidence from Aphasia*. Ph.D. thesis, University of London.

HOWARD D, ORCHARD-LISLE V (1984) On the origin of semantic errors in naming: evidence from the case of a global aphasic. *Cognitive Neuropsychology*, **1**, 163–190.

JAMES CT (1975) The role of semantic information in lexical decisions. *Journal of Experimental Psychology: Human Perception and Performance*, **1**, 130–136.

JONES GV, MARTIN M (1985) Deep dyslexia and the right-hemisphere hypothesis for semantic paralexia: a reply to Marshall and Patterson. *Neuropsychologia*, **23**, 685–688.

KAY J, LESSER R, COLTHEART M (1989) *Psycholinguistic Assessments of Language Processing in Aphasia (PALPA)*. London: Lawrence Erlbaum. In press.

KREMIN H (1985) Routes and strategies in surface dyslexia and dysgraphia. In: *Surface Dyslexia: Neuropsychological and Cognitive Studies of Phonological Reading*. Edited by K. E. Patterson, J. C. Marshall and M. Coltheart. London: Lawrence Erlbaum, pp. 105–137.

KUČERA H, FRANCIS WN (1967) *Computational Analysis of Present-Day American English*. Providence, RI: Brown University Press.

LAMBERT AJ (1982) Right hemisphere language ability. 1. Clinical evidence. *Current Psychological Reviews*, **2**, 77–94.

LANDIS T, REGARD M, SERRAT A (1980) Iconic reading in a case of alexia without agraphia caused by a brain tumor: a tachistoscopic study. *Brain and Language*, **11**, 45–53.

LANDIS T, REGARD M, GRAVES R, GOODGLASS H (1983) Semantic paralexia: a release of right hemispheric function from left hemispheric control? *Neuropsychologia*, **21**, 359–364.

LENNEBERG EH (1967) *Biological Foundations of Language*. New York: John Wiley.
LEVY J (1983) Language, cognition, and the right hemisphere: a response to Gazzaniga. *American Psychologist*, **38**, 538–541.
LIBERMAN IY, SHANKWEILER D, FISCHER FW, CARTER B (1974) Explicit syllable and phoneme segmentation in the young child. *Journal of Experimental Child Psychology*, **18**, 201–212.
MARGOLIN DI, MARCEL AJ, CARLSON NR (1985) Common mechanisms in dysnomia and post-semantic surface dyslexia: processing deficits and selective attention. In: *Surface Dyslexia: Neuropsychological and Cognitive Studies of Phonological Reading*. Edited by K. E. Patterson, J. C. Marshall and M. Coltheart. London: Lawrence Erlbaum, pp. 139–173.
MARSHALL JC, NEWCOMBE F (1966) Syntactic and semantic errors in paralexia. *Neuropsychologia*, **4**, 169–176.
MARSHALL JC, NEWCOMBE F (1973) Patterns of paralexia: a psycholinguistic approach. *Journal of Psycholinguistic Research*, **2**, 175–199.
MARSHALL JC, PATTERSON KE (1985) Left is still left for semantic paralexias: a reply to Jones and Martin. *Neuropsychologia*, **23**, 689–690.
MASTERSON J (1985) On how we read non-words: data from different populations. In: *Surface Dyslexia: Neuropsychological and Cognitive Studies of Phonological Reading*. Edited by K. E. Patterson, J. C. Marshall and M. Coltheart. London: Lawrence Erlbaum, pp. 289–299.
MENKES JH (1980) *Textbook of Child Neurology*. Second edition. Philadelphia: Lea and Febiger.
MILNER B (1975) Psychological aspects of focal epilepsy and its neurosurgical management. *Advances in Neurology*, **8**, 299–321.
MORAIS J, CARY L, ALEGRIA J, BERTELSON P (1979) Does awareness of speech as a sequence of phones arise spontaneously? *Cognition*, **7**, 323–331.
MORTON J, PATTERSON K (1980a) A new attempt at an interpretation, or, an attempt at a new interpretation. In: *Deep Dyslexia*. Edited by M. Coltheart, K. Patterson and J. C. Marshall. London: Routledge and Kegan Paul, pp. 91–118.
MORTON J, PATTERSON K (1980b) 'Little words—No!' In: *Deep Dyslexia*. Edited by M. Coltheart, K. Patterson and J. C. Marshall. London: Routledge and Kegan Paul, pp. 270–285.
MYERS JJ (1984) Right hemisphere language: science or fiction? *American Psychologist*, **39**, 315–320.
NEALE MD (1966) *Neale Analysis of Reading Ability*. Second edition. London: Macmillan.
NEWCOMBE F, MARSHALL JC (1980) Response monitoring and response blocking in deep dyslexia. In: *Deep Dyslexia*. Edited by M. Coltheart, K. Patterson and J. C. Marshall. London: Routledge and Kegan Paul, pp. 160–175.
OLDFIELD RC, WINGFIELD A (1965) Response latencies in naming objects. *Quarterly Journal of Experimental Psychology*, **17**, 273–281.
PATTERSON KE (1978) Phonemic dyslexia: errors of meaning and the meaning of errors. *Quarterly Journal of Experimental Psychology*, **30**, 587–607.
PATTERSON KE (1979) What is right with 'deep' dyslexic patients? *Brain and Language*, **8**, 111–129.
PATTERSON K, BESNER D (1984) Is the right hemisphere literate? *Cognitive Neuropsychology*, **1**, 315–341.
PATTERSON KE, MARSHALL JC, COLTHEART M (editors) (1985) *Surface Dyslexia: Neuropsychological and Cognitive Studies of Phonological Reading*. London: Lawrence Erlbaum.
RABINOWICZ B, MOSCOVITCH M (1984) Right hemisphere literacy: a critique of some recent approaches. *Cognitive Neuropsychology*, **1**, 343–350.
RASMUSSEN T (1978) Further observations on the syndrome of chronic encephalitis and epilepsy. *Applied Neurophysiology*, **41**, 1–12.
RASMUSSEN T (1983) Hemispherectomy for seizures revisited. *Canadian Journal of Neurological Sciences*, **10**, 71–78.
RASMUSSEN T, MILNER B (1977) The role of early left-brain injury in determining lateralization of cerebral speech functions. *Annals of the New York Academy of Sciences*, **299**, 355–369.

RICKARD SJ (1987) *Deep Dyslexia*. Ph.D. thesis, University of London.
ROZIN P, GLEITMAN LR (1977) The structure and acquisition of reading. II. The reading process and the acquisition of the alphabetic principle. In: *Toward a Psychology of Reading*. Edited by A. S. Reber and D. L. Scarborough. Hillsdale, NJ: Lawrence Erlbaum, pp. 55-141.
SAFFRAN EM (1980) Reading in deep dyslexia is not ideographic. *Neuropsychologia*, **18**, 219-223.
SAFFRAN EM, BOGYO LC, SCHWARTZ MF, MARIN OSM (1980) Does deep dyslexia reflect right-hemisphere reading? In: *Deep Dyslexia*. Edited by M. Coltheart, K. Patterson and J. C. Marshall. London: Routledge and Kegan Paul, pp. 381-406.
ST JAMES-ROBERTS I (1981) A reinterpretation of hemispherectomy data without functional plasticity of the brain. I. Intellectual function. *Brain and Language*, **13**, 31-53.
SCHONELL FJ (1942) *Backwardness in the Basic Subjects*. Edinburgh: Oliver and Boyd.
SEARLEMAN A (1977) A review of right hemisphere linguistic capabilities. *Psychological Bulletin*, **84**, 503-528.
SHALLICE T, SAFFRAN E (1986) Lexical processing in the absence of explicit word identification: evidence from a letter-by-letter reader. *Cognitive Neuropsychology*, **3**, 429-458.
SHALLICE T, WARRINGTON EK (1980) Single and multiple component central dyslexic syndromes. In: *Deep Dyslexia*: Edited by M. Coltheart, K. Patterson and J. C. Marshall. London: Routledge and Kegan Paul, pp. 119-145.
SMITH A (1966) Speech and other functions after left (dominant) hemispherectomy. *Journal of Neurology, Neurosurgery and Psychiatry*, **29**, 467-471.
STRAUSS E, VERITY C (1983) Effects of hemispherectomy in infantile hemiplegics. *Brain and Language*, **20**, 1-11.
TEUBER H-L (1975) Recovery of function after brain injury in man. In: *Outcome of Severe Damage to the Nervous System. Ciba Foundation Symposium* 34. Edited by R. Porter and D. W. Fitzsimons. Amsterdam and Oxford: Elsevier, pp. 159-190.
WAPNER W, HAMBY S, GARDNER H (1981) The role of the right hemisphere in the apprehension of complex linguistic materials. *Brain and Language*, **14**, 15-33.
WARRINGTON EK (1981) Concrete word dyslexia. *British Journal of Psychology*, **72**, 175-196.
WECHSLER D (1976) *Wechsler Intelligence Scale for Children—Revised*. British edition. Windsor: NFER-Nelson.
YOUNG AW, ELLIS AW (1985) Different methods of lexical access for words presented in the left and right visual hemifields. *Brain and Language*, **24**, 326-358.
ZAIDEL E (1977) Unilateral auditory language comprehension on the Token Test following cerebral commissurotomy and hemispherectomy. *Neuropsychologia*, **15**, 1-17.
ZAIDEL E (1978a) Lexical organization in the right hemisphere. In: *Cerebral Correlates of Conscious Experience*. Edited by P. A. Buser and A. Rougeul-Buser. Amsterdam and Oxford: North-Holland, pp. 177-197.
ZAIDEL E (1978b) Auditory language comprehension in the right hemisphere following cerebral commissurotomy and hemispherectomy: a comparison with child language and aphasia. In: *Language Acquisition and Language Breakdown: Parallels and Divergencies*. Edited by A. Caramazza and E. B. Zurif. Baltimore and London: Johns Hopkins University Press, pp. 229-275.
ZAIDEL E (1980) The split and half brains as models of congenital language disability. In: *The Neurological Bases of Language Disorders in Children: Methods and Directions for Research. NINCDS Monograph* No. 22. Edited by C. L. Ludlow and M. E. Doran-Quine. Bethesda, MD: U.S. Department of Health, Education and Welfare, Public Health Service, National Institutes of Health, National Institute for Communicative Disorders and Stroke, pp. 55-86.
ZAIDEL E (1982) Reading by the disconnected right hemisphere: an aphasiological perspective. In: *Dyslexia: Neuronal, Cognitive and Linguistic Aspects*. Edited by Y. Zotterman. Oxford: Pergamon Press, pp. 67-91.

ZAIDEL E (1983) A response to Gazzaniga: language in the right hemisphere, convergent perspectives. *American Psychologist*, **38**, 542–546.

ZAIDEL E, PETERS AM (1981) Phonological encoding and ideographic reading by the disconnected right hemisphere: two case studies. *Brain and Language*, **14**, 205–234.

ZAIDEL E, SCHWEIGER A (1984) On wrong hypotheses about the right hemisphere: commentary on K. Patterson and D. Besner, 'Is the right hemisphere literate?' *Cognitive Neuropsychology*, **1**, 351–364.

(Received December 1, 1987. Revised February 23, 1988. Accepted March 4, 1988)

APPENDIX

The table below provides general background information and standardized test results for the 2 patients. The first three sections of the table are self-explanatory, but the individual background tests on language abilities may require brief description.

1. The picture naming test from Oldfield and Wingfield (1965) consists of 36 line drawings of objects.

2. The Token Test (De Renzi and Vignolo, 1962) assesses auditory language comprehension by asking the subject to arrange tokens of various colours, shapes and sizes in response to spoken sentences.

3. TROG (Test for the Reception of Grammar, Bishop, 1982) also measures sentence comprehension. The subject is asked to point to 1 of 4 pictures to match a spoken sentence. Each of the 20 blocks in this test assesses a different aspect of syntactic comprehension, and 'passing' a block requires correct picture matching on all 4 items within the block.

4. The British Picture Vocabulary Scale (Dunn and Dunn, 1982) measures comprehension of single words, again with four alternative forced choice picture matching. The test, consisting of 150 items, is stopped when the subject reaches ceiling (6 errors in 8 consecutive items), and the subject's score is calculated by subtracting the total number of errors from ceiling. Since items occur in graded order of difficulty, the item number at which the subject makes his or her first error is also informative.

5. Schonell (1942) constructed standardized tests of (a) reading single words aloud and (b) writing single words to dictation, yielding estimates of reading and spelling age.

6. The Neale test of text reading (Neale, 1966) provides reading-age estimates for the speed and accuracy of oral reading and also for reading comprehension based on questions about the content of the text that has just been read.

Background information

	N.I.	H.P.
General		
Hemisphere removed	L	R
Original handedness	R	R
Year of birth	1969	1970
Age at onset of symptoms (yrs)	13	10
Age at operation (yrs)	15	15
Age at testing (yrs)	17	17
Postoperative IQ (NI: WISC-R; HP: WAIS)		
Verbal IQ	55	80
Performance IQ	80	75
Full scale IQ	66	77

	N.I.	H.P.
Wechsler Memory Scale		
MQ (based on adjusted age correction)	(62)	(83)
Logical memory—immediate	29%	49%
Visual reproduction—immediate	61%	39%
Digit span—forwards	4	4
—backwards	3	2
Language tests		
Naming (Oldfield-Wingfield, n = 36)	39%	92%
Token Test (n = 62)	69%	95%
TROG		
Blocks passed (of 20)	12	18
Total errors (of 80)	20	3
BPVS with spoken words		
Raw score	71	108
Item number of first error	34	58
BPVS with written words		
Raw score	43	93
Item number of first error	16	62
Schonell oral reading		
Number of words read	10	69
Reading age equivalent	6:9	11:1
Schonell spelling		
Number of words written	3	57
Spelling age equivalent	5:4	10:8
Neale reading age equivalents for:		
Rate	< 6:6	9:1
Accuracy	< 6:0	10:10
Comprehension	< 6:3	11:4

Glossary

agraphia - an inability to write
alexia - an inability to read
anomia - an inability to name things
astrocytic - referring to astrocytes; astrocytes are central nervous system cells that provide physical support for neurons and help keep some toxic substances in the blood from entering central nervous system neurons
atrophy - to degenerate or waste away
circumlocution - talking around the topic; a long speech that is slow getting to the point
corticosteroids - steroid hormones that are released from the adrenal cortex
CT scans - X-ray photographs of the living human brain taken by computerized tomography
deep dyslexia - an acquired reading disorder in which lexical reading is intact and nonlexical reading is impaired; patients with deep dyslexia can read words that they have previously memorized (lexical reading), but they cannot sound out even the most simple new words or nonwords (nonlexical reading)
Dyslexia - any disorder characterized by disturbances of reading
dysphasia - any disorder characterized by disturbances of speech
dyspraxia - a disorder characterized by a difficulty in carrying out voluntary movements
focal motor seizure - any motor seizure that is initially restricted to one part of the body; a focal motor seizure may remain focused or it may spread to other parts of the body
global aphasia - a severe disruption of all language-related skills
hemiparesis - a partial paralysis of half the body
hemispherectomy - surgical removal of one cerebral hemisphere
homonymous hemianopia - blindness in the same half (right or left) of the visual fields of both eyes
infarct - an area of cell damage resulting from ischemia
ischemia - any lack of blood supply that causes damage to an area of the body (i.e., that causes an infarct); ischemia is frequently caused by obstructions in the circulatory system; ischaemia is the British spelling
lexicon - the words of a language

nonsense words - strings of letters that have the appearance of words but are not (e.g., dipple)
orthographic - referring to the sight of the written or printed word
palsy - paralysis
paralexia - errors in reading caused by transposing syllables or words
phonemes - individual speech sounds, which combine to form a spoken word
phonological - referring to the sound of spoken language
plantar response - the reflexive response elicited by scratching the bottom of the foot; it is normal in adults for the toes to flex (curl); extension is sometimes a symptom of neurological disorder
progressive - refers to any disorder that grows worse
Rasmussen's encephalitis - a disorder that typically begins in late childhood or early adolescence and is characterized by unilateral focal motor seizures and progressive hemiparesis; its cause is unknown
rolandic region - the area around the central fissure, which is also called the rolandic fissure or the fissure of Rolando
semantic - referring to the meaning of language
sodium amytal test - a test that involves anesthetizing first one hemisphere and then later the other with an injection of sodium amytal into the circulatory systems of the left and right hemispheres; the subjects' language skills are quickly assessed in the minute or two after each injection
split-brain patients - patients in whom the neural connections between the left and right hemispheres have been severed for the treatment of epilepsy
surface dyslexia - an acquired reading disorder in which nonlexical reading is intact but lexical reading is impaired; patients with surface dyslexia can follow normal rules of pronunciation, but they cannot draw on their previous knowledge to correctly pronounce words that do not follow normal rules of pronunciation (e.g., "yacht," "have," and "lose")
Turner's syndrome - a genetic disorder in which females are born with only one X chromosome rather than two
unilateral - one side of the body

Essay Study Questions

1. According to the Patterson, Vargha-Khadem, and Polkey, what have been the major shortcomings of previous studies of the reading abilities of hemispherectomized patients?

2. Compare deep and surface dyslexia (see the glossary).

3. The results of these two case studies are described under three headings: (1) orthography, (2) from orthography to phonology, and (3) from orthography to semantics. What do these headings mean?

4. Compare the postoperative reading abilities of N.I. and H.P.

5. N.I. and H.P. did not receive exactly the same battery of tests. Why?

6. Unlike patients with massive left hemisphere lesions, N.I. could speak quite well. What interpretations do Patterson, Vargha-Khadem, and Polkey offer for this apparent discrepancy?

7. What roles in reading do the authors hypothesize for the right hemisphere?

8. N.I.'s performance was poorer than that of most deep dyslexics. What interpretations do the authors offer for this difference?

Multiple-Choice Study Questions

1. The authors conclude that N.I.'s reading deficits were similar to those of
    a. surface dyslexia.
    b. deep dyslexia.
    c. split-brain patients using their left hemispheres.
    d. global aphasia.
    e. both a and c

2. Which of the following patients has great difficulty pronouncing nonwords?
    a. N.I.
    b. a patient with deep dyslexia
    c. H.P.
    d. both a and b
    e. both b and c

3. N.I. and H.P. both
    a. were diagnosed as suffering from Rasmussen's encephalitis.
    b. experienced homonymous hemianopia.
    c. suffered epileptic attacks.
    d. had hemiatrophy, which was visible in their CT scans.
    e. all of the above

4. On the letter-naming test, N.I. scored
    a. 0/17.
    b. 17/17 but was very slow.
    c. 17/17 but worked at a normal speed.
    d. 7/17 and was very slow.
    e. none of the above

5. N.I. could recognize one class of words with 95% accuracy, but she was virtually incapable of reading them aloud. These words were
    a. high-frequency function words.
    b. high-frequency concrete nouns.
    c. places.
    d. animal names.
    e. number words.

6. N.I.'s reading problem was characterized by the authors as an inability to convert
    a. phonology to orthography.
    b. semantics to phonology.
    c. orthography to phonology.
    d. semantics to orthography.
    e. both a and b

The answers to the preceding questions are on page 290.

Food-For-Thought Questions

1. It is important not to lose sight of the fact that N.I. and H.P. are two real people who are living their lives with only one hemisphere. Some people are opposed to the correction of behavioral disorders by brain surgery; do you think that hemispherectomy improved the lives of N.I. and H.P.?

2. Both N.I. and H.P. lost half their cerebral tissue. What do you think would have happened if the same amount of cerebral tissue had ben removed but in another pattern such as the front half of both hemispheres, the back half of both hemispheres, or the front half of one hemisphere and the back half of the other? What do you think would have happened if the same amount of tissue had been removed from the brainstem?

Science

AMERICAN
ASSOCIATION FOR THE
ADVANCEMENT OF
SCIENCE

13 JANUARY 1989　　　　$3.50
VOL. 243 ■ PAGES 141–272

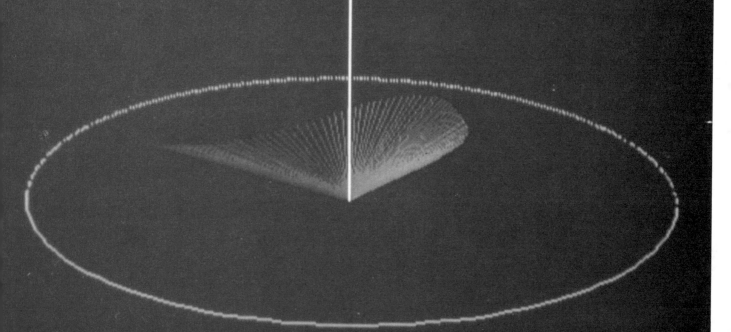

ARTICLE 24

Mental Rotation of the Neuronal Population Vector

A.P. Georgopoulos, J.T. Lurito, M. Petrides, A.B. Schwartz, and J.T. Massey
Reprinted from Science, 1989, volume 243, 234-236.

One of the major goals of biopsychology is to understand the neural events that underlie complex mental activity. Georgopoulos and his collaborators studied mental rotation in a rhesus monkey by adapting the human mental-rotation paradigm. The task of the monkey subject was to move a handle in a direction that was at a prescribed angle to the movement of a dot of light on a display screen. Correct performance was reinforced by a sweet-tasting liquid. The finding that the reaction times of the monkey increased with increases in the angle between the moving stimulus and the required handle movement was similar to the results of tests with human subjects engaging in mental rotation; this similarity suggested that the monkey was engaging in mental rotation (i.e., rotating the image of the stimulus in its mind) in order to figure out the correct angle of handle movement.

Georgopoulos et al. based their experiment on the fact that neurons in the motor cortex are directionally tuned; each motor cortex neuron becomes more active for movements in one direction than it does for movements in other directions. Because motor cortex neurons that are tuned to a particular direction become more active in the interval just before a movement takes place, it is possible, if one is recording from several of them and if one has determined the preferred direction of each, to calculate a so-called population vector, a kind of weighted average direction, that accurately predicts the direction of the subsequent movement. (You need not worry about the details of vector mathematics, they are not necessary for you to appreciate this research.)

Georgopoulos and his collaborators made a seemingly farfetched prediction. They predicted that there would be a rotation of their subject's motor neuron population vector, from the direction of stimulus movement to the direction of intended handle movement, in the brief interval between the presentation of the stimulus and the completion of the response. Amazingly, this is exactly what they found. On trials in which the monkey was required to make a response 90° counterclockwise to the direction of the stimulus, the population vector started off pointing in the direction of the stimulus, and over the ensuing 200 milliseconds, it rotated 90° counterclockwise so that it was pointing in the exact direction of the correct movement. Then the monkey responded. This finding suggests that some monkeys engage in mental rotation, and it illustrates an exciting new procedure for studying its neural basis.

Mental Rotation of the Neuronal Population Vector

Apostolos P. Georgopoulos,* Joseph T. Lurito,
Michael Petrides, Andrew B. Schwartz, Joe T. Massey

A rhesus monkey was trained to move its arm in a direction that was perpendicular to and counterclockwise from the direction of a target light that changed in position from trial to trial. Solution of this problem was hypothesized to involve the creation and mental rotation of an imagined movement vector from the direction of the light to the direction of the movement. This hypothesis was tested directly by recording the activity of cells in the motor cortex during performance of the task and computing the neuronal population vector in successive time intervals during the reaction time. The population vector rotated gradually counterclockwise from the direction of the light to the direction of the movement at an average rate of 732° per second. These results provide direct, neural evidence for the mental rotation hypothesis and indicate that the neuronal population vector is a useful tool for "reading out" and identifying cognitive operations of neuronal ensembles.

A FUNDAMENTAL PROBLEM IN COGnitive neuroscience is the identification and elucidation of brain events underlying cognitive operations (1). The technique of recording the activity of single cells in the brain of behaving animals (2) provides a direct tool for that purpose. Indeed, a wealth of knowledge has accumulated during the past 15 years concerning the activity of cells in several brain areas during performance by monkeys of complex tasks. A major finding of these studies has been that the activity of single cells in specific areas of the cerebral cortex changes during performance of particular tasks; these changes are thought to reflect the participation of the area under study in the cognitive function involved in the task (3). However, a direct visualization of a cognitive operation in terms of neuronal activation in the brain is lacking.

We chose as a test case for this problem the cognitive operation of mental rotation. Important work in experimental psychology during the past 20 years (4) has established the mental rotation paradigm as a standard in cognitive psychology and as a prime tool in investigating cognitive operations of the "analog" type. We adapted this procedure in a task that required movement of a handle in a direction that was at an angle with the direction of a stimulus. Under these conditions the reaction time increased with the angle, which suggests that the subject may solve this problem by a mental rotation of an imagined movement vector from the direction of the stimulus to the direction of the actual movement (5). Now, the direction of an upcoming movement in space seems to be represented in the motor cortex as the neuronal population vector (6), which is a weighted vector sum of contributions ("votes") of directionally tuned neurons: each neuron is assumed to vote in its own preferred direction with a strength that depends on how much the activity of the neuron changes for the movement under consideration. This vectorial analysis has proved useful in visualizing the directionality of the population in two- and three-dimensional space during the reaction time (7) and during an instructed delay period (8).

Given the mental rotation hypothesis above and the neuronal population vector as a neural representation of the movement direction, a strong test is as follows: if a monkey performs in the above-mentioned task and the neuronal activity in the motor cortex is recorded during performance, would the population vector rotate in time, as the hypothesis for a mental rotation of an imagined movement vector would predict? Because the appropriate movement direction can be arrived at by either a counterclockwise or a clockwise rotation, which of these two rotations would be realized by the population vector? Of course, there is no reason that the population vector should rotate at all, and if it rotates, there is no a

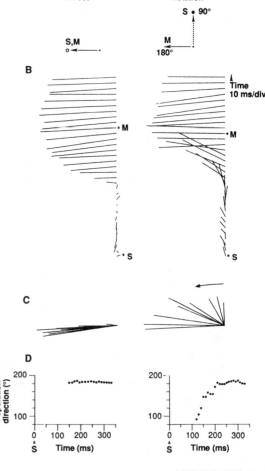

Fig. 1. Results from a direct (left) and rotation (right) movement. (**A**) Task. Unfilled and filled circles indicate dim and bright light, respectively. Interrupted and continuous lines with arrows indicate stimulus (S) and movement (M) direction, respectively. (**B**) Neuronal population vectors calculated every 10 ms from the onset of the stimulus (S) at positions shown in (A) until after the onset of the movement (M). When the population vector lengthens, for the direct case (left) it points in the direction of the movement, whereas for the rotation case it points initially in the direction of the stimulus and then rotates counterclockwise (from 12 o'clock to 9 o'clock) and points in the direction of the movement. (**C**) Ten successive population vectors from (B) are shown in a spatial plot, starting from the first population vector that increased significantly in length. Notice the counterclockwise rotation of the population vector (right panel). (**D**) Scatter plots of the direction of the population vector as a function of time, starting from the first population vector that increased significantly in length after stimulus onset (S). For the direct case (left panel) the direction of the population vector is in the direction of the movement (~180°); for the rotation case (right panel) the direction of the population vector rotates counterclockwise from the direction of the stimulus (~90°) to the direction of the movement (~180°).

A. P. Georgopoulos and J. T. Lurito, The Philip Bard Laboratories of Neurophysiology, Department of Neuroscience, The Johns Hopkins University School of Medicine, 725 North Wolfe Street, Baltimore, MD 21205.
M. Petrides, Department of Psychology, McGill University, 1205 Dr. Penfield Avenue, Montreal, Quebec, Canada H3A 1B1.
A. B. Schwartz, Division of Neurobiology, St. Joseph's Hospital and Medical Center, Barrow Neurological Institute, 350 West Thomas Road, Phoenix, AZ 85013.
J. T. Massey, Department of Neuroscience and Department of Biomedical Engineering, The Johns Hopkins University School of Medicine, 725 North Wolfe Street, Baltimore, MD 21205.

*To whom correspondence should be addressed.

priori reason that it should rotate in one or the other direction; for all we know, any of these alternatives is possible.

The activity of single cells in the motor cortex was recorded (9) while a rhesus monkey performed in the mental rotation task. In the beginning of a trial, a light appeared at the center of a plane in front of the animal, which moved its arm toward the light with a freely movable handle (10). After a variable period of time (0.75 to 2.25 s), the center light was turned off and turned on again, dim or bright, at one of eight positions on a circle of 2-cm radius (11). The monkey was trained to move the handle in the direction of the light when it came on dim (direct trials) or in a direction that was perpendicular (90°) to and counterclockwise from the direction of the light when it came on bright (rotation trials) (12). The movements of the animal were in the appropriate direction for both kinds of trials. The neuronal population vector was calculated every 10 ms starting from the onset of the peripheral light (that is, at the beginning of the reaction time). The preferred direction of each cell ($n = 102$ cells) was determined from the cell activity in the trials in which the animal moved toward the light (direct trials). For the calculation of the population vector, peristimulus time histograms (10-ms binwidth) were computed for each cell and each of the 16 combinations (classes) used [eight positions and two conditions (direct or rotation), see (11) above] with counts of fractional interspike intervals as a measure of the intensity of cell discharge. A square root transformation was applied to these counts to stabilize the variance (13). For a given time bin, each cell made a vectorial contribution in the direction of the cell's preferred direction and of magnitude equal to the change in cell activity from that observed during 0.5 s preceding the onset of the peripheral stimulus (control rate, that is, while the monkey was holding the handle at the center of the plane). The population vector **P** for the jth class and k^{th} time bin is

$$\mathbf{P}_{j,k} = \sum_{i}^{102} w_{i,j,k} \mathbf{C}_i$$

where $\mathbf{C}_i$ is the preferred direction of the ith cell and $w_{i,j,k}$ is a weighting function $w_{i,j,k} = (d_{i,j,k}) - a_{i,j}$ where $d_{i,j,k}$ is the square root–transformed (13) discharge rate of the ith cell for the jth class and k^{th} time bin, and $a_{i,j}$ is the similarly transformed control rate of the ith cell for the jth class.

Figure 1 illustrates the results obtained when the movement direction was the same (toward 9 o'clock) but the stimulus was either at 9 o'clock (direct trials, left panel) or at 12 o'clock (rotation trials, right panel). In the direct trials the population vector pointed in the direction of the movement (which coincided with the direction of the stimulus) (Fig. 1, left). However, in the rotation trials the population vector rotated in time counterclockwise from the direction of the stimulus to the direction of the movement (Fig. 1, right). Another example is shown in Fig. 2 and illustrated in the cover photograph. The working space is outlined in blue. The time axis is the white line directed upwards. The population vector is shown in green, as it rotates during the reaction time from the stimulus direction (between 1 and 2 o'clock) to the movement direction (between 10 and 11 o'clock). The population vector was calculated with a 20-ms bin sliding every 2 ms. The red lines are projections of the population vector onto the working space.

The rotation of the population vector was a linear function of time with an average slope (for the eight positions of the light used) of $732 \pm 456°/s$ (mean ± SD). The population vector began to change in length 125 ± 28 ms (mean ± SD, $n = 8$) after the stimulus onset. At this point its direction was close to the direction of the stimulus; the average angle between the direction of the population vector and that of the stimulus was 17° counterclockwise (the average absolute angle was 29°). The population vector stabilized in direction at 225 ± 50 ms after stimulus onset. At this point its direction was close to the direction of the movement; the average angle between the direction of the population vector and that of the movement was 0.5° clockwise (the average absolute angle was 8°). Finally, the movement began 260 ± 30 ms after stimulus onset, that is, 35 ms after the direction of the population vector became relatively stable; this difference was statistically significant ($P < 0.02$, paired t test).

These results support the hypothesis that the directional transformation required by the task was achieved by a counterclockwise rotation of an imagined movement vector. This process was reflected in the gradual change of activity of motor cortical cells, which led to the gradual rotation of the vectorial distribution of the neuronal ensemble and the population vector. The average slope of the rotation of the population vector (732°/s, see above) was comparable to but higher than that observed when human subjects performed a similar task (~400°/s) (5) and that observed in a task that involved mental rotation of two-dimensional images (~400°/s) (14). It is likely that all three experiments involved a process of mental rotation which, in the present case, was reflected in the motor cortical recordings of this study and identified by using the population vector analysis. Of course, other brain areas are probably involved in such complicated transformations; for example, recent experiments with measurements of regional cerebral blood flow (15) suggested that frontal and parietal areas seem to be involved in the mental rotation task of Shepard and Metzler (16), whereas frontal and central areas seem to be involved in a line orientation task (15); in both of these tasks there was a greater increase in blood flow in the right than in the left hemisphere.

The rotation of the neuronal population vector is of particular interest because there was no a priori reason for it to rotate at all. It is also interesting that the population vector rotated consistently in the counterclockwise direction: this suggests that the spatial-motor transformation imposed by the task was solved by a rotation through the shortest angular distance. Given that the mental rotation is time consuming, this solution was behaviorally meaningful, for it minimized both the time for the animal to get the reward and the computational effort which would have been longer if the rota-

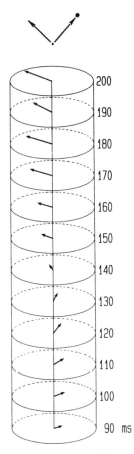

Fig. 2. Rotation of the population vector for a different set of rotation trials. The stimulus and movement directions are indicated by the interrupted and continuous lines at the top. The population vector in the two-dimensional space is shown for successive time frames beginning 90 ms after stimulus onset. Notice its rotation counterclockwise from the direction of the stimulus to the direction of the movement.

tion had been through 270° clockwise (17).

Finally, these results were obtained from one animal: because cognitive problems could be solved in different ways by different subjects, it is important that techniques for reading out brain operations be sensitive enough to be applied to single subjects. Indeed, the findings of our study indicate that the population vector is a sensitive tool by which an insight can be gained into the brain processes underlying cognitive operations in space.

REFERENCES AND NOTES

1. See, for example, V. B. Mountcastle, *Trends Neurosci.* **9**, 505 (1986); M. A. Arbib and M. B. Hesse, *The Construction of Reality* (Cambridge Univ. Press, New York, 1986); J. Z. Young, *Philosophy and the Brain* (Oxford Univ. Press, New York, 1987).
2. R. N. Lemon, *Methods for Neuronal Recording in Conscious Animals* (Wiley, Chisester, 1984).
3. See, for example, *Handbook of Physiology*, Section 1, *The Nervous System, Higher Function of the Brain*, parts 1 and 2, V. B. Mountcastle, F. Plum, S. R. Geiger, Eds. (American Physiological Society, Bethesda, MD, 1987), vol. 5.
4. R. N. Shepard and J. Metzler, *Science* **171**, 701 (1971); R. N. Shepard and L. A. Cooper, *Mental Images and Their Transformations* (MIT Press, Cambridge, MA, 1982).
5. A. P. Georgopoulos and J. T. Massey, *Exp. Brain Res.* **65**, 361 (1987).
6. A. P. Georgopoulos, P. Caminiti, J. F. Kalaska, J. T. Massey, *ibid. Suppl.* **7**, 327 (1983); A. P. Georgopoulos, A. B. Schwartz, R. E. Kettner, *Science* **233**, 1416 (1986); *J. Neurosci.* **8**, 2928 (1988).
7. A. P. Georgopoulos, J. F. Kalaska, M. D. Crutcher, R. Caminiti, J. T. Massey, in *Dynamic Aspects of Neocortical Function*, G. M. Edelman, W. E. Gall, W. M. Cowan, Eds. (Wiley, New York, 1984), pp. 501–524; A. P. Georgopoulos, A. B. Schwartz, R. E. Kettner, *J. Neurosci.* **8**, 2928 (1988).
8. A. P. Georgopoulos, M. D. Crutcher, A. B. Schwartz, *Exp. Brain Res.*, in press.
9. The electrical signs of activity of individual cells in the arm area of the motor cortex contralateral to the performing arm were recorded extracellularly [A. P. Georgopoulos, J. F. Kalaska, R. Caminiti, J. T. Massey, *J. Neurosci.* **2**, 1527 (1982)]. All surgical operations [A. P. Georgopoulos, J. F. Kalaska, R. Caminiti, J. T. Massey, *J. Neurosci.* **2**, 1527 (1982)] for the preparation of the animal for electrophysiological recordings were performed under general pentobarbital anesthesia. Behavioral control and data collection and analysis were performed with a laboratory minicomputer.
10. The apparatus was as described in A. P. Georgopoulos and J. T. Massey [*Exp. Brain Res.* **65**, 361 (1987)]. Briefly, it consisted of a 25 cm by 25 cm planar working surface made of frosted plexiglass onto which a He-Ne laser beam was back-projected with a system of mirrors and two galvanometers. The monkey (5 kg) sat comfortably on a primate chair and grasped a freely movable, articulated handle at its distal end, next to a 10-mm diameter transparent plexiglass circle within which the animal captured the center light.
11. The eight positions were equally spaced on the circle, that is, at angular intervals of 45°, and were the same throughout the experiment. The brightness condition (dim or bright) and the position of the light were mixed. The resulting 16 brightness-position combinations were randomized. Eight repetitions of these 16 combinations were presented in a randomized block design.
12. The term "counterclockwise" is simply descriptive; no counterclockwise or clockwise directions were indicated to the animal. The direction in which the animal was required to move can be described equivalently as either 90° counterclockwise or 270° clockwise. The animal received a liquid reward when its movement exceeded 3 cm and stayed within ±25° of the direction required. The average direction of the actual movement trajectories was within ±5° of the direction required. Performance was over 70% correct trials.
13. The square root transformation was used as a variance-stabilizing transformation for counts [G. W. Snedecor and W. G. Cochran, *Statistical Methods* (Iowa State Univ. Press, Ames, Iowa, ed. 7, 1980), pp. 288–290.] Although the results obtained without this transformation were similar, the transformation is more appropriate because of the small size of the time bins (10 ms), and, therefore, the small number of counts.
14. L. A. Cooper and R. N. Shepard, in *Visual Information Processing*, W. G. Chase, Ed. (Academic Press, New York, 1973), pp. 75–176; L. A. Cooper, *Cognitive Psychol.* **7**, 20 (1975).
15. G. Deutsch, W. T. Bourbon, A. C. Papanicolaou, H. M. Eisenberg *Neuropsychologia* **26**, 44 (1988).
16. R. N. Shepard and J. Metzler, *Science* **171**, 701 (1971).
17. The same principles of minimization of the time-to-reward and of reduction of computation load, even at the expense of mechanical work, were observed in strategies developed by human subjects and monkeys in a different task [J. T. Massey, A. P. Schwartz, A. P. Georgopoulos, *Exp. Brain Res. Suppl.* **15**, 242 (1986)].
18. We thank D. Brandt and N. Porter for help during some of the experiments. Supported by USPHS grants NS17413 and NS20868.

1 August 1988; accepted 1 November 1988

COVER Visualization of brain activity during mental rotation. The neuronal population vector (green) rotated gradually from 2 to 10 o'clock as a monkey was thinking (time upward, white line). The population vector was calculated from an ensemble of neurons recorded in the motor cortex. See page 234. [A. P. Georgopoulos, Department of Neuroscience, Johns Hopkins University School of Medicine, Baltimore, MD 21205]

Glossary

analog operation - any operation that is continuous; in contrast to digital operations that occur in discrete steps

cognitive - refers to any complex mental process; for example, thinking, remembering, or imagining

cognitive neuroscience - a branch of neuroscience that uses the methods of cognitive psychology to understand the relation between cognitive processes and neural activity; much of the research of cognitive neuroscientists is focused on the study of brain-damaged subjects

frontal - refers to the frontal lobes of the cerebral hemispheres; as their name implies, the frontal lobes are located at the front of the hemispheres

mental rotation - rotating an image of a visual stimulus in one's mind

motor cortex - an area of the frontal lobe that contains neurons whose axons descend into the motor circuits of the spinal cord

parietal - refers to the parietal lobes of the cerebral hemispheres; in humans, the parietal lobes are located at the top of the back half of the hemispheres

regional cerebral blood flow - the amount of blood flowing into a region of the brain; because more blood flows into areas of the brain that are active, increases in cerebral blood flow during various cognitive activities indicate the areas of the brain that are involved in the activities

vector - a mathematical representation of direction and strength; a vector is represented geometrically by a straight line of a particular length with an arrowhead at one end; the slope of the line indicates the angle of the vector, the length of the line indicates the strength of the vector, and the position of the arrowhead indicates the direction of the vector

Essay Study Questions

1. What was the hypothesis of Georgopoulos et al., and what was their ingenious idea for testing it?

2. Describe the task that was performed by the subject in this experiment. What were rotational trials and direct trials?

3. Describe how the population vector changed on rotational trials and on direct trials.

4. What evidence suggested that the subject performed the mental rotation in a counterclockwise direction?

5. What have measurements of increases in regional cerebral blood flow indicated about the neural mediation of mental rotation?

Multiple-Choice Study Questions

1. On rotational trials, the motor neuron population vector rotated
    a. 90°.
    b. 270°.
    c. 180°.
    d. 15°.
    e. either a and b

2. On direct trials, the motor neuron population vector rotated
    a. 0°.
    b. 15°.
    c. 90°.
    d. 180°.
    e. 270°.

3. On rotational trials, the population vector began to change
    a. about 1.5 seconds before the movement.
    b. immediately after the mental rotation.
    c. immediately before the stimulus onset.
    d. about 125 milliseconds before the mental rotation.
    e. about 125 milliseconds after the stimulus onset.

4. The movement began on each rotational trial
    a. before the population vector started to rotate.
    b. about 35 milliseconds after the population vector stopped rotating.
    c. while the population vector was rotating.
    d. only if the population vector did not rotate.
    e. only when the angle of required movement was the same as the angle of the stimulus.

5. Studies of regional blood flow in human subjects indicate that areas in the
    a. frontal lobes are involved in mental rotation.
    b. parietal lobes are involved in mental rotation.
    c. temporal lobes are involved in mental rotation.
    d. occipital lobes are involved in mental rotation.
    e. both a and b

6. The rhesus monkey learned that
    a. every other trial was a rotational trial.
    b. rotational trials occurred on alternate days.
    c. bright stimulus lights signalled rotational trials.
    d. only rotational trials were reinforced.
    e. both a and d

The answers to the preceding questions are on page 290.

Food-For-Thought Questions

1. In one sense, there was only one subject in this experiment, but in another sense, there were 102. Discuss.

2. It is usual to think of mental rotation as a cognitive process with a strong sensory component, yet the results of Georgopoulos and his collaborators suggest that many of the neurons involved in mental rotation are in the motor cortex. This illustrates the inappropriateness of labelling specific areas of the cortex as sensory, motor, or cognitive. Elaborate and discuss.

3. Does this experiment prove that monkeys can think?

Answers to the Multiple Choice Study Questions

Article 1
1. d
2. d
3. e
4. a
5. c
6. e
7. b

Article 2
1. e
2. b
3. c
4. e
5. a
6. e

Article 3
1. d
2. c
3. a
4. c
5. d
6. e

Article 4
1. d
2. e
3. e
4. d
5. d
6. a

Article 5
1. a
2. e
3. c
4. c
5. b
6. e
7. a

Article 6
1. c
2. e
3. a
4. e
5. c
6. e

Article 7
1. d
2. c
3. d
4. c
5. a
6. e

Article 8
1. e
2. e
3. d
4. b
5. b
6. c

Article 9
1. d
2. b
3. e
4. d
5. e
6. b
7. a

Article 10
1. a
2. e
3. c
4. b
5. d
6. e

Article 11
1. d
2. a
3. c
4. b
5. e
6. b
7. c

Article 12
1. a
2. e
3. b
4. c
5. b
6. a

Article 13
1. e
2. e
3. a
4. b
5. e
6. b

Article 14
1. e
2. d
3. c
4. a
5. e
6. c

Article 15
1. d
2. e
3. c
4. d
5. c
6. a

Article 16
1. e
2. a
3. c
4. b
5. e
6. e

Article 17
1. e
2. c
3. e
4. a
5. b
6. b

Article 18
1. b
2. b
3. e
4. c
5. c
6. d

Article 19
1. b
2. a
3. d
4. d
5. b
6. a

Article 20
1. b
2. e
3. e
4. e
5. d
6. b

Article 21
1. c
2. d
3. e
4. e
5. e
6. d

Article 22
1. e
2. b
3. c
4. b
5. e
6. c

Article 23
1. b
2. d
3. e
4. d
5. b
6. c

Article 24
1. a
2. a
3. e
4. b
5. e
6. c